Analysis and Modeling of Radio Wave Propagation

With this comprehensive guide you will understand the theory and learn the techniques needed to analyze and model radio wave propagation in complex environments. All of the essential topics are covered, from the fundamental concepts of radio systems to complex propagation phenomena. These topics include diffraction, ray tracing, scattering, atmospheric ducting, ionospheric ducting, scintillation and propagation through both urban and non-urban environments. Emphasis is placed on practical procedures, with detailed discussion of numerical and mathematical methods, providing you with the necessary skills to build your own propagation models and develop your own techniques. MATLAB functions illustrating key modeling ideas are available online.

This is an invaluable resource for anyone wanting to use propagation models to understand the performance of radio systems for navigation, radar, communications or broadcasting.

Christopher John Coleman is a Senior Visiting Research Fellow in the Department of Electronic and Electrical Engineering at the University of Bath, and a Visiting Research Fellow at the School of Electrical and Electronic Engineering at the University of Adelaide. From 1990 until 1999 he was a Principal Research Scientist on Australia's Jindalee Over the Horizon radar project. He is the author of the book *An Introduction to Radio Frequency Engineering* (Cambridge, 2004).

Analysis and Modeling of Radio Wave Propagation

CHRISTOPHER JOHN COLEMAN

University of Bath and University of Adelaide

CAMBRIDGE
UNIVERSITY PRESS

University Printing House, Cambridge CB2 8BS, United Kingdom

One Liberty Plaza, 20th Floor, New York, NY 10006, USA

477 Williamstown Road, Port Melbourne, VIC 3207, Australia

4843/24, 2nd Floor, Ansari Road, Daryaganj, Delhi – 110002, India

79 Anson Road, #06–04/06, Singapore 079906

Cambridge University Press is part of the University of Cambridge.

It furthers the University's mission by disseminating knowledge in the pursuit of education, learning and research at the highest international levels of excellence.

www.cambridge.org
Information on this title: www.cambridge.org/9781107175563

© Cambridge University Press 2017

This publication is in copyright. Subject to statutory exception
and to the provisions of relevant collective licensing agreements,
no reproduction of any part may take place without the written
permission of Cambridge University Press.

First published 2017

Printed in the United Kingdom by TJ International Ltd. Padstow Cornwall

A catalogue record for this publication is available from the British Library

Library of Congress Cataloging-in-Publication Data
Names: Coleman, Christopher, 1950– author.
Title: Analysis and modeling of radio wave propagation / Christopher John Coleman, University of Adelaide.
Description: Cambridge, United Kingdom ; New York, NY : Cambridge University Press, [2017] | Includes bibliographical references and index.
Identifiers: LCCN 2016045806| ISBN 9781107175563 (Hardback ; alk. paper)|
ISBN 1107175569 (Hardback ; alk. paper)
Subjects: LCSH: Radio wave propagation. | Radio wave propagation–Mathematical models. | Electromagnetic waves.
Classification: LCC TK6553 .C635 2017 | DDC 621.3841/1–dc23 LC record available at https://lccn.loc.gov/2016045806

ISBN 978-1-107-17556-3 Hardback

Additional resources for this publication at www.cambridge.org/coleman

Cambridge University Press has no responsibility for the persistence or accuracy of URLs for external or third-party Internet websites referred to in this publication, and does not guarantee that any content on such websites is, or will remain, accurate or appropriate.

Contents

	Preface		*page* ix
1	**Basic Concepts**		1
	1.1	Waves	1
	1.2	Electromagnetic Waves	2
	1.3	Communications Systems	5
	1.4	Cellular Radio	7
	1.5	Radar Systems	8
	1.6	Complex Propagation	9
2	**The Fundamentals of Electromagnetic Waves**		15
	2.1	Maxwell's Equations	15
	2.2	Plane Electromagnetic Waves	17
	2.3	Plane Waves in Anisotropic Media	22
	2.4	Boundary Conditions	27
	2.5	Transmission through an Interface	28
	2.6	Oblique Incidence	30
	2.7	Sources of Radio Waves	33
3	**The Reciprocity, Compensation and Extinction Theorems**		38
	3.1	The Reciprocity Theorem	38
	3.2	Reciprocity and Radio Systems	40
	3.3	Pseudo Reciprocity	42
	3.4	The Compensation Theorem	44
	3.5	The Extinction Theorem	46
4	**The Effect of Obstructions on Radio Wave Propagation**		50
	4.1	The Friis Equation	50
	4.2	Reflection by Irregular Terrain	53
	4.3	Diffraction	55
	4.4	Surface Waves	58
	4.5	The Geometric Theory of Diffraction	60

	4.6	Propagation in Urban Environments	64
	4.7	The Channel Impulse Response Function	68
5		**Geometric Optics**	70
	5.1	The Basic Equations	70
	5.2	Analytic Integration	78
	5.3	Geometric Optics in an Anisotropic Medium	80
	5.4	Weakly Anisotropic Medium	86
	5.5	Fermat's Principle for Anisotropic Media	87
6		**Propagation through Irregular Media**	94
	6.1	Scattering by Permittivity Anomalies	94
	6.2	The Rytov Approximation	98
	6.3	Mutual Coherence	102
	6.4	The Rytov Approximation and Irregular Media	103
	6.5	Parabolic Equations for the Average Field and MCF	110
	6.6	The Phase Screen Approximation	114
	6.7	Channel Simulation	118
	6.8	Rough Surface Scattering	121
7		**The Approximate Solution of Maxwell's Equations**	129
	7.1	The Two-Dimensional Approximation	129
	7.2	The Paraxial Approximation	135
	7.3	Kirchhoff Integral Approach	139
	7.4	Irregular Terrain	145
	7.5	3D Kirchhoff Integral Approach	148
	7.6	Time Domain Methods	151
8		**Propagation in the Ionospheric Duct**	159
	8.1	The Benign Ionosphere	159
	8.2	The Disturbed Ionosphere	165
	8.3	Vertical and Quasi-Vertical Propagation	167
	8.4	Oblique Propagation over Long Ranges	177
	8.5	Propagation Losses	187
	8.6	Fading	191
	8.7	Noise	192
	8.8	Full Wave Solutions	194
9		**Propagation in the Lower Atmosphere**	204
	9.1	Propagation in Tropospheric Ducts	204
	9.2	The Effect of Variations in Topography	209

	9.3	Surface Wave Propagation	211
	9.4	Propagation through Forest	216
	9.5	Propagation through Water	218
	9.6	Propagation through Rain	219

10 Transionospheric Propagation and Scintillation 222

10.1	Propagation through a Benign Ionosphere	222
10.2	Faraday Rotation and Doppler Shift	225
10.3	Small-Scale Irregularity	226
10.4	Scintillation	227

Appendix A Some Useful Mathematics 236

A.1	Vectors	236
A.2	Vector Operators	236
A.3	Cylindrical Polar Coordinates	237
A.4	Some Useful Integrals	238
A.5	Trigonometric Identities	239
A.6	Method of Stationary Phase	240
A.7	Some Expansions	240
A.8	The Airy Function	241
A.9	Hankel and Bessel Functions	242
A.10	Some Useful Series	245

Appendix B Numerical Methods 247

B.1	Numerical Differentiation and Integration	247
B.2	Zeros of a Function	248
B.3	Numerical Solution of Ordinary Differential Equations	249
B.4	Multidimensional Integration	251

Appendix C Variational Calculus 253

Appendix D The Fourier Transform 259

Appendix E Finding Stationary Values 263

E.1	Newton–Raphson Approach	263
E.2	Nelder–Mead Method	263

Appendix F Stratified Media 265

F.1	Two-Layer Medium	265
F.2	Three-Layer Medium	270

Appendix G	Useful Information	273
Appendix H	A Perfectly Matched Layer	274
Appendix I	Equations for TE and TM Fields	276
Appendix J	Canonical Solutions	278
	Index	284

Preface

The aim of this book is to provide the reader with the techniques and theory that are required for the analysis and modeling of radio wave propagation in complex environments. It is designed for the reader who might need to model propagation in order to understand the performance of radio systems for navigation, radar, communications or broadcasting. The book brings together a range of topics that are often treated separately, but all of which are important in the comprehensive modeling of a radio system. In particular, the book includes an extensive discussion of propagation through irregularity, of importance to systems that suffer from scintillation. The book is not intended to be just a cookbook of propagation formulae, but rather to provide readers with sufficient insight to enable them to produce their own specialized theory and techniques when required. It is my experience that many propagation problems are not amenable to off-the-shelf black box solutions. A black box will often only provide part of the solution and the modeler will need to modify and/or add capability. To do this successfully, the modeler will need to have some insight into the basis of the black box in order to effectively incorporate his/her own modifications. It is the intention of the author to provide the reader with such insight. The book leverages on my experience, over several decades, in the development of techniques for the analysis and modeling of propagation in a variety of radar and communication systems. In writing this book, I have been heavily influenced by the work of Professor James Wait, Dr. G.D. Monteath, Dr. Jenifer Haselgrove, and Dr. Kenneth Davies. In particular, the reciprocity ideas that have been developed by Dr. Monteath have proven invaluable in the development of many propagation modeling techniques.

The book is designed to take the reader from very basic ideas concerning radio systems to advanced propagation modeling. The first chapter will be useful to someone new to radio systems and provide them with an idea of the technology and the challenges that radio wave propagation imposes. Obviously, this chapter can be skipped by those readers who are already familiar with radio technology. Chapters 2 and 3 introduce some important electromagnetic ideas that are used in the rest of the book. Readers specifically interested in ionospheric propagation will find Chapters 5 and 8 of most use, while those interested in scintillation will find Chapters 6 and 10 of relevance. For those interested in propagation across terrain and through the lower atmosphere, Chapters 4, 7 and 9 are of greatest relevance. The appendices contain extensive notes on mathematics and numerical techniques that are used throughout the text. In addition, there are appendices in which important canonical solutions are derived.

I would like to thank Professor Christophe Fumeaux, Dr. L.J.Nickisch and Dr. Robert Watson for reading drafts of this book and providing useful feedback. I would also like to thank my wife, Marilyn, for her invaluable support and help in preparing this book.

1 Basic Concepts

This chapter introduces the fundamental ideas of radio waves and radio systems. It is designed to give a brief introduction to radio technology for those without a background in this area. The chapter includes an introduction to a variety of propagation phenomena as a motivation for the more detailed analysis in later chapters.

1.1 Waves

The concept of a wave is something for which it is very difficult to find a clear definition in the literature. Before proceeding, however, it is important that we have a good understanding of what we mean by a wave. In this regard, it is instructive to start with the surface water wave, a phenomenon that gives us one of the best practical illustrations of wave phenomena in general. Water waves are something that most of us have experienced and that exhibit many of the important features of waves and their propagation. As children, we have nearly all generated waves by throwing stones into a pond. Before the stone lands, the surface of the pond (the propagation medium) is calm. After impact, however, there is a ripple that travels radially outward from the point of impact. The ripple forms a circular band of disturbance that expands at a finite speed. Within the band the ripple maintains its shape but with amplitude that reduces as the radius of the band increases. As the ripple travels outward, it might encounter a floating object and then cause it to bob up and down. This motion can be used to extract energy from the wave, energy that was originally supplied by the stone's impact (the wave source). Further, the vertical motion of the object provides a means of detecting the passage of a wave.

Water waves illustrate several important features that are common to all wave phenomena. First, the wave can transport energy from one point (the source) to another (the detector), the energy being transported at a finite speed. Second, after the passage of the wave, the medium returns to its undisturbed state. This last point leads on to another important property of wave phenomena, the ability to make arbitrarily shaped waves. Instead of causing the wave by casting a single stone, we could simply drive the water up and down in an arbitrary fashion (by a sequence of impacts of varying force). The ripples would now constitute a record of the driving sequence and these would then be reproduced some distance away in a detector (our floating object bobbing up and down in sympathy with the wave as it passes). In this way, we could transfer information between the source and the detector by use of a suitable code. Not only can information be transmitted in such a fashion by water waves but, just as importantly, by sound and radio waves. For the generation of arbitrarily shaped waves, the important property is that disturbances of the propagation medium should allow for discontinuous behaviour across a surface in space and time, the wavefront. In the case of a uniform medium, such a surface is of the form $R - ct =$ constant where R is the distance from the source, c is the speed of wave propagation and t is time. The partial differential equations that admit solutions with the requisite property are known as hyperbolic equations, and wave phenomena satisfy equations of this form.

1.2 Electromagnetic Waves

Electromagnetic fields satisfy hyperbolic partial differential equations and therefore exhibit wave phenomena. Electromagnetic waves are generated when charge (the source of electromagnetic fields) accelerates. Static isolated charges cause electric fields that decay at least as fast as $1/R^2$ where R is the distance from the charge. When the charges accelerate, however, they cause a field that only decays as $1/R$. This is a wavelike field that can carry energy and information over vast distances. For example, consider a rearrangement of a system of charge that takes place over a short period of time. Before and after the rearrangement, the charge is static and its field falls off as $1/R^2$. During the rearrangement, however, the field will only fall off as $1/R$. Like the ripples in the pond, the effect of the rearrangement (the $1/R$ field) will travel outward as a ripple in the electromagnetic field (see Figure 1.1). The speed of propagation for this ripple is the speed of light ($c = 3 \times 10^8$ m/s). (Indeed, light is an example of electromagnetic waves.) Such a ripple could be detected through the motions that it induces in a second system of charge.

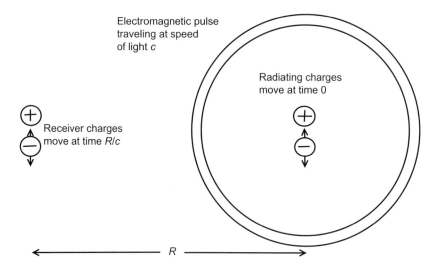

Figure 1.1 Propagation of an electromagnetic pulse.

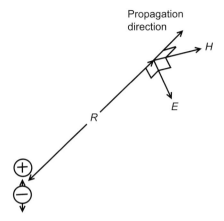

Figure 1.2 The electromagnetic fields caused by accelerating charge.

Unlike the case of static fields, there is always a magnetic wave field H associated with the electric wave field E. The associated magnetic field is proportional to the electric field

$$H = \frac{E}{\eta_o} \tag{1.1}$$

where η_o is the impedance of the free space (377 Ω). The magnetic field is perpendicular to the electric field and also to the direction of wave propagation (see Figure 1.2). Further, the electric wave is also perpendicular to the direction of propagation. Ignoring the static field (this falls off much faster than the wave field), the electric field behaves as

$$E = \frac{K(t - R/c)}{4\pi R} \tag{1.2}$$

where $K(t)$ is a function that depends on the motion of the source charge during rearrangement. Consequently, $K(t)$ will only be non-zero over the period of charge rearrangement, i.e the effect of this rearrangement will be a pulse of electromagnetic field that travels radially outward. Importantly, by suitably controlling the acceleration of the source charge, we can create an arbitrary $K(t)$ and hence transfer information to a distant observer due to the slow fall off in the field of the accelerating charge. This is the basis of radio communication.

If the charges in a system oscillate, they will nearly always be accelerating and so there will nearly always be a wave field. For an oscillation frequency ω (radians per second or $f = \omega/2\pi$ in terms of hertz)

$$K(t) = A \cos(\omega t + \phi) \tag{1.3}$$

where A determines the amplitude of the field and ϕ its phase. The waves produced by such a system will contain virtually no information and so the system needs *modulation* if it is to be used for transferring information. Modulation is achieved by arbitrary variations of A, ϕ and ω. Variations in A are known as amplitude modulation (AM), ω as frequency modulation (FM) and ϕ as phase modulation (PM). Many other forms of modulation can be created by suitable combinations of these basic forms of modulation. Modulated sinusoidal signals are the basis of most radio communications, but modulation causes the signal to spread in frequency around that of the unmodulated sinusoidal *carrier* signal. It is possible, however, for many communications systems to coexist by operating at different frequencies with the variations in modulation limited in order to prevent the signals overlapping. Detection systems are designed to select a particular band of frequencies (the extent of the band being known as the *bandwidth* of the system) when the signal is *modulated*.

In a practical communications system, radio waves will be produced by an electronic source that drives varying current into a structure (usually metal) known as an antenna. Variations in current constitute the accelerating charge that will cause radio waves. A simple antenna, known as a *dipole*, consists of a rod that is driven at its center. Waves travel radially outward from the antenna, and the power density in the waves ($E^2/2\eta_0$ where η_0 is the impedance of free space) will fall away as $1/R^2$. Furthermore, the power will not be uniform in all directions.

The effectiveness of the antenna in a particular direction can be described by its *directivity* in that direction:

$$\text{Directivity} = \frac{\text{power radiated in a particular direction}}{\text{average of power radiated in all directions}} \quad (1.4)$$

Antennas, however, lose some of their energy as heat (in their structure and its surroundings) and so an important quantity is the *antenna efficiency*

$$\text{efficiency } \eta = \frac{\text{total power radiated}}{\text{total power supplied}} \quad (1.5)$$

As a consequence, a more realistic measure of effectiveness is the *gain*

$$\text{gain} = \eta \times \text{directivity} \quad (1.6)$$

No antenna exists that has a uniform gain in all directions and so we quite often represent this variation as a gain pattern. This is a 3D surface whose distance from the origin is the gain of the antenna in that direction. For the dipole, the pattern will be rotationally symmetric about the axis along the dipole's length and so it is sufficient to represent the pattern as a 2D slice through the axis. Figure 1.3 shows a dipole together with its gain pattern and from which it will be noted that there is no gain along the dipole axis.

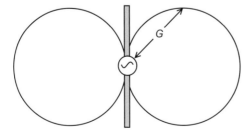

Figure 1.3 A dipole antenna and its gain pattern.

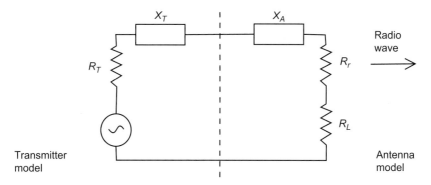

Figure 1.4 Circuit model of a transmitter system.

Radiation is maximum in directions perpendicular to the axis and, for a dipole with an efficiency of 1, this gain will be about 3/2.

An antenna will act as a load to the electronics that drives it; Figure 1.4 shows a circuit model of the transmit antenna and its driver. A dipole is resonant (no reactance X_A in its impedance) when its length is approximately 0.48λ (λ is the wavelength c/f at the operating frequency f). For this reason, a resonant dipole is often known as a half wave dipole. At resonance, the radiation resistance R_r is about 73 Ω and the loss resistance R_L is usually negligible. Dipoles that are much shorter than a wavelength, however, have a much smaller radiation resistance and a large capacitive reactance. For such dipoles, R_L and R_r can often be comparable and this can make them very inefficient.

In general, the amplitude of the electric field of an antenna that is sinusoidally excited has the form

$$E = \frac{\omega \mu_0 I h_{\mathit{eff}}}{4 \pi R} \tag{1.7}$$

where I is the current in the feed, μ_0 is the permeability of free space and h_{eff} is the *effective antenna length*. In the case of a resonant dipole, $h_{\mathit{eff}} \approx 0.64 l \sin \theta$ where l is the dipole length and θ is the angle that the propagation direction makes with the dipole axis.

An antenna can be used to extract energy from an electric field since an incident electromagnetic wave will set charges in motion and hence cause current to flow on the antenna structure. In this receive mode, an electric field E will induce an open circuit voltage $h_{\mathit{eff}} E$ in the antenna terminals. An antenna will act as a source to the electronics that acts as the radio receiver; Figure 1.5 shows a circuit model of the receive antenna and its receiver load. It will be noted that the antenna exhibits the same impedance $R_r + R_L + jX_A$ in both receive and transmit functions.

1.3 Communications Systems

Radio waves are used to communicate information over both large and small distances without the use of wires, hence the term *wireless*. The effectiveness of a communication

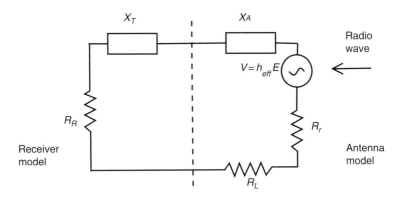

Figure 1.5 Circuit model of a receiver system.

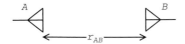

Figure 1.6 Communication system with two stations.

system is often calculated using the *Friis* equation. If the transmitter and a receiver are distance r_{AB} apart, the received power P_B is related to the transmitted power P_A by

$$P_B = P_A G_A G_B \left(\frac{\lambda}{4\pi r_{AB}} \right)^2 \qquad (1.8)$$

where G_A and G_B are the gains of the receive and transmit antennas, respectively. The last term represents the decay in the power of the wave as it spreads out from the source, and this *spreading loss* is defined to be $L_{sprd} = (4\pi r_{AB}/\lambda)^2$ (i.e. $P_B = P_A G_A G_B / L_{sprd}$). It should be noted that this loss is often quoted in terms of decibels, i.e $10 \log_{10} (L_{sprd})$.

In a radio system, the intentional radio signals must compete with *noise* (unwanted interfering signals) that is generated within the electronics of the system and within the propagation environment. Even a simple resistor will create a noise voltage v_n due to the random thermal motion of its electrons. Such noise has rms voltage

$$\overline{v_n^2} = 4kTBR \qquad (1.9)$$

where T (in Kelvin) is the absolute temperature, B (in hertz) is the bandwidth of the radio channel, R (in ohms) is the resistance and k is the Boltzmann constant (1.38×10^{-23} joules per Kelvin). Semiconductors are the source of many different sorts of noise, and so radio receivers will have a variety of contributions to their *internal noise*. In a real radio receiving system, *external noise* is just as important as the internal variety. This can arise from manmade sources (ignition interference, for example) and natural sources (lightning, for example). Consequently, the input signal will normally need to be at a level above that of the combined internal and external noise. In the case of an antenna, the external noise can be considered as that due to resistance $R_r + R_L$, but at a temperature T_A (the *antenna temperature*) that is possibly different from the ambient temperature (around 290 K). External noise is the ultimate constraint, and, for best performance, a radio receiver should be *externally noise limited* (i.e. the internal noise is below the level of external noise).

The crucial quantity in calculating radio system performance is the signal to noise ratio (SNR) that is required for detection. SNR is defined by

$$\text{SNR} = \frac{S}{N} = \frac{\text{signal power}}{\text{noise power}} \qquad (1.10)$$

It is necessary to make sure that the strength of the received is sufficient to make the SNR greater than that required for detection. Obviously, this will require the correct choice of transmit power and/or antenna gains.

1.4 Cellular Radio

Although we can accommodate many users by dividing the radio spectrum into many channels (slices of spectrum with limited bandwidth), this still fails to satisfy the modern demands for high volume personnel communications (including video and internet). A solution has been found by limiting the range coverage of a channel so its frequency can be reused at some other location. The full communication region is divided into small cells within which the transmit power is limited so that the users within any one cell can only communicate with the radio base station (RBS) for that cell. In this manner the same set of frequencies can be used in many different cells, provided that they are sufficiently isolated. The RBSs within the network are all connected to each other, and the fixed network, through a mobile switching center (MSC). When a user passes from one cell to another, control is passed to the RBS associated with the new cell and the frequency will appropriately change. Such a system is known as *cellular radio*.

A cellular radio system is designed around a *cluster* of cells, each cell in the cluster having a unique frequency set. A typical cluster consists of seven cells; Figure 1.7 shows a cellular system based on such a cluster (frequency sets are labeled A to G). It can be seen that the cluster topology allows reuse of the frequency sets within separate clusters. The reuse, however, has the potential to cause intercellular interference

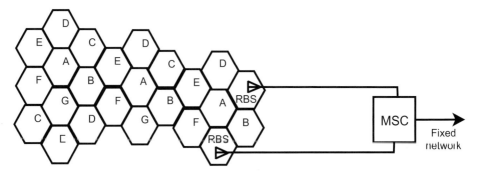

Figure 1.7 A cellular radio system.

(the dominant source of unwanted interfering signals in cellular systems). From Figure 1.7, it will be noted that the minimum distance between cells with the same frequency set is approximately $4.583R$ where R is the cell radius. On the assumption that all transmitters use the same power level, and that the power decays as $(1/distance)^n$, the *signal to interference ratio* (SIR) is given by

$$\text{SIR} = \frac{\text{minimum power within a cell}}{6 \times \text{maximum power between cells}} = \frac{R^{-n}}{6(4.583R)^{-n}} = \frac{4.583^n}{6} \quad (1.11)$$

From the Friis equation, it will be noted that a system in free space will have a value of 2 for n. There are, however, many factors that can modify free space propagation, and it is found, in practice, that a value of n around 4 is more appropriate. For the configuration of Figure 1.7 this would suggest an SIR of greater than 10 dB.

1.5 Radar Systems

One of the major non-communications applications of radio waves is *radar* (radio detection and ranging). In classical radar, the transmitted signal is interrupted by a *target* from which a small amount of energy is re-radiated back to a receiver. The receiver will normally ascertain the direction of the target using a steerable beam (mechanical or electronic steering) and the time of flight of the signal will then provide the target range. For radar systems, the power returned P_R is related to that transmitted P_T by the *radar equation*,

$$P_R = P_T G_R G_T \left(\frac{\lambda}{4\pi R_T}\right)^2 \left(\frac{\lambda}{4\pi R_R}\right)^2 \frac{4\pi \sigma}{\lambda^2} \quad (1.12)$$

where R_T and R_R are the ranges of the target from the transmitter and receiver, respectively, and G_T and G_R are the gains of the transmit and receive antennas, respectively (see Figure 1.8). σ is the *radar cross section* of the target and represents the amount of power reflected when a field with unit power per unit area is incident. The radar equation can be regarded as the double application of the Friis equation with the target acting as both receiver and transmitter. The term $4\pi\sigma/\lambda^2$ is effectively the product of the target receive and transmit gains. Radar cross sections can be quite complex, often

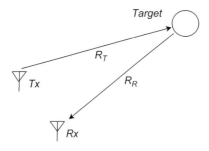

Figure 1.8 A general radar configuration.

depending on both the direction of the illuminator and the direction of the receiver. For a metallic sphere the back-scatter cross section is approximately direction independent with a value πa^2 when the radius a is greater than a wavelength.

One of the major problems with a radar system is that surfaces such as a rough sea and ground can also return a considerable amount of power, known as *clutter*, which can mask a radar target. This interference will be in addition to the noise we have discussed for communications systems. Consequently, the signal to clutter ratio (SCR) can be just as important as SNR in determining radar performance. Fortunately, the motion of the target will cause a shift in frequency for the return signal from the target and so the target can be discriminated from the clutter in the frequency domain. The frequency shift is known as the *Doppler shift* and is related to the target dynamics through

$$\Delta f = -\frac{f}{c}\left(\frac{dR_T}{dt} + \frac{dR_R}{dt}\right) \quad (1.13)$$

Radar systems can effectively be regarded as radio systems in which the environment modulates the signal, hence allowing a radar operator to glean information about the environment. The most obvious application of radar is in the detection of ships and aircraft. However, we increasingly use radar to monitor things such as weather (wind profiling radar, for example) and underground conditions (ground penetrating radar, for example).

1.6 Complex Propagation

The Friis equation assumes there to be no interaction of the radio waves with their environment. At a minimum, however, the ground will reflect radio waves and these reflected waves provide additional paths for the waves between the antennas of a communication system (see Figure 1.8). The waves on the direct and reflected paths will travel different distances and so there will be the potential for interference at the receiver. If the antennas are at heights H_A and H_B above the reflection point, the Friis equation will require modification to

$$P_B = P_A G_A G_B \left(\frac{\lambda}{4\pi r_{AB}}\right)^2 \left|1 + R\exp\left(-2j\frac{\beta H_A H_B}{r_{AB}}\right)\right|^2 \quad (1.14)$$

where R is the reflection coefficient of the ground (approximately -1 at frequencies above a few hundred MHz). From this expression it can be seen that there can be constructive, or destructive, interference between direct and reflected signals that will

Figure 1.9 Propagation with ground reflections.

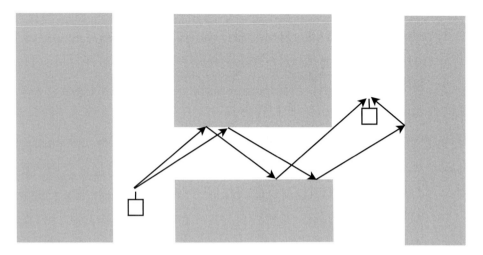

Figure 1.10 Propagation through an urban environment.

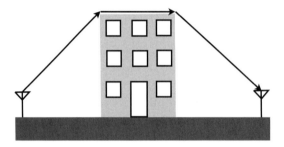

Figure 1.11 Diffractive propagation over a building.

vary with distance between transmitter and receiver. The effect is often pronounced in mobile communications where it can cause the signal to experience *flutter* as the vehicle moves. This can get worse in urban environments where the paths can be more tortuous and numerous (see Figure 1.10). The reflected signals, however, can have a positive effect in that they can allow communication into a *shadow* region (a region where no line-of-sight path is available).

Even without reflections, non-line-of-sight propagation is still possible through a mechanism known as *diffraction*. Consider the propagation of radio waves over a building, as illustrated in Figure 1.11. A small amount of energy will be diffracted into the *shadow* region at the top of the building and then a small amount of this power will be diffracted into the shadow region over the other side of the building. Such power can sometimes be sufficient for communication purposes.

Diffraction, and many other propagation phenomena, can be explained through *Huygens' principle*:

Each point on the wavefront of a general wave can be considered as the source of a secondary spherical wave. A subsequent wavefront of the general wave can then be constructed as the envelope of secondary wavefronts.

1.6 Complex Propagation

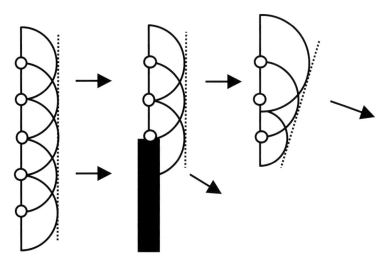

Figure 1.12 Huygens' principle applied to diffraction and refraction.

We will look at diffraction over a screen, as illustrated in Figure 1.12. To the left of the screen there is a plane wavefront that, according to Huygens' principle, can considered as a series of sources from which spherical waves emanate. After time δt, the wavefront (the envelope of the spherical wavefronts) will have moved distance $c\delta t$ from the sources. Eventually, this wavefront reaches the screen and here there will be a series of sources above the screen that act as sources for waves to the right. On the screen, however, there will be no sources for waves to the right (potential sources have been blocked by the screen). Above the screen, the new wavefront to the right will consist of an envelope that is a plane. Below the top of the screen, however, the wavefront to the right will be that of the source just above the screen. This is the wavefront of the diffracted wave that penetrates into the shadow region behind the screen.

Diffraction and reflection are not the only mechanisms whereby radio waves can access regions that are not available through line-of-sight propagation. Propagation media, other than a vacuum, can have propagation speeds that vary with location and this can cause radio signals to bend around obstacles through a process known as *refraction*. This phenomena can also be understood through Huygens' principle. In Figure 1.12, the effect of a spatially varying wave speed is illustrated to the right of the screen. The effect of a vertically increasing speed is to cause the spherical wavefronts to increase in radius, thus causing the wavefront to tilt downward. Such a mechanism can also cause propagation into a shadow region. The variation in propagation direction is described by *Snell's law* (see Figure 1.13) according to which

$$\frac{\sin\theta}{\sin\theta_A} = \frac{c}{c_A} \qquad (1.15)$$

where θ is the angle of this direction to the vertical. Angle θ_A is the angle of propagation at some reference level and c_A is the propagation speed at this level.

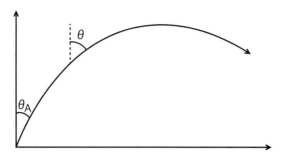

Figure 1.13 Variation of propagation direction caused by a continuously varying propagation speed.

Figure 1.14 Ducting propagation in the atmosphere (wave speed increasing with height).

An example of refraction occurs in the atmosphere near Earth's surface where, under certain meteorological conditions, the atmosphere can present a propagation speed that increases with height. Normally communications would be limited by the horizon but, in this case, the radio waves can be bent back to the ground beyond the horizon. The wave can then be reflected at the ground and bent back at even further ranges (see Figure 1.14). In this way propagation can be *ducted* over great distances. A further example of such ducting propagation is caused by the *ionosphere*. The ionosphere is an ionized layer several hundred kilometers above the Earth that is caused by solar radiation (see Figure 1.15). For frequencies below about 30 MHz, this layer can present a medium with wave speed increasing with height and hence causing radio waves to be refracted back toward the ground. Such propagation, together with reflections at the ground, can thus provide reliable radio communications on a global scale.

Another important propagation mechanism is that of scatter. Consider an electromagnetic wave normally incident upon a plane screen (see Figure 1.16). There will be some diffraction into the shadow region behind the plate and direct reflections at the surface of the screen. By Huygens' principle, however, we can represent the reflected wave by sources on the screen. Just in front of the screen the wavefronts will be plane but, above the screen, wavefront will be that of the topmost source on the screen. Consequently, energy will be scattered backward into the region above the plate. This type of scatter is sometimes known as *back scatter* (diffraction is sometimes known as *forward scatter*).

Propagation media can often be subject to irregularity in their properties and this can generate a large number of scattering surfaces. Even in line-of-sight communication,

1.6 Complex Propagation

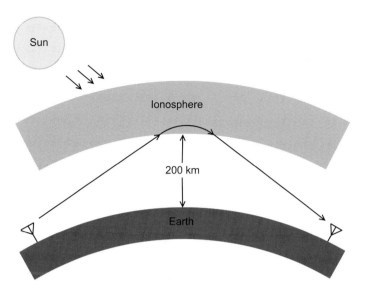

Figure 1.15 Propagation in the ionosphere.

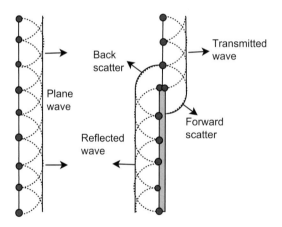

Figure 1.16 The scattering caused by a flat plate.

these scatterers can have a significant effect on the quality of communication. As shown in Figure 1.17, there will be scatter signal paths besides the direct path and this will cause a spread in transmission delay. The additional signals can also interfere with the main signal and hence cause fluctuations in signal level when there is relative motion between the irregularity and the line of sight. This phenomenon, known as *scintillation*, is most evident in satellite communications and astronomical observations (the twinkle of stars). Irregularity can be caused by turbulence in the ionosphere as well as the atmosphere. In addition, weather phenomena (such as rain, hail and snow) can also cause fluctuations in atmospheric properties and this can greatly affect propagation at frequencies in the high GHz range resulting in severe signal loss.

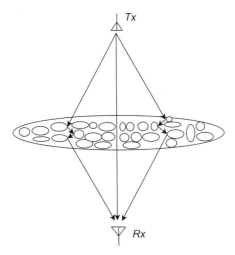

Figure 1.17 Scintillation caused by permittivity irregularity.

As can be seen from the above considerations, the understanding of radio wave propagation is of fundamental importance to the analysis and modeling of radio system performance. For example, cellular radio systems depend on the isolation of cells for their effective operation and the available propagation between cells is an important issue for the design of a cellular network. In the case of over-the-horizon radio systems, these can only work through complex propagation mechanisms, and the understanding of such mechanisms is of paramount importance in their design. The following chapters will consider, in detail, the techniques for analyzing the complexities of radio wave propagation and their application to the prediction of radio system performance. The problem of predicting performance is usually split into the analysis of the hardware (transmitters, receivers and antennas) and the analysis of radio wave propagation between antennas. In reality, the antennas and propagation are part of one grand electromagnetic system. However, the antennas and propagation can usually be treated in isolation, and so, in the next chapter, we will introduce some important antenna concepts in order to relate them to the propagation that they produce.

1.7 References

L. Barclay (editor), *Propagation of Radio Waves*, Institution of Electrical Engineers, London, 2003.

N. Blaunstein, *Radio Propagation in Cellular Networks*, Artech House, Boston, 2000.

C.J. Coleman, *An Introduction to Radio Frequency Engineering*, Cambridge University Press, Cambridge, 2004.

S.R. Saunders, *Antennas and Propagation for Wireless Communication Systems*, John Wiley, Chichester, 1999.

2 The Fundamentals of Electromagnetic Waves

The equations that describe electromagnetic fields are due to James Clerk Maxwell and were the culmination of the discoveries of many different scientists, particularly during the eighteenth and nineteenth centuries. The final form of the equations required the modification of previously existing laws, but led to a total description of electromagnetism that has stood the test of time. Indeed, Maxwell's equations were some of the few that came through the revolution of relativity unscathed. Further attesting to their correctness, they predicted radio waves and showed light to be an electromagnetic phenomenon. The following chapter develops some of the fundamental ideas of electromagnetic waves, starting with Maxwell's equations. In addition, it introduces some ideas concerning antennas that are of importance to the development of the propagation aspect of radio waves.

2.1 Maxwell's Equations

The more popular form of Maxwell's equations is

$$\nabla \cdot \mathcal{B} = 0 \tag{2.1}$$

$$\nabla \cdot \mathcal{D} = \rho \tag{2.2}$$

$$\nabla \times \mathcal{E} = -\frac{\partial \mathcal{B}}{\partial t} \tag{2.3}$$

$$\nabla \times \mathcal{H} = \frac{\partial \mathcal{D}}{\partial t} + \mathcal{J} \tag{2.4}$$

where $\mathcal{E}, \mathcal{D}, \mathcal{B}, \mathcal{H}$ are the electric intensity (V/m), electric flux density (C/m^2), magnetic flux density (W/m^2) and magnetic intensity (A/m), respectively. The field sources are the electric charge distribution ρ (C/m^3) and the current distribution $\mathcal{J}$ ($\mathcal{J} = \rho \mathbf{v}$ where $\mathbf{v}$ is the velocity field of the charge) given in A/m^2. For an isotropic dielectric material,

$$\mathcal{B} = \mu \mathcal{H} \tag{2.5}$$

$$\mathcal{D} = \epsilon \mathcal{E} \tag{2.6}$$

where μ and ϵ are permeability and permittivity (sometimes known as the *dielectric constant*) of the material. These quantities can be tensor in nature if the propagation medium is anisotropic, but we shall initially assume the medium to be isotropic and hence these quantities are scalars. If ϵ_0 is the permittivity of free space (8.854×10^{-12} F/m), we define the relative permittivity ϵ_r by $\epsilon = \epsilon_r \epsilon_0$ and, if μ_0 is the permeability of free space ($4\pi \times 10^{-7}$ H/m), we define the relative permeability μ_r by $\mu = \mu_r \mu_0$. The field sources will satisfy the charge continuity equation

$$\frac{\partial \rho}{\partial t} + \nabla \cdot \mathcal{J} = 0 \qquad (2.7)$$

since charge cannot be created or destroyed.

In a radio system, the sources are usually currents generated electronically and fed into a metallic structure known as an antenna. Accelerating charge in the currents that travel across the antenna then cause electromagnetic waves that carry energy away from the antenna. Current consists of charge in motion and the effect of the electromagnetic field on an individual charge q will be the *Lorentz force*

$$\mathcal{F} = q(\mathcal{E} + \mathbf{v} \times \mathcal{B}) \qquad (2.8)$$

when the charge travels with velocity **v**. Consequently, since antennas are usually constructed of highly conducting material, the free electrons will be set in motion by the radio wave that is incident on the antenna. In this way, an antenna can extract energy from an incident electromagnetic wave and act as a source to the electronics of a radio receiver. In general, some current will flow in most materials when subjected to an electromagnetic field. This current is usually well approximated by the generalized *Ohm's law*. This law states that an electric field $\mathcal{E}$ will cause a current density $\mathcal{J} = \sigma \mathcal{E}$, where σ (units of S/m) is the *conductivity* of the material.

When current is driven into an antenna, some of the energy will be dissipated in the antenna structure, some will be stored in the field close to the antenna and some will be carried away as electromagnetic waves. From Maxwell's equations, and the divergence theorem, it can be shown that the power $\mathcal{P}_S$ delivered by an antenna within a volume V is related to these different mechanisms through

$$\mathcal{P}_S = \int_V \mathcal{J} \cdot \mathcal{E} dV \qquad (2.9)$$

$$= \int_V \sigma \mathcal{E} \cdot \mathcal{E} dV + \frac{\partial}{\partial t} \int_V \left(\frac{\mathcal{D} \cdot \mathcal{E}}{2} + \frac{\mathcal{H} \cdot \mathcal{B}}{2} \right) dV + \oint_S (\mathcal{E} \times \mathcal{H}) \cdot d\mathbf{S} \qquad (2.10)$$

where S is the surface of volume V. In order, the terms on the right represent the energy dissipated in the antenna and its surrounds (i.e. *Ohmic loss*), the rate of change of field energy $\mathcal{W} \left(= \frac{1}{2} \int_V (\epsilon \mathcal{E} \cdot \mathcal{E} + \mu \mathcal{H} \cdot \mathcal{H}) dV \right)$ and the rate of flow of wave energy out of the volume.

The vector

$$\mathcal{P} = \mathcal{E} \times \mathcal{H} \qquad (2.11)$$

is known as the *Poynting vector* and is the instantaneous rate of energy flow across a unit area that is perpendicular to $\mathcal{P}$.

2.2 Plane Electromagnetic Waves

The above form of Maxwell's equations does not immediately indicate that they will support wavelike phenomena. However, we will now show that electromagnetic fields satisfy the wave equation, the form of hyperbolic equation that is normally associated with waves. Consider an isotropic homogenous medium, then Maxwell's equations imply that

$$\nabla \times \mathcal{E} = -\mu \frac{\partial \mathcal{H}}{\partial t} \tag{2.12}$$

$$\nabla \cdot \mathcal{E} = 0 \tag{2.13}$$

$$\nabla \times \mathcal{H} = \epsilon \frac{\partial \mathcal{E}}{\partial t} \tag{2.14}$$

and

$$\nabla \cdot \mathcal{H} = 0 \tag{2.15}$$

for points outside the field sources. Taking the curl of Equation 2.12 and noting Equation 2.14,

$$\nabla \times \nabla \times \mathcal{E} = -\mu \frac{\partial}{\partial t} \nabla \times \mathcal{H}$$

$$= -\mu \epsilon \frac{\partial^2 \mathcal{E}}{\partial t^2} \tag{2.16}$$

Furthermore, noting the vector identity $\nabla \times (\nabla \times \mathbf{G}) = \nabla(\nabla \cdot \mathbf{G}) - \nabla^2 \mathbf{G}$ and Equation 2.13, we obtain that

$$\nabla^2 \mathcal{E} - \frac{1}{c^2} \frac{\partial^2 \mathcal{E}}{\partial t^2} = 0 \tag{2.17}$$

where $c = (\mu \epsilon)^{-\frac{1}{2}}$. The quantity c is the speed of wave propagation and was found to be exactly that of light. This discovery was important evidence that the phenomena of light was electromagnetic in nature. In a similar fashion to the electric field, the magnetic field can also be shown to satisfy a wave equation

$$\nabla^2 \mathcal{H} - \frac{1}{c^2} \frac{\partial^2 \mathcal{H}}{\partial t^2} = 0 \tag{2.18}$$

Although fields $\mathcal{E}$ and $\mathcal{H}$ must satisfy wave equations, these equations alone do not uniquely specify them as they do not incorporate the coupling between these fields.

The above wave equations have what is known as plane wave solutions (solutions that are constant in any plane perpendicular to the propagation direction). Although these waves cannot exist in reality, they do form an effective approximation at great distances from sources where the curvature of the wavefront is small. Plane waves have the form

$$\mathcal{E} = \mathcal{E}\left(t - \frac{\mathbf{r} \cdot \hat{\mathbf{p}}}{c}\right) \tag{2.19}$$

and
$$\mathcal{H} = \mathcal{H}\left(t - \frac{\mathbf{r} \cdot \hat{\mathbf{p}}}{c}\right) \tag{2.20}$$

where $\hat{\mathbf{p}}$ is a unit vector in the propagation direction. From the Maxwell Equations 2.12 and 2.14,

$$-\frac{1}{c}\hat{\mathbf{p}} \times \frac{\partial \mathcal{E}}{\partial t} = -\mu \frac{\partial \mathcal{H}}{\partial t} \tag{2.21}$$

and

$$-\frac{1}{c}\hat{\mathbf{p}} \times \frac{\partial \mathcal{H}}{\partial t} = \epsilon \frac{\partial \mathcal{E}}{\partial t} \tag{2.22}$$

Then, on integrating these equations with respect to time,

$$\mathcal{H} = \frac{\hat{\mathbf{p}} \times \mathcal{E}}{\mu c} \tag{2.23}$$

and

$$\mathcal{E} = -\frac{\hat{\mathbf{p}} \times \mathcal{H}}{\epsilon c} \tag{2.24}$$

From the Maxwell Equations 2.13 and 2.15, we will also have

$$\hat{\mathbf{p}} \cdot \mathcal{E} = \hat{\mathbf{p}} \cdot \mathcal{H} = 0 \tag{2.25}$$

Bringing these results together, we see that $\mathcal{E}$ can be an arbitrary vector function of $t - \mathbf{r} \cdot \hat{\mathbf{p}}/c$ providing that $\mathcal{E}$ is orthogonal to the propagation direction $\hat{\mathbf{p}}$. The corresponding magnetic field is then given by

$$\mathcal{H} = \frac{\hat{\mathbf{p}} \times \mathcal{E}}{\eta} \tag{2.26}$$

where $\eta = \sqrt{\mu/\epsilon}$ is the *impedance of the propagation medium*. The Poynting vector for the above field will take the form

$$\frac{\hat{\mathbf{p}}}{\eta} \mathcal{E} \cdot \mathcal{E} \tag{2.27}$$

from which it can be seen that the plane wave will transport energy in the propagation direction (Figure 2.1).

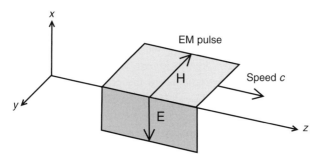

Figure 2.1 A plane wave consisting of an electromagnetic pulse.

2.2 Plane Electromagnetic Waves

An important specialization of the electromagnetic field is the case where it oscillates in time on a single frequency (a *time harmonic* field), i.e.

$$\mathcal{H}(\mathbf{r},t) = \Re\left\{\mathbf{H}(\mathbf{r})\exp(j\omega t)\right\} \tag{2.28}$$

and

$$\mathcal{E}(\mathbf{r},t) = \Re\left\{\mathbf{E}(\mathbf{r})\exp(j\omega t)\right\} \tag{2.29}$$

with nonitalic capital letters representing the fields with the $\exp(j\omega t)$ term factored out. (In theory, through Fourier transform techniques, a general electromagnetic field can be analyzed in terms of time harmonic fields.) The time harmonic fields $\mathbf{H}(\mathbf{r})$ and $\mathbf{E}(\mathbf{r})$ will satisfy the equations

$$\nabla \times \mathbf{E} = -j\omega\mu\mathbf{H} \tag{2.30}$$

and

$$\nabla \times \mathbf{H} = j\omega\epsilon\mathbf{E} + \mathbf{J} \tag{2.31}$$

where $\mathbf{J} = \Re\left\{\mathbf{J}e^{j\omega t}\right\}$. (It should be noted that the divergence Maxwell equations are automatically satisfied providing the time harmonic continuity equation $j\omega\rho + \nabla \cdot \mathbf{J} = 0$ is satisfied.) For a time harmonic field, we measure the energy flow in terms of the average flow over a wave period, giving the *time harmonic Poynting vector*

$$\mathbf{P} = \frac{1}{2}\Re\left\{\mathbf{E} \times \mathbf{H}^*\right\} \tag{2.32}$$

A time harmonic plane wave has the form

$$\mathcal{E} = \Re\left\{\mathbf{E}_0 \exp\left(j\omega\left(t - \frac{\hat{\mathbf{p}} \cdot \mathbf{r}}{c}\right)\right)\right\} \tag{2.33}$$

and

$$\mathcal{H} = \Re\left\{\mathbf{H}_0 \exp\left(j\omega\left(t - \frac{\hat{\mathbf{p}} \cdot \mathbf{r}}{c}\right)\right)\right\} \tag{2.34}$$

where $\hat{\mathbf{p}}$ is the unit vector in the direction of propagation. Fields $\mathbf{E}_0$ and $\mathbf{H}_0$ will be related through

$$\mathbf{H}_0 = \frac{\hat{\mathbf{p}} \times \mathbf{E}_0}{\eta} \tag{2.35}$$

and $\mathbf{E}_0$ will satisfy $\hat{\mathbf{p}} \cdot \mathbf{E}_0 = 0$. From these relations, it is obvious that $\mathbf{H}_0$, $\mathbf{E}_0$ and $\hat{\mathbf{p}}$ form a mutually orthogonal triad. For a plane wave, the time harmonic Poynting vector simplifies to

$$\mathbf{P} = \frac{\hat{\mathbf{p}}}{2\eta}\mathbf{E} \cdot \mathbf{E}^* \tag{2.36}$$

Without loss of generality, we now consider propagation in the z direction ($\hat{\mathbf{p}} = \hat{\mathbf{z}}$), then

$$\mathbf{E} = \mathbf{E}_0 \exp(-j\beta z) \tag{2.37}$$

where $\beta = \omega/c$ is the propagation constant ($\beta = 2\pi/\lambda$ in terms of wavelength λ) and $\mathbf{E}_0$ has no component in the $\hat{\mathbf{z}}$ direction. We can express $\mathbf{E}$, up to an arbitrary phase, as

$$\mathbf{E} = E_0 \left(\cos\gamma \hat{\mathbf{x}} + \sin\gamma \exp(j\delta)\hat{\mathbf{y}} \right) \exp(-j\beta z) \tag{2.38}$$

For the case that $\delta = 0$, the $\hat{\mathbf{x}}$ and $\hat{\mathbf{y}}$ components will be in phase and we have what is known as *linear polarization*. In this case, the electric field $\Re\{\mathbf{E}e^{j\omega t}\}$ will trace out a finite length line at an angle γ to the $\hat{\mathbf{x}}$ direction. For the case that $\delta \neq 0$, the electric field vector will rotate about the propagation direction. When $\gamma = 45°$ and either $\delta = -90°$ or $\delta = 90°$, we have the important special case of *circular polarization*. In this case,

$$\mathcal{E}_{\pm}(\mathbf{r},t) = \frac{E_0}{\sqrt{2}} \left(\cos(\omega t - \beta z)\hat{\mathbf{x}} \pm \sin(\omega(t - \beta z)\hat{\mathbf{y}} \right) \tag{2.39}$$

and the electric field traces out a circle about the propagation direction. For the circle traced in the anticlockwise direction ($\delta = -90°$) we have right-hand circular polarization and in the clockwise direction ($\delta = 90°$) we have left-hand circular polarization. Any linear polarization can be expressed as a combination of left- and right-hand circular polarizations. For a wave propagating in the $\hat{\mathbf{z}}$ direction

$$\mathbf{E} = E_0 \hat{\mathbf{x}} e^{-j\beta z} \tag{2.40}$$

$$= \mathbf{E}_+ + \mathbf{E}_- \tag{2.41}$$

$$= \frac{E_0}{2}(\hat{\mathbf{x}} - j\hat{\mathbf{y}})\exp(-j\beta z) + \frac{E_0}{2}(\hat{\mathbf{x}} + j\hat{\mathbf{y}})\exp(-j\beta z) \tag{2.42}$$

as illustrated in Figure 2.2. In between the cases of circular and linear polarization (i.e. δ not a multiple of $90°$), the electric field vector will trace out an ellipse and so we will have *elliptical polarization*. The electric field is now given by

$$\mathcal{E}(\mathbf{r},t) = \mathcal{E}_x \hat{\mathbf{x}} + \mathcal{E}_y \hat{\mathbf{y}} = \Re \left\{ E_0(\cos\gamma \cos(\omega t - \beta z)\hat{\mathbf{x}} + \sin\gamma \cos(\omega(t - \beta z + \delta))\hat{\mathbf{y}}) \right\} \tag{2.43}$$

and so the ellipse will have the equation

$$\tan^2\gamma \mathcal{E}_x^2 + \mathcal{E}_y^2 - 2\tan\gamma \cos\delta \mathcal{E}_x \mathcal{E}_y = 2E_0^2 \sin^2\gamma \sin^2\delta \tag{2.44}$$

For a medium with finite conductivity σ, the time harmonic electric field $\mathbf{E}$ will cause a time harmonic current $\mathbf{J} = \sigma \mathbf{E}$ to flow and the time harmonic Maxwell equations can now be expressed as

$$\nabla \times \mathbf{E} = -j\omega\mu\mathbf{H} \tag{2.45}$$

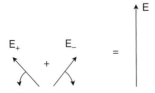

Figure 2.2 The decomposition of linear polarization into circularly polarized modes.

and
$$\nabla \times \mathbf{H} = j\omega \epsilon_{\text{eff}} \mathbf{E} \tag{2.46}$$

where $\epsilon_{\text{eff}} = \epsilon - j\sigma/\omega$ is known as the *effective permittivity* of the medium. In the case of a plane wave, the impedance and propagation constant will now have the form

$$\eta = \sqrt{\frac{\mu}{\epsilon_{\text{eff}}}} = \sqrt{\frac{\mu}{\epsilon}} \frac{1}{\sqrt{1 - \frac{j\sigma}{\omega\epsilon}}} \tag{2.47}$$

and

$$\beta = \omega\sqrt{\mu\epsilon_{\text{eff}}} = \frac{\omega}{c}\sqrt{1 - \frac{j\sigma}{\omega\epsilon}} \tag{2.48}$$

from which we see that both η and β are complex. If the wave is traveling in the z direction, we will now have

$$\mathcal{E} = \exp(\Im\{\beta\}z) \, \Re\{\mathbf{E}_0 \exp j(\omega t - \Re\{\beta\}z)\} \tag{2.49}$$

and

$$\mathcal{H} = \exp(\Im\{\beta\}z) \, \Re\left\{\frac{\hat{\mathbf{z}} \times \mathbf{E}_0}{\eta} \exp j(\omega t - \Re\{\beta\}z)\right\} \tag{2.50}$$

where $\mathbf{E}_0$ is the vector amplitude of the electric field. In the case that the frequency is high, or the conductivity low, $\Im\{\beta\} \approx -\sigma\eta/2$ and it is clear that wave suffers attenuation as it propagates (note that $\alpha = -\Im\{\beta\}$ is sometimes known as the *attenuation constant*). It should also be noted that the real part of β is frequency dependent and so the speed of the phase front $v_p = \omega/\Re\{\beta\}$ is also frequency dependent. When the phase speed is frequency dependent, the medium is said to be *dispersive*. Dispersive media can have serious consequences for communications systems. In particular, a modulated signal will have components over a range of frequencies and these can propagate at different speeds. As a consequence, the modulation can lose its integrity after the signal has propagated a sufficient distance (as illustrated in Figure 2.3).

For a modulated signal in a dispersive medium, the concept of speed becomes more complex and we need to introduce the concept of group speed (the speed at which the modulation propagates). In the case of a pulse, this can be interpreted as the speed of the

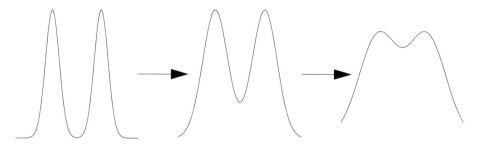

Figure 2.3 Loss of integrity due to dispersion.

pulse peak. Consider the case of a general modulating signal $m(t)$ that has bandwidth B. The complex spectrum of this *baseband* signal will be given by

$$M(f) = \frac{1}{2\pi} \int_{-B}^{B} m(t) \exp(-2\pi j f t) dt \qquad (2.51)$$

Then, if we modulate a carrier signal of frequency f_c, the complex transmitted signal will take the form

$$s_T(t) = \exp(2\pi j f_c t) m(t) \qquad (2.52)$$

$$= \int_{-B}^{B} M(f) \exp(2\pi j (f + f_c) t) df \qquad (2.53)$$

Consider the propagation of a modulated signal from the transmit location T to the receive location R (a distance D). This signal will be made up of a combination of sinusoidal signals on different frequencies. On frequency $f + f_c$, let $\tau_p = D/v_p$ be the time a phase front takes to travel between T and R (this is known as the *phase delay*). The phase delay, however, will be frequency dependent in a dispersive medium (i.e. $\tau_p = \tau_p(f + f_c)$) and so the received signal will take the form

$$s_R(t) = \int_{-B}^{B} M(f) \exp(2\pi j (f + f_c)(t - \tau_p(f + f_c))) df \qquad (2.54)$$

Assume that $f_c \gg B$, then

$$\tau_p(f + f_c) \approx \tau_p + f \frac{\partial \tau_p}{\partial f} \qquad (2.55)$$

with τ_p and $\partial \tau_p / \partial f$ evaluated at frequency f_c. Consequently

$$(f + f_c)(t - \tau_p(f + f_c)) \approx f_c t - f_c \tau_p + f(t - \tau_g) \qquad (2.56)$$

where $\tau_g = \tau_p + f_c \partial \tau_p / \partial f$. We therefore have that

$$s_R(t) \approx \int_{-B}^{B} M(f) \exp(2\pi j (f_c t - f_c \tau_p + f(t - \tau_g))) df \qquad (2.57)$$

$$= \exp(2\pi j f_c (t - \tau_p)) \int_{-B}^{B} M(f) \exp(2\pi j f(t - \tau_g)) df \qquad (2.58)$$

i.e.

$$s_R(t) = \exp(2\pi j f_c (t - \tau_p)) m(t - \tau_g) \qquad (2.59)$$

The quantity τ_p is known as the *group delay* since this is the delay in the modulation during transit from T to R. It should be noted that the quantities $c_0 \tau_p$ and $c_0 \tau_g$ are often referred to as the *phase distance* and *group distance*, respectively.

2.3 Plane Waves in Anisotropic Media

As mentioned previously, electromagnetic media can exhibit anisotropic behavior and this will manifest itself as a relative permittivity that is tensor in nature. A particularly important example of this is Earth's ionosphere, a plasma layer hundreds of kilometers

2.3 Plane Waves in Anisotropic Media

above Earth that is caused by the sun's radiation. Within the ionospheric plasma, the electrons are free to move and so, if an electromagnetic field is present, they will experience a Lorentz force. This force causes the electrons to move according to

$$m \frac{d\mathbf{v}}{dt} = e\mathcal{E} + e\mathbf{v} \times \mathbf{B}_0 - m\nu\mathbf{v} \tag{2.60}$$

where $\mathbf{v}$ is the velocity of the electron, m is its mass, e is its charge, ν is the electron collision frequency and $\mathbf{B}_0$ is Earth's magnetic field (assumed to be the dominant magnetic field). We will assume the fields to be time harmonic and so the electron motion will also be time harmonic. Consequently, the derivative with respect to time can be replaced by the factor $j\omega$. Further, since the electron current density will be given by $\mathcal{J} = eN_e\mathbf{v}$ where N_e is the electron density, Equation 2.60 will imply

$$-j\omega X \epsilon_0 \mathbf{E} = U\mathbf{J} + j\mathbf{Y} \times \mathbf{J} \tag{2.61}$$

where $U = 1 - \frac{j\nu}{\omega}$, $X = \omega_p^2/\omega^2$, $\mathbf{Y} = (\omega_H/\omega)(\mathbf{B}_0/|\mathbf{B}_0|)$, $\omega_H = |e\mathbf{B}_0/m|$ is the gyro frequency and $\omega_p = \sqrt{N_e e^2/\epsilon_0 m}$ is the plasma frequency. Taking the dot and vector products of $\mathbf{Y}$ with Equation 2.61 we obtain

$$-j\omega\epsilon_0 X \mathbf{Y} \cdot \mathbf{E} = U\mathbf{Y} \cdot \mathbf{J} \tag{2.62}$$

and

$$-j\omega X \epsilon_0 \mathbf{Y} \times \mathbf{E} = U\mathbf{Y} \times \mathbf{J} + j\mathbf{YY} \cdot \mathbf{J} - jY^2 \mathbf{J} \tag{2.63}$$

respectively. We can eliminate $\mathbf{Y} \times \mathbf{J}$ from Equation 2.63 using Equation 2.61 to yield

$$\omega X \epsilon_0 \mathbf{Y} \times \mathbf{E} = -U^2 \mathbf{J} - j\omega U X \epsilon_0 \mathbf{E} - \mathbf{YY} \cdot \mathbf{J} + Y^2 \mathbf{J} \tag{2.64}$$

and then eliminate $\mathbf{Y} \cdot \mathbf{J}$ using Equation 2.62 to yield

$$\left(Y^2 - U^2\right) \mathbf{J} = \omega X \epsilon_0 \mathbf{Y} \times \mathbf{E} + j\omega U X \epsilon_0 \mathbf{E} - j\frac{\omega \epsilon_0 X}{U} \mathbf{YY} \cdot \mathbf{E} \tag{2.65}$$

where $Y^2 = \mathbf{Y} \cdot \mathbf{Y}$.

Since we assume the electric field to be time harmonic, Equations 2.30 and 2.31 will imply

$$\nabla \times \nabla \times \mathbf{E} - \omega^2 \mu_0 \epsilon_0 \mathbf{E} = -j\omega \mu_0 \mathbf{J} \tag{2.66}$$

where the current density $\mathbf{J}$ consists of that due to the plasma alone, i.e. that given by Equation 2.65, and that due to any sources (antennas, for example) denoted by $\mathbf{J}_s$. Consequently, we can eliminate $\mathbf{J}$ from Equation 2.66 and obtain

$$\nabla \times \nabla \times \mathbf{E} - \beta_0^2 \left(1 - \frac{UX}{U^2 - Y^2}\right) \mathbf{E}$$
$$= -j\frac{\beta_0^2 X}{U^2 - Y^2} \left(-\mathbf{Y} \times \mathbf{E} + \frac{j}{U} \mathbf{YY} \cdot \mathbf{E}\right) + \mathbf{J}_s \tag{2.67}$$

where $\beta_0^2 = \omega^2 \mu_0 \epsilon_0$. This equation describes the electric field within the ionospheric plasma. This equation can be reinterpreted as

$$\nabla \times \nabla \times \mathbf{E} - \omega^2 \mu_0 \epsilon_0 \epsilon_r \mathbf{E} = \mathbf{J}_s \qquad (2.68)$$

where the relative permittivity ϵ_r is now a tensor of the form

$$\epsilon_r = \begin{pmatrix} 1 - \frac{XU}{U^2-Y^2} + \frac{XY_x^2}{U(U^2-Y^2)} & \frac{XY_xY_y}{U(U^2-Y^2)} - \frac{jXY_z}{U^2-Y^2} & \frac{XY_xY_z}{U(U^2-Y^2)} + \frac{jXY_y}{U^2-Y^2} \\ \frac{XY_yY_x}{U(U^2-Y^2)} + \frac{jXY_z}{U^2-Y^2} & 1 - \frac{XU}{U^2-Y^2} + \frac{XY_y^2}{U(U^2-Y^2)} & \frac{XY_yY_z}{U(U^2-Y^2)} - \frac{jXY_x}{U^2-Y^2} \\ \frac{XY_zY_x}{U(U^2-Y^2)} - \frac{jXY_y}{U^2-Y^2} & \frac{XY_zY_y}{U(U^2-Y^2)} + \frac{jXY_x}{U^2-Y^2} & 1 - \frac{XU}{U^2-Y^2} + \frac{XY_z^2}{U(U^2-Y^2)} \end{pmatrix} \qquad (2.69)$$

There are two important properties of ϵ_r worth mentioning, first that $\epsilon_r^T = \epsilon_r^*$ and second that ϵ_r^* can be obtained from ϵ_r by reversing the direction of the magnetic field. These properties will exploited in the next chapter.

We now consider the propagation of a plane harmonic wave in a source free ($\mathbf{J}_s = 0$) anisotropic medium. The electric and current fields will take the form

$$\mathbf{E} = \mathbf{E}_0 \exp(-j\beta \hat{\mathbf{p}} \cdot \mathbf{r}) \qquad (2.70)$$

and

$$\mathbf{J} = \mathbf{J}_0 \exp(-j\beta \hat{\mathbf{p}} \cdot \mathbf{r}) \qquad (2.71)$$

From Equation 2.66

$$(\beta_r^2 - 1)\mathbf{E}_0 - \beta_r^2 \hat{\mathbf{p}}\hat{\mathbf{p}} \cdot \mathbf{E}_0 = -\frac{j}{\epsilon_0 \omega}\mathbf{J}_0 \qquad (2.72)$$

where $\beta_r = \beta/\beta_0$. Without loss of generality, we now choose the z axis to be the propagation direction ($\hat{p}_x = \hat{p}_y = 0$ and $\hat{p}_z = 1$) and choose the y axis such that the magnetic field direction lies in the yz plane ($Y_x = 0$). Equation 2.72 will then yield

$$(\beta_r^2 - 1)E_0^x = -\frac{j}{\omega\epsilon_0}J_0^x \qquad (2.73)$$

$$(\beta_r^2 - 1)E_0^y = -\frac{j}{\omega\epsilon_0}J_0^y \qquad (2.74)$$

and

$$-E_0^z = -\frac{j}{\omega\epsilon_0}J_0^z \qquad (2.75)$$

Further, from Equation 2.61,

$$-j\omega X\epsilon_0 E_0^x = UJ_0^x + jY_y J_0^z - jY_z J_0^y \qquad (2.76)$$

$$-j\omega X\epsilon_0 E_0^y = UJ_0^y + jY_z J_0^x \qquad (2.77)$$

and

$$-j\omega X\epsilon_0 E_0^z = UJ_0^z - jY_y J_0^x \qquad (2.78)$$

2.3 Plane Waves in Anisotropic Media

From Equations 2.75 and 2.78, we obtain

$$(U - X)J_0^z = jY_y J_0^x \tag{2.79}$$

Then, from Equations 2.76 and 2.79,

$$-j\omega X \epsilon_0 E_0^x = \left(U - \frac{Y_y^2}{U - X}\right) J_0^x - jY_z J_0^y \tag{2.80}$$

and, substituting from Equations 2.73 and 2.74,

$$\left(U + \frac{X}{\beta_r^2 - 1} - \frac{Y_y^2}{U - X}\right) E_0^x - jY_z E_0^y = 0 \tag{2.81}$$

The important thing to note is that the x and y components of the electric field are out of phase and so the polarization is elliptical. From Equations 2.75, 2.73 and 2.79

$$E_0^z = -jY_y \frac{\beta_r^2 - 1}{U - X} E_0^x \tag{2.82}$$

Consequently, unlike the propagation in an isotropic medium, there is a component of electric field in the propagation direction.

From Equation 2.77 we note that there is a further equation connecting the x and y components of the electric field. On substituting Equations 2.73 and 2.74 into Equation 2.77, we obtain

$$jY_z E_0^x + \left(\frac{X}{\beta_r^2 - 1} + U\right) E_0^y = 0 \tag{2.83}$$

Then, for Equations 2.81 and 2.83 to be compatible, we will need the matrix

$$\begin{pmatrix} U + \frac{X}{\beta_r^2 - 1} + \frac{Y_y^2}{X - U} & -jY_z \\ jY_z & \frac{X}{\beta_r^2 - 1} + U \end{pmatrix} \tag{2.84}$$

to have a zero determinant. This results in a quadratic expression for β_r^2, which provides an equation for determining the possible values of β_r and hence the wave modes. Expanding the determinant, we obtain an equation of the form

$$((X - U)U^2 + UY_T^2 - (X - U)Y_L^2) + Q(2(X - U)U + Y_T^2) + Q^2(X - U) = 0 \tag{2.85}$$

where we have expressed the equation as a polynomial in $Q = X/(\beta_r^2 - 1)$. Further, we have denoted the component of Earth's magnetic field that is transverse to the propagation by Y_T and the component that is parallel by Y_L (in line with conventional notation). We can solve the above equation to obtain

$$Q = \frac{2U(U - X) - Y_T^2 \pm \left(Y_T^4 + 4(U - X)^2 Y_L^2\right)^{\frac{1}{2}}}{2(X - U)} \tag{2.86}$$

from which follows the Appleton–Hartree formula (Budden, 1985) for β_r

$$\beta_r^2 = 1 - \frac{2X(U-X)}{2U(U-X) - Y_T^2 \pm \left(Y_T^4 + 4(U-X)^2 Y_L^2\right)^{\frac{1}{2}}} \quad (2.87)$$

The important thing to note is that there are two distinct solutions for β_r^2 and therefore two distinct elliptically polarized modes.

The complex Poynting vector

$$\mathbf{P} = \frac{1}{2}\Re\{\mathbf{E} \times \mathbf{H}^*\} \quad (2.88)$$

is parallel to the propagation direction in the case of an isotropic medium, but things are not quite so simple in the anisotropic case. On noting that $\mathbf{H} = (j\beta/\omega\mu)\hat{\mathbf{p}} \times \mathbf{E}$, we will have that

$$\mathbf{P} = \frac{\beta_r}{2\eta_0}\Re\{\hat{\mathbf{p}}\mathbf{E} \cdot \mathbf{E}^* - \mathbf{E}^*\hat{\mathbf{p}} \cdot \mathbf{E}\} \quad (2.89)$$

where $\mathbf{E} \cdot \mathbf{p}$ is not necessarily zero. From the above considerations, the vector magnitude of the electric field will be given by

$$\mathbf{E}_0 = E_0^x \hat{\mathbf{x}} - \frac{jY_z q}{1+Uq} E_0^x \hat{\mathbf{y}} - \frac{jqY_y X}{U-X} E_0^x \hat{\mathbf{z}} \quad (2.90)$$

where $q = (\beta_r^2 - 1)/X$. From Equation 2.90, $\hat{\mathbf{p}} \cdot \mathbf{E}_0 = -jqY_y X E_0^x/(U-X)$ and from which it is evident that the direction of energy flow is not totally in the propagation direction. Further, we will have that

$$\mathbf{E} \cdot \mathbf{E}^* = E_0^{x2}\left(1 + \frac{Y_z^2 q^2}{(1+Uq)^2} + \frac{q^2 Y_y^2 X^2}{(U-X)^2}\right) \quad (2.91)$$

and so, from Equation 2.89,

$$\mathbf{P} = \frac{E_0^{x2} \beta_r}{2\eta_0}\left(\frac{-Y_z Y_y X q^2}{(1+Uq)(U-X)}\hat{\mathbf{y}} + \left(1 + \frac{Y_z^2 q^2}{(1+Uq)^2}\right)\hat{\mathbf{z}}\right) \quad (2.92)$$

In the case that $U = 1$, we can simplify this to

$$\mathbf{P} = \frac{E_0^{x2} \beta_r}{2\eta_0 Q(1+Q)(1-X)}\left(-Y_z Y_y X \hat{\mathbf{y}} + (1-X)Q\left(1 + Q + \frac{Y_z^2}{1+Q}\right)\hat{\mathbf{z}}\right) \quad (2.93)$$

From Equation 2.86,

$$Q + 1 = \frac{Y_y^2 \mp \sqrt{Y_y^4 + 4(1-X)^2 Y_z^2}}{2(1-X)} \quad (2.94)$$

and, after some algebra, Equation 2.93 simplifies to

$$\mathbf{P} = \frac{E_0^{x2} \beta_r}{2\eta_0 Q(1+Q)(1-X)}\left(-Y_z Y_y X \hat{\mathbf{y}} \mp Q\sqrt{Y_y^4 + 4(1-X)^2 Y_z^2}\,\hat{\mathbf{z}}\right) \quad (2.95)$$

We will use this last expression when we come to consider ray tracing in Chapter 5.

2.4 Boundary Conditions

Boundary conditions pick out the unique solution to the Maxwell equations that describes a particular physical situation. Some important conditions are:

(1) The tangential components of magnetic intensity and electric intensity are continuous across the interface between two dielectrics (i.e. $\mathbf{H} \times \mathbf{n}$ and $\mathbf{E} \times \mathbf{n}$ are continuous where $\mathbf{n}$ is the unit normal at the interface).

(2) A surface with $\sigma = \infty$ is said to be perfectly electrically conducting (PEC). For such a surface, the tangential components of electric intensity are zero (i.e. $\mathbf{E} \times \mathbf{n} = 0$). The current on a perfect conductor will concentrate at the surface with a density per unit area $\mathbf{J}_s$ that adjusts itself to be $\mathbf{J}_s = \mathbf{n} \times \mathbf{H}$.

Due to the high conductivity of metals (5.7×10^7 S/m for copper), it is frequently a good approximation to treat them as PEC. Seawater (conductivity about 5 S/m) is sometimes also treated in this manner. Freshwater (conductivity about 5.5×10^{-6} S/m), however, is certainly not PEC.

(3) Another useful boundary condition is given by $\mathbf{H} \times \mathbf{n} = 0$, and a material that satisfies this on its boundary is known as a perfectly magnetically conducting (PMC) material. Such a material does not exist in practice, but it is a good approximation to a corrugated metallic surface. In addition, it is often used as a boundary condition at an aperture in order to avoid the complex radiation problem that would normally arise in this situation.

(4) All fields fall away to zero far away from a bounded source.

(5) Bounded sources will only produce outgoing waves (the *radiation condition*). In the case of time harmonic fields, this requires that field quantities ψ satisfy

$$\frac{\partial \psi}{\partial r} + j\beta\psi = 0 \qquad (2.96)$$

as the distance from the source r tends to infinity.

The first boundary condition can be difficult to implement in general and so some approximate conditions have been developed that are more convenient (Senior and Volakis, 1995). Of particular importance is the *impedance boundary condition* that can be applied at the interface between two media when one of them has much lower impedance than the other (due to high conductivity and/or large relative permittivity). The condition takes the form

$$\mathbf{n} \times (\mathbf{n} \times \mathbf{E}) = -Z\mathbf{n} \times \mathbf{H} \qquad (2.97)$$

or

$$\mathbf{n} \times \mathbf{E} = Z\mathbf{n} \times (\mathbf{n} \times \mathbf{H}) \qquad (2.98)$$

where $\mathbf{n}$ is the unit normal pointing away from the lower impedance medium and Z is the *surface impedance* (we usually take Z to be the impedance η for this medium). This condition is usually applied when propagation in the medium of low impedance is of little interest. (A limiting case of this situation occurs at the surface of a perfectly conducting material where $\eta = 0$.) The condition can also be used to model the effects of a rough surface (Senior, 1995). Consider a perfectly conducting surface (the x–y plane) with fluctuations ξ about a mean level and with statistics described by

$$\langle \xi(x,y)\xi(x',y') \rangle = \xi_0^2 F(\rho) \tag{2.99}$$

where $\rho = \sqrt{(x-x')^2 + (y-y')^2}$ and $F(\rho)$ is a suitable autocorrelation function. The effective surface impedance will be given by (Senior and Volakis, 1995)

$$Z = j\eta_0 \beta_0 \xi_0^2 C \tag{2.100}$$

where

$$C = \frac{1}{4} \int_0^\infty \left(\frac{\partial^2}{\partial \rho^2} + \frac{1}{\rho} \frac{\partial}{\partial \rho} + 2\beta_0^2 \right) \left(FJ_0 \left(\frac{\beta_0 \rho}{\sqrt{2}} \right) \right) \exp(-j\beta_0 \rho) d\rho$$
$$- \frac{1}{4} \int_0^\infty \frac{\beta_0}{\sqrt{2}} \frac{\partial F}{\partial \rho} J_1 \left(\frac{\beta_0 \rho}{\sqrt{2}} \right) \exp(-j\beta_0 \rho) d\rho \tag{2.101}$$

For a Gaussian autocorrelation $F(\rho) = \exp(-4\rho^2/l^2)$, and small roughness ($\beta_0 l \ll 1$), $C \simeq -\sqrt{\pi}/2l$.

The fifth boundary condition can also be difficult to implement in the numerical solution of propagation problems since we must approximate infinity by a surface at a finite distance from the sources. An alternative here is the *absorbing boundary condition* in which we make the medium lossy as we approach this boundary. In this case, we can apply any convenient boundary condition at the boundary since any artificial reflections it causes will be absorbed by the lossy medium. In this way no radiation will be reflected back to the source and the radiation condition will be satisfied. Fields in the artificial region, however, will need to be discounted.

2.5 Transmission through an Interface

A simple example that illustrates the importance of boundary conditions is provided by the problem of a plane wave that is normally incident upon a plane interface between two different media (Figure 2.4). Some of the energy will be reflected at the interface

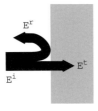

Figure 2.4 Reflection and transmission at a plane interface.

2.5 Transmission through an Interface

and some transmitted. We assume the incident wave to be plane harmonic and traveling in the z direction, i.e.

$$\mathbf{E}^i(z) = E_0^i e^{-j\beta_1 z} \hat{\mathbf{x}} \tag{2.102}$$

$$\mathbf{H}^i(z) = H_0^i e^{-j\beta_1 z} \hat{\mathbf{y}} \tag{2.103}$$

and then the transmitted wave will be of the form

$$\mathbf{E}^t(z) = E_0^t e^{-j\beta_2 z} \hat{\mathbf{x}} \tag{2.104}$$

$$\mathbf{H}^t(z) = H_0^t e^{-j\beta_2 z} \hat{\mathbf{y}} \tag{2.105}$$

and the reflected wave of the form

$$\mathbf{E}^r(z) = E_0^r e^{+j\beta_1 z} \hat{\mathbf{x}} \tag{2.106}$$

$$\mathbf{H}^r(z) = H_0^r e^{+j\beta_1 z} \hat{\mathbf{y}} \tag{2.107}$$

(Note that β_1 is the propagation constant for the medium of the incident wave and β_2 is the propagation constant for the medium of the transmitted wave.) The electric and magnetic fields will be related through

$$E_0^i = \eta_1 H_0^i \tag{2.108}$$

$$E_0^r = -\eta_1 H_0^r \tag{2.109}$$

and

$$E_0^t = \eta_2 H_0^t \tag{2.110}$$

where η_1 and η_2 are the impedances of media 1 and 2, respectively. At the interface, the boundary conditions will require the tangential components of the electric and magnetic fields (components in the $\hat{\mathbf{x}}$ and $\hat{\mathbf{y}}$ directions) to be continuous across the interface. This implies the relations

$$E_0^i + E_0^r = E_0^t \tag{2.111}$$

and

$$H_0^i + H_0^r = H_0^t \tag{2.112}$$

We can replace the magnetic fields in Equation 2.112 using Equations 2.108 to 2.110 and obtain

$$\frac{E_0^i}{\eta_1} - \frac{E_0^r}{\eta_1} = \frac{E_0^t}{\eta_2} \tag{2.113}$$

From Equations 2.113 and 2.111 we now have

$$E_0^t = \frac{2\eta_2}{\eta_1 + \eta_2} E_0^i \tag{2.114}$$

$$E_0^r = \frac{\eta_2 - \eta_1}{\eta_1 + \eta_2} E_0^i \tag{2.115}$$

which relates the transmitted and reflected waves to the incoming wave.

The Fundamentals of Electromagnetic Waves

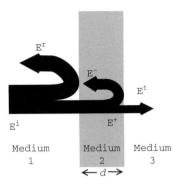

Figure 2.5 Transmission through a screen.

Consider now the case of a wave incident upon a screen of finite thickness (Figure 2.5). It is assumed that medium 2 (impedance η and propagation constant β) is strongly attenuating and that media 1 and 3 consist of free space (impedance η_0). Due to the strong attenuation in medium 2, we can ignore E^- (the reflection at the second interface) and hence treat the interfaces separately. The transmitted field that leaves the first interface will be

$$E_0^+ = \frac{2\eta}{\eta_0 + \eta} E_0^i \qquad (2.116)$$

and will suffer attenuation $\exp(-|\Im\{\beta\}|d)$ through the screen. At the second interface, the incident field will now have magnitude $E_0^+ \exp(-|\Im\{\beta\}|d)$. Consequently, at this interface, the transmitted field will have magnitude

$$\begin{aligned} E^t &= \frac{2\eta_0}{\eta_0 + \eta} E_0^+ e^{-|\Im\{\beta\}|d} \\ &= \frac{4\eta_0 \eta}{(\eta_0 + \eta)^2} E_0^i e^{-|\Im\{\beta\}|d} \end{aligned} \qquad (2.117)$$

This expression allows us to calculate the attenuation caused by a strongly conducting screen.

2.6 Oblique Incidence

Consider a wave, traveling through free space, that is incident upon an imperfectly conducting half space (the interface is a plane that we assume to go through the origin and so has the equation $\mathbf{r} \cdot \mathbf{n} = 0$). Given an incident plane wave field ($\mathbf{H}^i, \mathbf{E}^i$), with propagation direction $\hat{\mathbf{r}}^i$, we would like to determine the transmitted ($\mathbf{H}^t, \mathbf{E}^t$) and reflected ($\mathbf{H}^r, \mathbf{E}^r$) wave fields, together with their respective propagation directions $\hat{\mathbf{r}}^t$ and $\hat{\mathbf{r}}^r$ (Figure 2.6). The incident, transmitted and reflected electric fields will take the form

$$\mathbf{E}^i = \mathbf{E}_0^i \exp(-j\beta \, \hat{\mathbf{r}}^i \cdot \mathbf{r}) \qquad (2.118)$$
$$\mathbf{E}^t = \mathbf{E}_0^t \exp(-j\beta \, \hat{\mathbf{r}}^t \cdot \mathbf{r}) \qquad (2.119)$$
$$\mathbf{E}^r = \mathbf{E}_0^r \exp(-j\beta \, \hat{\mathbf{r}}^r \cdot \mathbf{r}) \qquad (2.120)$$

2.6 Oblique Incidence

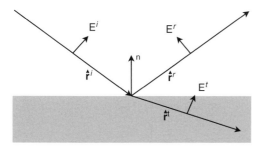

Figure 2.6 Oblique incidence of plane waves.

where $\mathbf{E}_0^r$, $\mathbf{E}_0^t$ and $\mathbf{E}_0^i$ are constant vectors perpendicular to the propagation directions $\hat{\mathbf{r}}^r$, $\hat{\mathbf{r}}^t$ and $\hat{\mathbf{r}}^i$, respectively. (It should be noted that we only need to calculate the electric fields since $\mathbf{H}^t = \hat{\mathbf{r}}^t \times \mathbf{E}^t / \eta_0 \eta_r$ and $\mathbf{H}^r = \hat{\mathbf{r}}^r \times \mathbf{E}^r / \eta_0$.)

We first assume the impedance boundary condition to be valid, so, at the interface,

$$\mathbf{n} \times (\mathbf{n} \times \mathbf{E}) = -\eta \mathbf{n} \times \mathbf{H} \qquad (2.121)$$

where $\mathbf{n}$ is the unit normal. From this condition, we will have

$$\mathbf{n} \times \left(\mathbf{n} \times \mathbf{E}_0^i \exp(-j\beta \hat{\mathbf{r}}^i \cdot \mathbf{r}) + \mathbf{n} \times \mathbf{E}_0^r \exp(-j\beta \hat{\mathbf{r}}^r \cdot \mathbf{r})\right)$$
$$= -\eta \mathbf{n} \times \left(\mathbf{H}_0^i \exp(-j\beta \hat{\mathbf{r}}^i \cdot \mathbf{r}) + \mathbf{H}_0^r \exp(-j\beta \hat{\mathbf{r}}^r \cdot \mathbf{r})\right) \qquad (2.122)$$

for points satisfying $\mathbf{r} \cdot \mathbf{n} = 0$ (which is the equation of the interface). The relationship between electric and magnetic fields will imply that

$$\mathbf{n} \times \left(\mathbf{n} \times \mathbf{E}_0^i + \mathbf{n} \times \mathbf{E}_0^r\right) = -\frac{\eta}{\eta_0} \mathbf{n} \times \left(\hat{\mathbf{r}}^i \times \mathbf{E}_0^i + \hat{\mathbf{r}}^r \times \mathbf{E}_0^r\right) \qquad (2.123)$$

and

$$\hat{\mathbf{r}}^i \cdot \mathbf{r} = \hat{\mathbf{r}}^r \cdot \mathbf{r} \qquad (2.124)$$

for points $\mathbf{r}$ on the interface. Condition 2.124 will be satisfied if

$$\hat{\mathbf{r}}^r = \hat{\mathbf{r}}^i - 2\mathbf{n}(\mathbf{n} \cdot \hat{\mathbf{r}}^i) \qquad (2.125)$$

We now consider Equation 2.123 for two separate cases. First, when the incident electric field is parallel to the interface and, second, when magnetic field is parallel to the interface (i.e. the electric field is in the plane that is normal to the interface and contains the propagation direction). For obvious reasons, the first of these fields is known as *transverse electric* (TE) and the second as *transverse magnetic* (TM). A general incident field can be considered as a combination of two such fields. Equation 2.123 can be expanded using the vector identity $\mathbf{A} \times (\mathbf{B} \times \mathbf{C}) = (\mathbf{A} \cdot \mathbf{C})\mathbf{B} - (\mathbf{A} \cdot \mathbf{B})\mathbf{C}$ and, noting that $\mathbf{E}_0^i \cdot \hat{\mathbf{r}}^i = 0$ and $\mathbf{E}_0^r \cdot \hat{\mathbf{r}}^r = 0$, Equation 2.123 reduces to

$$\mathbf{n} \cdot \mathbf{E}_0^r \mathbf{n} - \mathbf{E}_0^r + \frac{\eta}{\eta_0} \mathbf{n} \cdot \mathbf{E}_0^r \hat{\mathbf{r}}^r - \frac{\eta}{\eta_0} \hat{\mathbf{r}}^r \cdot \mathbf{n} \mathbf{E}_0^r = -\mathbf{n} \cdot \mathbf{E}_0^i \mathbf{n} + \mathbf{E}_0^i - \frac{\eta}{\eta_0} \mathbf{n} \cdot \mathbf{E}_0^i \hat{\mathbf{r}}^i + \frac{\eta}{\eta_0} \hat{\mathbf{r}}^i \cdot \mathbf{n} \mathbf{E}_0^i \qquad (2.126)$$

For a TE field, we have $\mathbf{n} \cdot \mathbf{E}_0^i = 0$ and $\mathbf{n} \cdot \mathbf{E}_0^r = 0$, so that Equation 2.126 reduces to

$$-\mathbf{E}_0^r + \frac{\eta}{\eta_0}\hat{\mathbf{r}}^i \cdot \mathbf{n}\mathbf{E}_0^r = \mathbf{E}_0^i + \frac{\eta}{\eta_0}\hat{\mathbf{r}}^i \cdot \mathbf{n}\mathbf{E}_0^i \tag{2.127}$$

where we have replaced $\hat{\mathbf{r}}^r$ using Equation 2.125. The reflected electric field will now have the form

$$\mathbf{E}_0^r = R_H \mathbf{E}_0^i \tag{2.128}$$

for which

$$R_H = \frac{C\eta_r - 1}{C\eta_r + 1} \tag{2.129}$$

where $\eta_r \left(= \frac{\eta}{\eta_0}\right)$ is the relative impedance of the reflecting medium and $C = -\mathbf{n} \cdot \hat{\mathbf{r}}^i$ is the cosine of the angle between the incoming propagation direction and the normal to the interface.

For the TM case, the reflected electric field will need to have the form

$$\mathbf{E}_0^r = R_V \left(2\mathbf{n}(\mathbf{n} \cdot \mathbf{E}_0^i) - \mathbf{E}_0^i\right) \tag{2.130}$$

in order for it to be orthogonal to the propagation direction (given by Equation 2.125). Equation 2.126 is identically zero in the normal direction ($\mathbf{n}$) and, due to the direction of the electric field, will be satisfied in the direction orthogonal to the plane of propagation and the normal. Consequently, Equation 2.126 will only need to be satisfied in an additional independent direction, which we take to be the propagation direction of the incident wave. Taking the dot product of $\hat{\mathbf{r}}^i$ with Equation 2.126, we obtain

$$-\mathbf{n} \cdot \mathbf{E}_0^r \hat{\mathbf{r}}^i \cdot \mathbf{n} + \eta_r \mathbf{n} \cdot \mathbf{E}_0^r = -\mathbf{n} \cdot \mathbf{E}_0^i \hat{\mathbf{r}}^i \cdot \mathbf{n} - \eta_r \mathbf{n} \cdot \mathbf{E}_0^i \tag{2.131}$$

from which, on replacing $\mathbf{E}_0$ using Equation 2.130, we obtain

$$R_V = \frac{C - \eta_r}{C + \eta_r} \tag{2.132}$$

Coefficients R_V and R_H are known as the *reflection coefficients* for vertical and horizontal polarization, respectively. The above expressions are valid when η_r is small, but, for general values, it can be shown that

$$R_V = \frac{C\eta_r^{-2} - \sqrt{\eta_r^{-2} - S^2}}{C\eta_r^{-2} + \sqrt{\eta_r^{-2} - S^2}} \tag{2.133}$$

$$R_H = \frac{C - \sqrt{\eta_r^{-2} - S^2}}{C + \sqrt{\eta_r^{-2} - S^2}} \tag{2.134}$$

where $S^2 = 1 - C^2$. (Note that we have implicitly assumed that $\mu = \mu_0$ since this is a very good approximation for the media of interest in the current text.) The above expressions reduce to the results 2.129 and 2.132 in the limit of a low impedance for the reflecting surface. What is clear from the above considerations is that reflections can occur when there is a discontinuity in the impedance of the medium. If there is a discontinuity, however, this does not necessarily mean that there will be reflections.

From the expression for the reflection coefficient of a vertically polarized wave, there can be an angle of incidence for which the coefficient is zero. This angle is known as the *Brewster angle* and satisfies $S = 1/\sqrt{1+\eta_r^2}$. If the medium is lossy, η_r is complex and the reflection coefficient does not go through a zero as the angle of incidence changes. There will, however, be an angle at which the reflection coefficient is minimum and this is sometimes known as the *pseudo Brewster angle*.

In the situation that the lower medium has low conductivity, the transmitted wave can be significant. If $\hat{\mathbf{r}}^t$ is the direction of transmission, the continuity of the tangential electric field at the interface implies that

$$\eta_r \hat{\mathbf{r}}^i \cdot \mathbf{r} = \hat{\mathbf{r}}^t \cdot \mathbf{r} \qquad (2.135)$$

and this is satisfied when

$$\hat{\mathbf{r}}^t = \eta_r \left(\hat{\mathbf{r}}^i + C\mathbf{n} \right) - \eta_r \sqrt{\eta_r^{-2} - S^2} \mathbf{n} \qquad (2.136)$$

From the boundary conditions, the transmitted wave is

$$\mathbf{E}_0^t = (1 + R_H)\mathbf{E}_0^i = \frac{2C}{C + \sqrt{\eta_r^{-2} - S^2}} \mathbf{E}_0^i \qquad (2.137)$$

for an incident wave with electric field parallel to the interface. For an incident electric field in the plane containing the normal and the propagation direction, the transmitted wave is

$$\mathbf{E}_0^t = (1 - R_V) \left(\mathbf{E}_0^i - \mathbf{n}\mathbf{n} \cdot \mathbf{E}_0^i + \frac{C}{\sqrt{\eta_r^{-2} - S^2}} \mathbf{n}\mathbf{n} \cdot \mathbf{E}_0^i \right) \qquad (2.138)$$

2.7 Sources of Radio Waves

Although plane waves are a useful idealization, the radio waves that are produced by real sources will be spherical in nature. Nevertheless, at large distances from the sources, the waves will be well approximated locally by a plane wave. The sources will usually consist of a metallic structure (an antenna) into which currents are driven by a radio transmitter. The currents on the antenna will constitute a current density $\mathbf{J}$ that, in turn, will give rise to the electric field

$$\mathbf{E}(\mathbf{r}) = \frac{1}{j\omega\epsilon} \nabla \times \nabla \times \int_V \frac{\mathbf{J}(\mathbf{r}') \exp(-j\beta |\mathbf{r} - \mathbf{r}'|)}{4\pi |\mathbf{r} - \mathbf{r}'|} dV' \qquad (2.139)$$

where dV' is a volume element and V is a volume that contains the sources. The corresponding magnetic field will be given by

$$\mathbf{H}(\mathbf{r}) = \nabla \times \int_V \frac{\mathbf{J}(\mathbf{r}') \exp(-j\beta |\mathbf{r} - \mathbf{r}'|)}{4\pi |\mathbf{r} - \mathbf{r}'|} dV' \qquad (2.140)$$

Noting that

$$|\mathbf{r} - \mathbf{r}'| = \sqrt{(\mathbf{r} - \mathbf{r}') \cdot (\mathbf{r} - \mathbf{r}')} = r\sqrt{1 - 2\frac{\mathbf{r} \cdot \mathbf{r}'}{r^2} + \frac{|\mathbf{r}'|^2}{r^2}} \qquad (2.141)$$

where $r = \sqrt{\mathbf{r} \cdot \mathbf{r}}$, it can be seen that $|\mathbf{r} - \mathbf{r}'| \approx r - \mathbf{r} \cdot \mathbf{r}'/r$ at large distances from the source. Consequently, to the leading order in $1/r$, the electric field can be approximated by

$$\mathbf{E} = j\omega\mu \frac{\exp(-j\beta r)}{4\pi r} \int_V \hat{\mathbf{r}} \times \hat{\mathbf{r}} \times \mathbf{J}(\mathbf{r}') \exp\left(j\beta \hat{\mathbf{r}} \cdot \mathbf{r}'\right) dV' \qquad (2.142)$$

where $\hat{\mathbf{r}}$ is a unit vector in the radiation direction. Locally, the field has the character of a plane wave moving in the radial direction $\hat{\mathbf{r}}$ and so the magnetic field is given by

$$\mathbf{H} = \frac{1}{\eta}\hat{\mathbf{r}} \times \mathbf{E} \qquad (2.143)$$

The region over which Equation 2.142 is valid is known as the *radiation (or Fraunhofer) zone* and, for a source with outer dimension D, this is the region at distances greater than $r_{ff} = 2D^2/\lambda$ from the source. In this region, the phase error in approximation 2.142 is less than $\pi/8$. The radiation zone is the region that we study in propagation theory, the focus of the present book. The region where the radiation still dominates, but Equation 2.142 is invalid, is known as the *radiating near-field (or Fresnel) zone* (Figure 2.7). This zone starts at a distance of about $\sqrt{D^3/\lambda}$ and ends at distance r_{ff}. In this region the curvature of the wavefront is too large to be ignored. Inside the Fresnel zone, the region is known as the *reactive near-field* and, in this region, all terms in the electromagnetic field are important.

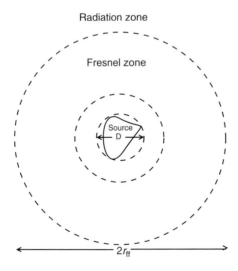

Figure 2.7 Field zones in relation to the radiating sources.

2.7 Sources of Radio Waves

In general, the electric field of an antenna in the radiation zone will have the form

$$E = \frac{j\omega\mu I_m}{4\pi} \mathbf{h}_{\text{eff}} \frac{\exp(-j\beta r)}{r} \quad (2.144)$$

where I_m is the current that feeds the antenna and $\mathbf{h}_{\text{eff}}$ is the *vector effective length*. The effective antenna length is a function of the direction of radiation, and it is important to note that there is no such thing as an antenna that radiates uniformly in all directions. In terms of the current distribution on the antenna, the effective antenna length is given by

$$\mathbf{h}_{\text{eff}} = I_m^{-1} \int_V \left((\hat{\mathbf{r}} \cdot \mathbf{J}(\mathbf{r}'))\hat{\mathbf{r}} - \mathbf{J}(\mathbf{r}') \right) \exp(j\beta \hat{\mathbf{r}} \cdot \mathbf{r}') \, dV' \quad (2.145)$$

where V is a volume that contains the antenna. Although the concept of effective antenna length has been introduced from the point of view of a transmitting antenna, it is also a relevant concept in the case of a receiving antenna. When immersed in electromagnetic wave field $\mathbf{E}$, an antenna will exhibit an open circuit voltage of the form

$$V_{OC} = \mathbf{h}_{\text{eff}} \cdot \mathbf{E} \quad (2.146)$$

where $\mathbf{h}_{\text{eff}}$ is the effective antenna length in the direction of the source (we will prove this in the next chapter). It will be noted that the polarization of a receiving antenna needs to be matched to the polarization of the incoming wave in order that the antenna extract maximum energy from this wave.

The concept of effective antenna length is very useful for describing the radiation field of an antenna, but the properties of an antenna are more often than not described in terms of the alternative concept of directivity or its related concept of gain. For an antenna that radiates a total power P_{rad}, we define the directivity $D(\hat{\mathbf{r}})$ in a particular direction to be the power radiated into a unit solid angle in that direction when scaled on $P_{\text{rad}}/4\pi$, that is,

$$D(\hat{\mathbf{r}}) = \frac{r^2 \mathbf{E} \cdot \mathbf{E}^*/2\eta}{P_{\text{rad}}/4\pi} \quad (2.147)$$

where P_{rad} is calculated from

$$P_{\text{rad}} = \int_S \frac{\mathbf{E} \cdot \mathbf{E}^*}{2\eta} dS \quad (2.148)$$

and S is a surface, within the radiation zone, that surrounds the antenna. Consequently, in terms of effective antenna length,

$$D(\theta,\phi) = \frac{4\pi \mathbf{h}_{\text{eff}} \cdot \mathbf{h}_{\text{eff}}^*}{\int_0^{2\pi} \int_0^{\pi} \mathbf{h}_{\text{eff}} \cdot \mathbf{h}_{\text{eff}}^* \sin\theta \, d\theta \, d\phi} \quad (2.149)$$

where θ and ϕ are spherical polar angular coordinates.

The simplest practical antenna is a dipole and this consists of a metallic rod that is driven at a suitable point by a radio transmitter. When the time variation is harmonic, the distribution of current will be sinusoidal with zero current at the ends of the rod. For a dipole orientated along the z axis with center at the origin, the corresponding harmonic current density will be approximately given by $\mathbf{J} = I_m \left(\sin(\beta(l - |z|))/\sin\beta l \right) \delta(x)\delta(y)\hat{\mathbf{z}}$

for $-l \leq z \leq l$ with $\mathbf{J} = 0$ otherwise. For this current distribution, we will have an effective antenna length of the form

$$\mathbf{h}_{\text{eff}} = I_m^{-1} \left(\cos\theta\hat{\mathbf{r}} - \mathbf{z}\right) \int_{-l}^{l} \frac{\sin(\beta(l - |z|))}{\sin\beta l} \exp\left(j\beta\cos\theta z'\right) dz'$$

$$= 2I_m^{-1} \left(\hat{\mathbf{z}} - \cos\theta\hat{\mathbf{r}}\right) \frac{\cos(\beta l \cos\theta) - \cos\beta l}{\sin^2\theta \sin\beta l} \tag{2.150}$$

where θ is the angle between the z axis and the radial direction $\hat{\mathbf{r}}$. As a consequence, the directivity of a half wave dipole ($l = \lambda/4$) is

$$D(\theta, \phi) = \frac{5}{3} \left(\frac{\cos(\frac{\pi}{2}\cos\theta)}{\sin\theta}\right)^2 \tag{2.151}$$

and, in the case of a short dipole ($l \ll \lambda$),

$$D(\theta, \phi) = \frac{3}{2} \sin^2\theta \tag{2.152}$$

In practice, we are more interested in the *gain* G of the antenna, rather than its directivity. This is the ratio defined by the power radiated in a particular direction divided by the total power P_a accepted by the antenna, i.e.

$$G = \frac{r^2 \mathbf{E} \cdot \mathbf{E}^*/2\eta}{P_a/4\pi} \tag{2.153}$$

where G is a function of radiation direction (i.e. $\hat{\mathbf{r}}$). Gain is proportional to directivity, i.e. $G(\theta, \phi) = \eta_A D(\theta, \phi)$ where η_A is a constant known as the efficiency of the antenna. The gain of an antenna is usually represented by a gain surface, a surface that is drawn around the origin with its distance from the origin in a particular direction being the gain in that particular direction. For the above short dipole, the gain surface is given by Figure 2.8.

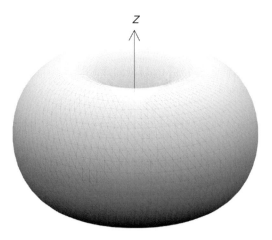

Figure 2.8 Gain surface of a short dipole.

Another important concept is that of *effective antenna area*. If a power density P_D (power per unit area) is incident upon an antenna, and causes a power P_{Rx} to appear in the antenna terminals, then the effective area is defined by $A_{\text{eff}} = P_{Rx}/P_D$. A_{eff} is effectively the area over which the antenna collects power. For a lossless isotropic antenna, $A_{\text{eff}} = \lambda^2/4\pi$ and so the relationship between effective aperture and gain is given by $G = 4\pi A_{\text{eff}}/\lambda^2$. For an isotropic antenna, the power collected is effectively that collected on a sphere of radius $\lambda/4\pi$. We will see in the next chapter that there is reciprocity between the behavior of antennas in the receive and transmit modes.

2.8 References

K.G. Budden, *The Propagation of Radio Waves*, Cambridge University Press, Cambridge, 1985.

C.J. Coleman, *An Introduction to Radio Frequency Engineering*, Cambridge University Press, Cambridge, 2004.

R.F. Harrington, *Time Harmonic Electromagnetic Fields*, McGraw-Hill, New York, 1961.

Z. Popovic and B.D. Popovic, *Introductory Electromagnetics*, Prentice Hall, Englewood Cliffs, NJ, 1999.

D.M. Pozar, *Microwave Engineering*, 3rd edition, John Wiley and Sons, Hoboken, NJ, 2005.

T.B.A. Senior and J.L. Volakis, *Approximate Boundary Conditions*, IEE Electromagnetic Waves Series 41, IEE, London, 1995.

W.L. Stutzman and G.A. Thiele, *Antenna Theory and Design*, John Wiley and Sons, New York, 1981.

3 The Reciprocity, Compensation and Extinction Theorems

In this chapter, we will discuss some of the important integral theorems of electromagnetism. These theorems can provide an alternative formulation of electromagnetic boundary value problems and open up an alternative approach to solving problems that are difficult when tackled as a direct solution of Maxwell's equations. In particular, many problems in propagation can be effectively solved through an integral formulation (Monteath, 1973). The important theorems of reciprocity, compensation and extinction are considered, theorems that are all intimately connected. We derive some extensions of the reciprocity theorem that are used in later chapters.

3.1 The Reciprocity Theorem

Consider an electromagnetic field $(\mathbf{H}_A, \mathbf{E}_A)$ generated by current distribution $\mathbf{J}_A$ and another $(\mathbf{H}_B, \mathbf{E}_B)$ generated by current distribution $\mathbf{J}_B$. From Maxwell's equations

$$\nabla \times \mathbf{H}_A = j\omega\epsilon \mathbf{E}_A + \mathbf{J}_A \tag{3.1}$$

$$\nabla \times \mathbf{E}_A = -j\omega\mu \mathbf{H}_A \tag{3.2}$$

$$\nabla \times \mathbf{H}_B = j\omega\epsilon \mathbf{E}_B + \mathbf{J}_B \tag{3.3}$$

$$\nabla \times \mathbf{E}_B = -j\omega\mu \mathbf{H}_B \tag{3.4}$$

Consider the identity

$$\nabla \cdot (\mathbf{X} \times \mathbf{Y}) = \mathbf{Y} \cdot \nabla \times \mathbf{X} - \mathbf{X} \cdot \nabla \times \mathbf{Y} \tag{3.5}$$

and then

$$\nabla \cdot (\mathbf{E}_A \times \mathbf{H}_B - \mathbf{E}_B \times \mathbf{H}_A) = \mathbf{H}_B \cdot \nabla \times \mathbf{E}_A - \mathbf{E}_A \cdot \nabla \times \mathbf{H}_B - \mathbf{H}_A \cdot \nabla \times \mathbf{E}_B + \mathbf{E}_B \cdot \nabla \times \mathbf{H}_A \tag{3.6}$$

Substituting from Maxwell's equations, we obtain

$$\nabla \cdot (\mathbf{E}_A \times \mathbf{H}_B - \mathbf{E}_B \times \mathbf{H}_A) = \mathbf{J}_A \cdot \mathbf{E}_B - \mathbf{J}_B \cdot \mathbf{E}_A \tag{3.7}$$

Now consider a volume V with surface S (see Figure 3.1). Integrating Equation 3.7 over V and applying the divergence theorem (see Appendix A)

$$\int_S (\mathbf{E}_A \times \mathbf{H}_B - \mathbf{E}_B \times \mathbf{H}_A) \cdot \mathbf{n}\, dS = \int_V (\mathbf{J}_A \cdot \mathbf{E}_B - \mathbf{J}_B \cdot \mathbf{E}_A)\, dV \tag{3.8}$$

3.1 The Reciprocity Theorem

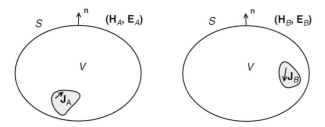

Figure 3.1 Fields $(\mathbf{H}_A, \mathbf{E}_A)$ and $(\mathbf{H}_B, \mathbf{E}_B)$ in the reciprocity theorem.

where **n** is the unit normal on the surface S. If the sources of the fields are bounded, and surface S surrounds all sources, the right-hand side of Equation 3.8 will yield the same value for all such S. If S is a sphere with sufficiently large radius, the fields will be in their radiation zone and will cancel. Consequently, the integral on the left-hand side of Equation 3.8 will be zero and, as a consequence, so will the volume integral on the right-hand side. This will mean that

$$\int_S (\mathbf{E}_A \times \mathbf{H}_B - \mathbf{E}_B \times \mathbf{H}_A) \cdot \mathbf{n}\, dS = 0 \qquad (3.9)$$

for any surface S that contains all sources and

$$\int_V \mathbf{J}_A \cdot \mathbf{E}_B\, dV = \int_V \mathbf{J}_B \cdot \mathbf{E}_A\, dV \qquad (3.10)$$

for volume a V that contains all sources. These results are known as the *Lorentz reciprocity theorem* for electromagnetic fields with current sources.

When volume V only contains the sources of field A, Equation 3.8 is modified to read

$$\int_V \mathbf{J}_A \cdot \mathbf{E}_B\, dV = \int_S (\mathbf{E}_A \times \mathbf{H}_B - \mathbf{E}_B \times \mathbf{H}_A) \cdot \mathbf{n}\, dS \qquad (3.11)$$

In the case that field $(\mathbf{H}_A, \mathbf{E}_A)$ is generated by the magnetic current distribution $\mathbf{M}_A$, i.e.

$$\nabla \times \mathbf{H}_A = j\omega\epsilon \mathbf{E}_A \qquad (3.12)$$
$$\nabla \times \mathbf{E}_A = -j\omega\mu \mathbf{H}_A - \mathbf{M}_A \qquad (3.13)$$

we have the related result

$$-\int_V \mathbf{H}_B \cdot \mathbf{M}_A\, dV = \int_S (\mathbf{E}_A \times \mathbf{H}_B - \mathbf{E}_B \times \mathbf{H}_A) \cdot \mathbf{n}\, dS \qquad (3.14)$$

We consider $(\mathbf{H}_A, \mathbf{E}_A)$ to be generated by electric and magnetic dipoles in turn (see Figure 3.2). Let $(\mathbf{H}_A^e, \mathbf{E}_A^e)$ be the field generated by a unit electric dipole at point A $(\mathbf{J}_A = \delta(\mathbf{r} - \mathbf{r}_A)\hat{\mathbf{s}}$, where $\hat{\mathbf{s}}$ is an arbitrary unit vector), then Equation 3.11 implies

$$\hat{\mathbf{s}} \cdot \mathbf{E}_B = \int_S (\mathbf{E}_A^e \times \mathbf{H}_B - \mathbf{E}_B \times \mathbf{H}_A^e) \cdot \mathbf{n}\, dS \qquad (3.15)$$

The Reciprocity, Compensation and Extinction Theorems

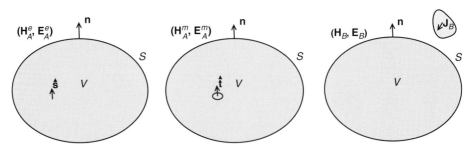

Figure 3.2 Sources and fields in the reciprocity related integral relations.

In a similar manner, let $(\mathbf{H}_A^m, \mathbf{E}_A^m)$ be the field generated by a unit magnetic dipole at point A ($\mathbf{M}_A = \delta(\mathbf{r} - \mathbf{r}_A)\hat{\mathbf{t}}$ where $\hat{\mathbf{t}}$ is an arbitrary unit vector), then Equation 3.14 implies

$$-\hat{\mathbf{t}} \cdot \mathbf{H}_B = \int_S (\mathbf{E}_A^m \times \mathbf{H}_B - \mathbf{E}_B \times \mathbf{H}_A^m) \cdot \mathbf{n}\, dS \qquad (3.16)$$

If the electromagnetic field $(\mathbf{H}_B, \mathbf{E}_B)$ is given on surface S, Equations 3.15 and 3.16 allow us to calculate the field on a new surface to the fore of surface S. We will find these equations useful in developing propagation algorithms. In a sense, the equations can be regarded as a mathematical expression of Huygens' principle since the fields on the new surface are represented as a sum of point source fields on the old surface S.

3.2 Reciprocity and Radio Systems

The most common practical expression of reciprocity comes from antenna theory. Let the field $(\mathbf{H}_A, \mathbf{E}_A)$ result when an antenna A is driven and another antenna B is open circuit. Conversely, let field $(\mathbf{H}_B, \mathbf{E}_B)$ result when antenna B is driven and antenna A is open circuit. Since the electric field will be zero inside the metal of the antennas, the only contribution to the volume integrals of Equation 3.10 will come from the region of the antenna feeds. Let the antenna feeds be represented by cylinders of length L and cross sectional area S. There will be a current flow $I = \int_S \mathbf{J} \cdot d\mathbf{S}$ through the feed and a voltage drop $V = -\int_0^L \mathbf{E} \cdot d\mathbf{r}$ across the feed. From Equation 3.10

$$\int_S \mathbf{J}_A \cdot \mathbf{n}\, dS \int_0^L \mathbf{E}_B \cdot d\mathbf{r} = \int_S \mathbf{J}_B \cdot \mathbf{n}\, dS \int_0^L \mathbf{E}_A \cdot d\mathbf{r} \qquad (3.17)$$

Then, from Equation 3.17,

$$I_A V_{AB} = I_B V_{BA} \qquad (3.18)$$

where V_{AB} is the open circuit voltage across the terminals of A due to a current I_B in the feed of B and V_{BA} is the voltage drop across the terminals of B due to a current I_A in the feed of A. If we drive the antennas with the same current ($I = I_A = I_B$), Equation 3.18 will then imply that the open circuit antennas will see the same voltage ($V = V_{AB} = V_{BA}$). This is the standard reciprocity result for antennas (see Figure 3.3).

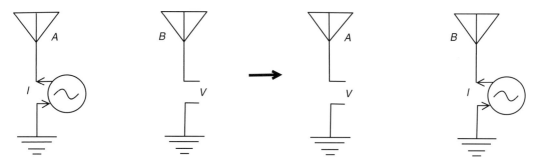

Figure 3.3 Antenna reciprocity.

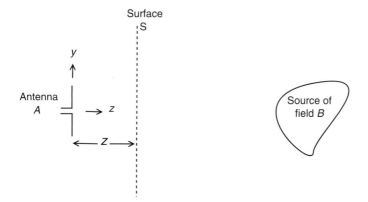

Figure 3.4 Voltage in a receiving antenna.

We now consider two antennas (A and B) located on the z axis and assume that B is sufficiently isolated from A for its field to be regarded as a plane wave at A, (Figure 3.4) i.e.

$$\mathbf{E}_B = \mathbf{E}_B^A \exp(j\beta z) \tag{3.19}$$

where $\mathbf{E}_B^A$ is the value of $\mathbf{E}_B$ at the location of A. In the radiation zone, the field of antenna A will have the form

$$\mathbf{E}_A = \frac{j\omega\mu I_A}{4\pi} \mathbf{h}_{\text{eff}}^A \frac{\exp(-j\beta r)}{r} \tag{3.20}$$

where $\mathbf{h}_{\text{eff}}^A$ is its effective antenna length. Now consider a plane surface R^2 that is orthogonal to the z axis and a distance Z from A. We form a closed surface S by adding a hemisphere at infinity. Consider Equation 3.11 and note that the integrand vanishes on the hemisphere at infinity, then

$$-I_A V_{AB} = \int_{R^2} (\mathbf{E}_A \times \mathbf{H}_B - \mathbf{E}_B \times \mathbf{H}_A) \cdot \hat{\mathbf{z}} \, dS \tag{3.21}$$

It turns out that the major contribution to the above integral will come from points around the z axis and so we can approximate the field of antenna A as

$$\mathbf{E}_A = \frac{j\omega\mu I_A}{4\pi} \mathbf{h}_{\text{eff}}^A \frac{\exp\left(-j\beta\left(Z + \frac{x^2+y^2}{2Z}\right)\right)}{Z} \qquad (3.22)$$

and use the relations $\mathbf{H}_A \approx \hat{\mathbf{z}} \times \mathbf{E}_A/\eta$ and $\mathbf{H}_B \approx -\hat{\mathbf{z}} \times \mathbf{E}_B/\eta$ to remove dependence on the magnetic fields. Equation 3.21 will now reduce to

$$I_A V_{AB} \approx \frac{2}{\eta} \int_{R^2} \mathbf{E}_A \cdot \mathbf{E}_B \, dS \qquad (3.23)$$

On substituting Equations 3.19 and 3.22 into Equation 3.23, we obtain

$$V_{AB} \approx \frac{j\omega\mu}{2\pi\eta} \mathbf{h}_{\text{eff}}^A \cdot \mathbf{E}_B^A \int_{-\infty}^{\infty} \int_{-\infty}^{\infty} \frac{\exp(-\frac{j\beta}{2Z}(x^2+y^2))}{Z} \, dx \, dy \qquad (3.24)$$

The above integrals can be evaluated analytically (see Appendix A) and this results in the relation

$$V_{AB} = \mathbf{h}_{\text{eff}}^A \cdot \mathbf{E}_B^A \qquad (3.25)$$

If the field $\mathbf{E}_B$ results from antenna with effective antenna length $\mathbf{h}_{\text{eff}}^B$, then

$$\mathbf{E}_B^A = \frac{j\omega\mu I_B}{4\pi} \mathbf{h}_{\text{eff}}^B \frac{\exp(-j\beta R_{AB})}{R_{AB}} \qquad (3.26)$$

where R_{AB} is the distance between the antennas and I_B is the current in antenna B. From this, we obtain that the mutual impedance between the antennas ($Z_{AB} = V_{AB}/I_B$) is given by

$$Z_{AB} = \frac{j\omega\mu \, \mathbf{h}_{\text{eff}}^A \cdot \mathbf{h}_{\text{eff}}^B}{4\pi} \frac{\exp(-j\beta R_{AB})}{R_{AB}} \qquad (3.27)$$

This is an important result that we will use in later chapters. Further, from a modeling viewpoint, it allows us to treat propagation aspects in a radio system as part of a grand circuit model.

3.3 Pseudo Reciprocity

Up to now we have assumed that the medium is isotropic, i.e. that μ is a simple scalar quantity. In fact, this was crucial in deriving Equation 3.7 on which the reciprocity theorem depended. However, as we have seen, the ionosphere is an example of an important medium that is not isotropic. Because of its usefulness, we would like to see whether anything can be rescued from the reciprocity theorem in the case of a non-isotropic medium. As it turns out, media such as the ionosphere have a property that we will term *pseudo reciprocity*. In the case of the ionospheric medium, such properties have been investigated by Budden (1988). However, more general results can be found in the book by De Hoop (1995).

3.3 Pseudo Reciprocity

As before, we start with fields $(\mathbf{H}_A, \mathbf{E}_A)$ and $(\mathbf{H}_B, \mathbf{E}_B)$ that satisfy the time harmonic Maxwell equations

$$\nabla \times \mathbf{H}_A = j\omega\epsilon_A \mathbf{E}_A + \mathbf{J}_A \tag{3.28}$$

$$\nabla \times \mathbf{E}_A = -j\omega\mu_A \mathbf{H}_A - \mathbf{M}_A \tag{3.29}$$

$$\nabla \times \mathbf{H}_B = j\omega\epsilon_B \mathbf{E}_B + \mathbf{J}_B \tag{3.30}$$

$$\nabla \times \mathbf{E}_B = -j\omega\mu_B \mathbf{H}_B - \mathbf{M}_B \tag{3.31}$$

where $(\mathbf{M}_A, \mathbf{J}_A)$ are the magnetic and electric sources of field A and $(\mathbf{M}_B, \mathbf{J}_B)$ are those for field B. Quantities μ_A and μ_B are tensors representing the non-isotropic permeability of medium A and B, respectively, and ϵ_A and ϵ_B are tensors representing the respective non-isotropic permittivity of these media. We first note that fields A and B will satisfy the identity

$$\nabla \cdot (\mathbf{E}_A \times \mathbf{H}_B - \mathbf{E}_B \times \mathbf{H}_A)$$
$$= \mathbf{H}_B \cdot \nabla \times \mathbf{E}_A - \mathbf{E}_A \cdot \nabla \times \mathbf{H}_B - \mathbf{H}_A \cdot \nabla \times \mathbf{E}_B + \mathbf{E}_B \cdot \nabla \times \mathbf{H}_A \tag{3.32}$$

On substituting Equation 3.28 into Equation 3.32,

$$\nabla \cdot (\mathbf{E}_A \times \mathbf{H}_B - \mathbf{E}_B \times \mathbf{H}_A)$$
$$= -j\omega\mathbf{H}_B \cdot \mu_A \mathbf{H}_A - \mathbf{H}_B \cdot \mathbf{M}_A - j\omega\mathbf{E}_A \cdot \epsilon_B \mathbf{E}_B - \mathbf{E}_A \cdot \mathbf{J}_B$$
$$+ j\omega\mathbf{H}_A \cdot \mu_B \mathbf{H}_B + \mathbf{H}_A \cdot \mathbf{M}_B + j\omega\mathbf{E}_B \cdot \epsilon_A \mathbf{E}_A + \mathbf{E}_B \cdot \mathbf{J}_A \tag{3.33}$$

We now assume that $\epsilon_B = \epsilon_A^T$ and $\mu_B = \mu_A^T$, which implies that $\mathbf{H}_B \cdot \mu_A \mathbf{H}_A = \mathbf{H}_A \cdot \mu_B \mathbf{H}_B$ and $\mathbf{E}_B \cdot \epsilon_A \mathbf{E}_A = \mathbf{E}_A \cdot \epsilon_B \mathbf{E}_B$. As a consequence

$$\nabla \cdot (\mathbf{E}_A \times \mathbf{H}_B - \mathbf{E}_B \times \mathbf{H}_A)$$
$$= -\mathbf{H}_B \cdot \mathbf{M}_A - \mathbf{E}_A \cdot \mathbf{J}_B + \mathbf{H}_A \cdot \mathbf{M}_B + \mathbf{E}_B \cdot \mathbf{J}_A \tag{3.34}$$

Integrating Equation 3.34 over a volume V that contains all the field sources, we obtain

$$\int_S (\mathbf{E}_A \times \mathbf{H}_B - \mathbf{E}_B \times \mathbf{H}_A) \cdot \mathbf{n} dS$$
$$= \int_V (\mathbf{E}_B \cdot \mathbf{J}_A - \mathbf{E}_A \cdot \mathbf{J}_B - \mathbf{H}_B \cdot \mathbf{M}_A + \mathbf{H}_A \cdot \mathbf{M}_B) dV \tag{3.35}$$

where S is the surface of volume V and $\mathbf{n}$ is the unit normal on S. We will now assume that all sources are bounded, that $\mathbf{M}_A = \mathbf{M}_B = 0$ and that the electric properties of both media become isotropic, and identical, at infinity. As surface S tends to infinity, the integrand on the left-hand side will tend to zero and so

$$\int_V \mathbf{E}_B \cdot \mathbf{J}_A dV = \int_V \mathbf{E}_A \cdot \mathbf{J}_B dV \tag{3.36}$$

for any volume V that contains all the sources. In addition,

$$\int_S (\mathbf{E}_A \times \mathbf{H}_B - \mathbf{E}_B \times \mathbf{H}_A) \cdot \mathbf{n} dS = 0 \tag{3.37}$$

for any surface that contains all the sources. Although this looks like the standard reciprocity relation, it must be noted that media A and B must be related according to the relations $\epsilon_B = \epsilon_A^T$ and $\mu_B = \mu_A^T$. Budden (1990) has considered such an extension for the non-isotropic medium that is generated by the effect of Earth's magnetic field on the ionosphere. Here the appropriate change in media properties (ϵ to ϵ^T) is achieved by reversing the direction of Earth's magnetic field. It is important to note that true reciprocity does not hold for the ionospheric medium and this is an important consideration for ionospheric communications (Budden, 1961).

We now perturb medium B so that $\epsilon_B = \epsilon_A^T + \delta\epsilon$ and $\mu_B = \mu_B^T + \delta\mu$ where $\delta\mu$ and $\delta\epsilon$ are tensor perturbations to the permittivity and permeability. The pseudo reciprocity result will now take the form

$$\int_V \mathbf{E}_B \cdot \mathbf{J}_A dV = \int_V \mathbf{E}_A \cdot \mathbf{J}_B dV - j\omega \int_V (\mathbf{H}_A \cdot \delta\mu \mathbf{H}_B - \mathbf{E}_A \cdot \delta\varepsilon \mathbf{E}_B) dV \tag{3.38}$$

We will consider the case of perturbations in permittivity alone (i.e. $\delta\mu = 0$ and $\delta\epsilon \neq 0$) and take $\mathbf{J}_A$ to be a unit electric dipole at point A ($\mathbf{J}_A = \delta(\mathbf{r}-\mathbf{r}_A)\hat{\mathbf{s}}$ where $\hat{\mathbf{s}}$ is an arbitrary unit vector). Then from Equations 3.36 and 3.38, we have

$$\delta\mathbf{E}_B(\mathbf{r}_A) \cdot \hat{\mathbf{s}} = j\omega \int_V \mathbf{E}_A \cdot \delta\varepsilon \mathbf{E}_B dV \tag{3.39}$$

where $\delta\mathbf{E}_B$ is the deviation in electric field caused by a perturbation $\delta\epsilon$ in permittivity (Coleman, 2007). If the perturbation is small, an estimate for $\delta\mathbf{E}_B$ can be found by replacing $\mathbf{E}_B$ in the right-hand side of Equation 3.39 with its unperturbed value.

3.4 The Compensation Theorem

We consider vector identity 3.6 for the case where the fields $(\mathbf{H}_A, \mathbf{E}_A)$ and $(\mathbf{H}_B, \mathbf{E}_B)$ satisfy Maxwell's equations with different material properties; i.e. we will take field B to have its permeability perturbed by amount $\delta\mu$ and permittivity by amount $\delta\epsilon$. Assuming the materials to be isotropic, Maxwell's equations will imply

$$\begin{aligned}\nabla \cdot (\mathbf{E}_A \times \mathbf{H}_B &- \mathbf{E}_B \times \mathbf{H}_A) \\ = \mathbf{J}_A \cdot \mathbf{E}_B &- \mathbf{J}_B \cdot \mathbf{E}_A - \mathbf{H}_B \cdot \mathbf{M}_A + \mathbf{H}_A \cdot \mathbf{M}_B \\ &+ j\omega\delta\mu \mathbf{H}_A \cdot \mathbf{H}_B - j\omega\delta\epsilon \mathbf{E}_A \cdot \mathbf{E}_B\end{aligned} \tag{3.40}$$

We consider the perturbations in material properties to be confined to a volume V with surface S. (The sources of fields A and B are assumed to have finite extent and to be exterior to V.) If we integrate Equation 3.40 over the whole of space, the divergence

3.4 The Compensation Theorem

term will become a surface integral at infinity and, as we have bounded sources, this term will vanish. Consequently, we now have

$$\int_{R^3} (\mathbf{J}_A \cdot \mathbf{E}_B - \mathbf{J}_B \cdot \mathbf{E}_A - \mathbf{H}_B \cdot \mathbf{M}_A + \mathbf{H}_A \cdot \mathbf{M}_B)\, dV$$
$$= \int_V (j\omega\delta\epsilon \mathbf{E}_A \cdot \mathbf{E}_B - j\omega\delta\mu \mathbf{H}_A \cdot \mathbf{H}_B)\, dV \quad (3.41)$$

From this, it is clear that the effect of changing the properties of the medium can be mimicked by keeping the original medium, but compensating by adding electric current distribution $\delta\mathbf{J}_B = j\omega\delta\epsilon\mathbf{E}_B$ and magnetic current distribution $\delta\mathbf{M}_B = j\omega\delta\mu\mathbf{H}_B$. Further, provided the perturbations in material properties are small, we can use the $\delta\mu = \delta\epsilon = 0$ values of $\mathbf{E}_B$ and $\mathbf{H}_B$ in this expressions.

We now integrate Equation 3.40 over the volume V alone and obtain

$$\int_V (j\omega\delta\epsilon\mathbf{E}_A \cdot \mathbf{E}_B - j\omega\delta\mu\mathbf{H}_A \cdot \mathbf{H}_B)\, dV = -\int_S (\mathbf{E}_A \times \mathbf{H}_B - \mathbf{E}_B \times \mathbf{H}_A) \cdot \mathbf{n}\, dS \quad (3.42)$$

and, as a consequence, can eliminate the term in $\delta\mu$ and $\delta\epsilon$ between Equations 3.41 and 3.42 to obtain

$$\int_{R^3} (\mathbf{J}_A \cdot \mathbf{E}_B - \mathbf{J}_B \cdot \mathbf{E}_A - \mathbf{H}_B \cdot \mathbf{M}_A + \mathbf{H}_A \cdot \mathbf{M}_B)\, dV$$
$$= -\int_S (\mathbf{E}_A \times \mathbf{H}_B - \mathbf{E}_B \times \mathbf{H}_A) \cdot \mathbf{n}\, dS \quad (3.43)$$

For the special case where the sources of fields A and B consist of antennas that are driven by currents I_A and I_B, respectively, Equation 3.41 will imply

$$\delta Z_{AB} I_A I_B = -\int_V (j\omega\delta\epsilon\mathbf{E}_A \cdot \mathbf{E}_B - j\omega\delta\mu\mathbf{H}_A \cdot \mathbf{H}_B)\, dV \quad (3.44)$$

where δZ_{AB} is the change in mutual impedance between the antennas caused by the perturbation in material properties ($\delta\mu$ and $\delta\epsilon$). This is another form of the *compensation theorem* for fields (Monteath, 1973).

We can derive another form of the compensation theorem that is extremely useful in the study of surface wave propagation. Let S be a surface on which the material properties change (the permittivity and permeability are constant either side of the surface). Assume the change can be modeled using impedance boundary condition 2.97,

$$(\mathbf{E}_A \times \mathbf{H}_B) \cdot \mathbf{n} = -\eta \mathbf{H}_A^{tan} \cdot \mathbf{H}_B^{tan} \quad (3.45)$$

and

$$(\mathbf{E}_B \times \mathbf{H}_A) \cdot \mathbf{n} = -(\eta + \delta\eta)\mathbf{H}_A^{tan} \cdot \mathbf{H}_B^{tan} \quad (3.46)$$

where $\mathbf{H}_A^{tan} = \mathbf{H}_A - \mathbf{n}\mathbf{n} \cdot \mathbf{H}_A$ and $\mathbf{H}_B^{tan} = \mathbf{H}_B - \mathbf{n}\mathbf{n} \cdot \mathbf{H}_B$ are the tangential components of their respective fields. Substituting into Equation 3.43, we obtain

$$\int_V (\mathbf{J}_A \cdot \mathbf{E}_B - \mathbf{J}_B \cdot \mathbf{E}_A - \mathbf{H}_B \cdot \mathbf{M}_A + \mathbf{H}_A \cdot \mathbf{M}_B)\, dV = -\int_S \delta\eta \mathbf{H}_A^{tan} \cdot \mathbf{H}_B^{tan}\, dS \quad (3.47)$$

When the sources of fields A and B consist of antennas that are driven by currents I_A and I_B, respectively, Equation 3.47 reduces to (Monteath, 1973)

$$\delta Z_{AB} I_A I_B = \int_S \delta \eta \mathbf{H}_A^{tan} \cdot \mathbf{H}_B^{tan} \, dS \qquad (3.48)$$

3.5 The Extinction Theorem

We consider two fields, $(\mathbf{H}, \mathbf{E})$ and $(\mathbf{H}_T, \mathbf{E}_T)$, that satisfy the time harmonic Maxwell's equations

$$\nabla \times \mathbf{H} = j\omega\epsilon \mathbf{E} + \mathbf{J}_S \qquad (3.49)$$
$$\nabla \times \mathbf{E} = -j\omega\mu \mathbf{H} \qquad (3.50)$$
$$\nabla \times \mathbf{H}_T = j\omega\epsilon \mathbf{E}_T + \hat{\mathbf{s}}\delta(\mathbf{r} - \mathbf{a}) \qquad (3.51)$$
$$\nabla \times \mathbf{E}_T = -j\omega\mu \mathbf{H}_T \qquad (3.52)$$

Field $(\mathbf{H}, \mathbf{E})$ is caused by an isolated source with current density $\mathbf{J}_S$ and $(\mathbf{H}_T, \mathbf{E}_T)$ is caused by an infinitesimal current element $\mathbf{J}_T = \delta(\mathbf{r} - \mathbf{a})\hat{\mathbf{s}}$ at location $\mathbf{a}$ where $\hat{\mathbf{s}}$ is an arbitrary unit vector. Note that, from Equations 2.139 and 2.140, that the test field $(\mathbf{H}_T, \mathbf{E}_T)$ is given by

$$\mathbf{E}_T(\mathbf{r}) = \frac{1}{j\omega\epsilon} \nabla \times \nabla \times \frac{\hat{\mathbf{s}} \exp(-j\beta |\mathbf{r} - \mathbf{a}|)}{4\pi |\mathbf{r} - \mathbf{a}|} \qquad (3.53)$$

and

$$\mathbf{H}_T(\mathbf{r}) = \nabla \times \frac{\hat{\mathbf{s}} \exp(-j\beta |\mathbf{r} - \mathbf{a}|)}{4\pi |\mathbf{r} - \mathbf{a}|} \qquad (3.54)$$

We now consider a bounded surface S and a volume V that is the region outside this surface. The source with current vector $\mathbf{J}_S$ lies within V and its field illuminates the surface S (see Figure 3.5). From Equations 3.53 and 3.11, we obtain

$$-\int_S (\mathbf{E}(\mathbf{r}) \times \mathbf{H}_T(\mathbf{r}) - \mathbf{E}_T(\mathbf{r}) \times \mathbf{H}(\mathbf{r})) \cdot \mathbf{n}(\mathbf{r}) \, dS = -C\hat{\mathbf{s}} \cdot \mathbf{E}(\mathbf{a}) + \hat{\mathbf{s}} \cdot \mathbf{E}^i(\mathbf{a}) \qquad (3.55)$$

where $C = 1$ for points $\mathbf{a}$ outside S and $C = 0$ for points inside (note that the unit vector $\mathbf{n}$ is normal to S and points into volume V). Field $(\mathbf{H}^i, \mathbf{E}^i)$ is that caused by the sources of $(\mathbf{H}, \mathbf{E})$ acting in a homogeneous space that is devoid of other sources. Rearranging the surface integral, we obtain

$$\int_S (\mathbf{H}_T(\mathbf{r}) \cdot (\mathbf{E}(\mathbf{r}) \times \mathbf{n}(\mathbf{r})) + \mathbf{E}_T(\mathbf{r}) \cdot (\mathbf{H}(\mathbf{r}) \times \mathbf{n}(\mathbf{r}))) \, dS = -C\hat{\mathbf{s}} \cdot \mathbf{E}(\mathbf{a}) + \hat{\mathbf{s}} \cdot \mathbf{E}^i(\mathbf{a}) \qquad (3.56)$$

We now define a surface current density $\mathbf{J}$ by $\mathbf{J} = \mathbf{n} \times \mathbf{H}$ and assume S is the surface of a body whose properties can be represented by the impedance boundary condition

3.5 The Extinction Theorem

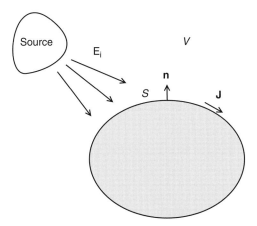

Figure 3.5 Geometry for the extinction theorem.

$\mathbf{n} \times \mathbf{E} = \eta_r \eta_0 \mathbf{n} \times (\mathbf{n} \times \mathbf{H})$ (η_r is the relative surface impedance). Consequently, Equation 3.56 can be recast as

$$-\int_S (\eta_r \eta_0 \mathbf{H}_T(\mathbf{r}) \cdot (\mathbf{n}(\mathbf{r}) \times \mathbf{J}(\mathbf{r})) + \mathbf{E}_T(\mathbf{r}) \cdot \mathbf{J}(\mathbf{r}))\, dS = -C\hat{\mathbf{s}} \cdot \mathbf{E}(\mathbf{a}) + \hat{\mathbf{s}} \cdot \mathbf{E}^i(\mathbf{a}) \quad (3.57)$$

Substituting for the field ($\mathbf{H}_T, \mathbf{E}_T$), and noting that $\hat{\mathbf{s}}$ is an arbitrary unit vector, we obtain

$$\begin{aligned}\mathbf{E}^i(\mathbf{a}) - \eta_0 \nabla \times \int_S \eta_r \frac{\exp(-j\beta |\mathbf{r} - \mathbf{a}|)}{4\pi |\mathbf{r} - \mathbf{a}|} \mathbf{J}(\mathbf{r}) \times \mathbf{n}(\mathbf{r})\, dS \\ + \frac{\eta_0}{j\beta} \nabla \times \nabla \times \int_S \frac{\exp(-j\beta |\mathbf{r} - \mathbf{a}|)}{4\pi |\mathbf{r} - \mathbf{a}|} \mathbf{J}(\mathbf{r})\, dS = C\mathbf{E}(\mathbf{a})\end{aligned} \quad (3.58)$$

where ∇ is now an operator in **a** coordinates. It will be noted that, for points **a** inside the body, the field **E** is extinguished and Equation 3.58 provides an integral equation for the surface current field **J**. For points **a** outside the body, Equation 3.58 enables the calculation of field **E** once **J** is known. This is a generalized form of the *extinction theorem* (Coleman, 1996). If we set $\eta_r = 0$, we obtain the standard form of the Ewald–Oseen extinction theorem (Born and Wolf, 1980) for a body with perfectly conducting surface.

A useful application of the generalized extinction theorem is in the calculation of the effect of a non-perfectly conducting ground on an incident field (Coleman, 1996). We assume that the ground is sufficiently absorbing and that the region of induced current is effectively bounded. Then, in the limit that $a = |\mathbf{a}| \to \infty$, 3.58 becomes

$$\begin{aligned}\mathbf{E}^i(\mathbf{a}) + \frac{j\beta \eta_0 \exp(-j\beta a)}{4\pi a} \hat{\mathbf{a}} \times \int_S \eta_r \exp(j\beta \mathbf{r} \cdot \hat{\mathbf{a}}) \mathbf{J}(\mathbf{r}) \times \mathbf{n}(\mathbf{r})\, dS \\ + \frac{j\beta \eta_0 \exp(-j\beta a)}{4\pi a} \hat{\mathbf{a}} \times \hat{\mathbf{a}} \times \int_S \exp(j\beta \mathbf{r} \cdot \hat{\mathbf{a}}) \mathbf{J}(\mathbf{r})\, dS = C\mathbf{E}(\mathbf{a})\end{aligned} \quad (3.59)$$

where $\hat{\mathbf{a}} = \mathbf{a}/a$. We consider a surface that is the $z = 0$ plane with constant relative impedance η_r and define a vector field $\mathbf{K}$ by

$$\mathbf{K}(\mathbf{a}) = \frac{j\beta\eta_0 \exp(-j\beta a)}{4\pi a} \int_{R^2} \exp(j\beta \, \mathbf{r} \cdot \hat{\mathbf{a}}) \mathbf{J}(\mathbf{r}) \, dxdy \qquad (3.60)$$

where it should be noted that $\mathbf{K}$ has no $\hat{\mathbf{z}}$ component. Equation 3.59 can be now be rewritten as

$$\mathbf{E}^i(\mathbf{a}) - \eta_r \hat{\mathbf{a}} \times \hat{\mathbf{z}} \times \mathbf{K} + \hat{\mathbf{a}} \times \hat{\mathbf{a}} \times \mathbf{K} = C\mathbf{E}(\mathbf{a}) \qquad (3.61)$$

or, alternatively, as

$$\mathbf{E}^i(\mathbf{a}) + \eta_r \hat{\mathbf{a}} \cdot \hat{\mathbf{z}} \mathbf{K} - \eta_r \hat{\mathbf{z}} \hat{\mathbf{a}} \cdot \mathbf{K} + \hat{\mathbf{a}} \hat{\mathbf{a}} \cdot \mathbf{K} - \mathbf{K} = \mathbf{E}(\mathbf{a}) \qquad (3.62)$$

for points $\mathbf{a}$ above the plane and

$$\mathbf{E}^i(\mathbf{a}) + \eta_r \hat{a}_z \mathbf{K} - \eta_r \hat{\mathbf{z}} \hat{\mathbf{a}} \cdot \mathbf{K} + \hat{\mathbf{a}} \hat{\mathbf{a}} \cdot \mathbf{K} - \mathbf{K} = 0 \qquad (3.63)$$

for points below the plane. We will take the point below the plane to be the conjugate point $\mathbf{a}_c$ of $\mathbf{a}$, i.e. $\mathbf{a}_c = \mathbf{a} - 2a_z\hat{\mathbf{z}}$, so that

$$\mathbf{E}^i(\mathbf{a}_c) - \eta_r \hat{a}_z \mathbf{K} - \eta_r \hat{\mathbf{z}} \hat{\mathbf{a}} \cdot \mathbf{K} + \hat{\mathbf{a}} \hat{\mathbf{a}} \cdot \mathbf{K} - 2\hat{a}_z \hat{\mathbf{z}} \hat{\mathbf{a}} \cdot \mathbf{K} - \mathbf{K} = 0 \qquad (3.64)$$

We now split $\mathbf{E}$ into vertical ($\hat{\mathbf{z}}$ direction) component $\mathbf{E}_v$ and horizontal component $\mathbf{E}_h = \mathbf{E} - \mathbf{E}_v$. Then, from Equations 3.62 and 3.64

$$\mathbf{E}_h(\mathbf{a}) = \mathbf{E}_h^i(\mathbf{a}) - \mathbf{E}_h^i(\mathbf{a}_c) + 2\eta_r \hat{a}_z \mathbf{K} \qquad (3.65)$$

and

$$\mathbf{E}_v(\mathbf{a}) = \mathbf{E}_h^i(\mathbf{a}) + \mathbf{E}_h^i(\mathbf{a}_c) - 2\eta_r \hat{\mathbf{a}} \cdot \mathbf{K}\hat{\mathbf{z}} \qquad (3.66)$$

Consider vector $\hat{\mathbf{a}}_h = \hat{\mathbf{a}} - \hat{a}_z\hat{\mathbf{z}}$ and take the component of Equation 3.63 in this direction, then

$$\hat{\mathbf{a}}_h \cdot \mathbf{E}^i(\mathbf{a}_c) - \eta_r \hat{a}_z \hat{\mathbf{a}} \cdot \mathbf{K} - \hat{a}_z^2 \hat{\mathbf{a}} \cdot \mathbf{K} = 0 \qquad (3.67)$$

on noting that $\hat{\mathbf{a}}_h \cdot \hat{\mathbf{a}}_h = 1 - \hat{a}_z^2$. Rearranging Equation 3.67, we obtain

$$\hat{\mathbf{a}}_h \cdot \mathbf{K} = \frac{\hat{\mathbf{a}}_h \cdot \mathbf{E}_h^i(\mathbf{a}_c)}{\hat{a}_z^2 + \eta_r \hat{a}_z} \qquad (3.68)$$

We can now replace $\hat{\mathbf{a}}_h \cdot \mathbf{K}$ in the horizontal component of Equation 3.68 and obtain an expression for $\mathbf{K}$ in terms of known quantities, i.e.

$$\mathbf{K} = \left(\mathbf{E}_h^i(\mathbf{a}_c) + \hat{\mathbf{a}}_h \frac{\hat{\mathbf{a}}_h \cdot \mathbf{E}_h^i(\mathbf{a}_c)}{\hat{a}_z^2 + \eta_r \hat{a}_z}\right) \frac{1}{1 + \eta_r \hat{a}_z} \qquad (3.69)$$

Expressions 3.65 and 3.66, together with Equations 3.68 and 3.69, will now yield the field $\mathbf{E}$ with the effect of non-perfectly conducting ground included.

3.6 References

M. Born and E. Wolf, *Principles of Optics*, 6th edition, Pergamon Press, Oxford, 1980.

K.G. Budden, *The Propagation of Radio Waves*, Cambridge University Press, Cambridge, 1988.

K.G. Budden, *Radio Waves in the Ionosphere*, Cambridge University Press, Cambridge, 1961.

C.J. Coleman, Application of the extinction theorem to some antenna problems, IEE Proc.-Microw. Antennas Propag., vol. 143, pp. 471–474, 1996.

C.J. Coleman, *An Introduction to Radio Frequency Engineering*, Cambridge University Press, Cambridge, 2004.

C.J. Coleman, Pseudo reciprocity in a disturbed non isotropic medium and its application to radio wave propagation, Radio Sci., 42, RS3013, doi:10.1029/2006RS003015, 2007.

A.T. De Hoop, *Handbook of Radiation and Scattering of Waves*, Academic Press, London, 1995.

L.B. Felsen and N. Marcuvitz, *Radiation and Scattering of Waves*, IEEE Press and Oxford University Press, New York and Oxford, 1994.

D.S. Jones, *Methods in Electromagnetic Wave Propagation*, 2nd edition, IEEE/OUP series in Electromagnetic Wave Theory, Oxford University Press, New York and Oxford, 1999.

G.D. Monteath, *Applications of the Electro-magnetic Reciprocity Principle*, Pergamon Press, Oxford, 1973.

4 The Effect of Obstructions on Radio Wave Propagation

In the following chapter we will consider the propagation medium to be *free space* ($\mu = \mu_0$ and $\epsilon = \epsilon_0$), but with the radio waves interrupted by obstructions such as hills and buildings. The Friis equation is extended to include the effects of reflection by a planar ground, and then Fermat's principle is introduced as a method for calculating reflections in the case of irregular ground. Reciprocity-based arguments are used to derive analytic expressions for the loss incurred in the propagation mechanism of diffraction, and this is followed by an introduction to the geometric theory of diffraction (GTD). The chapter ends with a discussion of propagation in urban environments in which important results concerning propagation over buildings are described.

4.1 The Friis Equation

The most basic tool in propagation modeling is the Friis equation for line-of-sight communications between two stations (A and B). This equation relates the power P_B received by the antenna of station B as a result of power P_A transmitted by the antenna of station A (Figure 4.1). Let the stations be separated by distance r_{AB} and have antennas with gains G_A and G_B, respectively. Due to spreading, the power flow per unit area at distance r_{AB} from B is given by $P = G_A P_A / 4\pi r_{AB}^2$. The power received by station B will be $P_B = P A_{\text{eff}}$ where A_{eff} is the effective area of the antenna of station B. Since $A_{\text{eff}} = G_B \lambda^2 / 4\pi$, we will therefore have the Friis equation

$$P_B = P_A G_A G_B \left(\frac{\lambda}{4\pi r_{AB}}\right)^2 \qquad (4.1)$$

where λ is the wavelength. It should be noted, however, that the above Friis equation assumes that the antennas are *polarization matched* (i.e. $\left|\mathbf{h}_{\text{eff}}^A \cdot \mathbf{h}_{\text{eff}}^B\right| = \left|\mathbf{h}_{\text{eff}}^A\right|\left|\mathbf{h}_{\text{eff}}^B\right|$). Consequently, when mismatch is present, the power will be reduced by the *polarization efficiency*

$$\eta_P = \frac{\left|\mathbf{h}_{\text{eff}}^A \cdot \mathbf{E}_B^A\right|^2}{\left|\mathbf{h}_{\text{eff}}^A\right|^2 \left|\mathbf{E}_B^A\right|^2} \qquad (4.2)$$

Besides polarization matching, the Friis equation also assumes there to be no interaction of the radio waves with the ground or obstacles. In reality, however, there will certainly

Figure 4.1 Communication system with two stations.

Figure 4.2 Propagation with ground reflections.

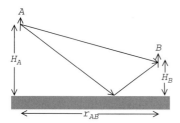

Figure 4.3 Propagation between two antennas.

be at least some ground reflections in addition to the direct wave, and the Friis equation will need further modification (Figure 4.2).

Consider two radio stations that have antennas at heights H_A and H_B above the ground (see Figure 4.3) with effective antenna lengths h_A and h_B (we assume the same polarization for both antennas). Further, assume the stations are separated by a distance r_{AB} such that $r_{AB} \gg H_A$ and $r_{AB} \gg H_B$. As a consequence, the propagation is approximately horizontal and the polarization approximately constant (i.e. the electric field can be represented by a scalar value). We now calculate the voltage V in one antenna due to a current I in the other (note that by reciprocity, it does not matter which antenna is chosen).

Since the antennas have the same polarization, the voltage in the terminals of the receive antenna A due to direct propagation is $E_D h_A$ and that due to the reflected propagation is $E_R h_A$. The direct field E_D is given by

$$E_D = \frac{j\omega\mu_0 I h_A}{4\pi} \frac{\exp(-j\beta l)}{l} \quad (4.3)$$

where l is the direct distance between the antennas ($l = \sqrt{r_{AB}^2 + (H_A - H_B)^2}$). The reflected electric field E_R is given by

$$E_R = R \frac{j\omega\mu_0 I h_A}{4\pi} \frac{\exp(-j\beta s)}{s} \quad (4.4)$$

where s is the distance over the reflected path between antennas ($s = \sqrt{r_{AB}^2 + (H_A + H_B)^2}$) and R is the reflection coefficient of the ground. Since $r_{AB} \gg H_A$ and $r_{AB} \gg H_B$ we can use the binomial approximation ($\sqrt{1+x} \approx 1 + x/2$ when $x \ll 1$) to obtain an approximate expression for l

$$l = r_{AB}\sqrt{1 + \frac{(H_A - H_B)^2}{r_{AB}^2}} \approx r_{AB} + \frac{(H_A - H_B)^2}{2r_{AB}} \quad (4.5)$$

and one for s

$$s = r_{AB}\sqrt{1 + \frac{(H_A + H_B)^2}{r_{AB}^2}} \approx r_{AB} + \frac{(H_A + H_B)^2}{2r_{AB}} \quad (4.6)$$

The fields can now be approximated by

$$E_D \approx \frac{j\omega\mu_0 I h_A}{4\pi} \frac{\exp(-j\beta r_{AB})}{r_{AB}} \exp\left(-j\beta \frac{(H_A - H_B)^2}{2r_{AB}}\right) \quad (4.7)$$

and

$$E_R \approx R \frac{j\omega\mu_0 I h_A}{4\pi} \frac{\exp(-j\beta r_{AB})}{r_{AB}} \exp\left(-j\beta \frac{(H_A + H_B)^2}{2r_{AB}}\right) \quad (4.8)$$

As a consequence, the total voltage V in antenna A due to current I in antenna B is given by

$$V = h_B E_D + h_B E_R$$

$$\approx \frac{j\omega\mu_0 I h_A h_B}{4\pi} \frac{\exp\left(-j\beta\left(r_{AB} + \frac{H_A^2 + H_B^2}{2r_{AB}}\right)\right)}{r_{AB}}$$

$$\times \left(\exp\left(j\beta \frac{H_A H_B}{r_{AB}}\right) + R\exp\left(-j\beta \frac{H_A H_B}{r_{AB}}\right)\right) \quad (4.9)$$

In order to calculate the reflection coefficient R we will need the polarization and the angle of reflection α ($\approx (H_A + H_B)/r_{AB}$). However, for high frequencies (above about 500 MHz), or for small α, $R = -1$ is usually a good approximation. Since $P_A \propto V^2$, the modified Friis equation will therefore take the form

$$P_B = P_A G_A G_B \left(\frac{\lambda}{4\pi r_{AB}}\right)^2 \left|1 + R\exp\left(-2j\frac{\beta H_A H_B}{r_{AB}}\right)\right|^2 \quad (4.10)$$

where R is the reflection coefficient of the ground. It is obvious that there can be constructive or destructive interference between direct and reflected signals and this will vary with distance between transmitter and receiver. The effect can often be pronounced in the case of mobile communications where the signal can experience *flutter* as the vehicle moves. This can get worse in urban environments where the propagation paths can be more tortuous and numerous.

4.2 Reflection by Irregular Terrain

In deriving the above extension to the Friis equation, the path of the reflected propagation was calculated on the basis of a plane ground. In the case of more complex topography, the path can be calculated using Fermat's principle:

The propagation path between two points is that for which the time of propagation is minimum with respect to small variations in path.

We will see later that this principle needs further generalization when the medium is nonhomogeneous, but this form is sufficient for present purposes. Fermat's principle can be illustrated through the following example. Consider the situation of Figure 4.4 with a transmitter A above ground defined by the equation $y = f(x)$ (A at $x = 0$) and a receiver B at ground range D. There are two propagation paths between A and B, the direct path and the one by means of reflection at the ground. It is a trivial matter to find the direct path, but the reflected path is more involved. For a reflection point at horizontal distance x, the total distance s of propagation for the reflected path is given by

$$s(x) = \sqrt{(H_A - f(x))^2 + x^2} + \sqrt{(D-x)^2 + (f(x) - H_B)^2} \qquad (4.11)$$

Then, according to Fermat's principle, we will need to find the value of x for which this is minimum. This will occur where $s'(x) = 0$, i.e.

$$\frac{-f'(x)(H_A - f(x)) + x}{\sqrt{(H_A - f(x))^2 + x^2}} + \frac{x - D + f'(x)(f(x) - H_B)}{\sqrt{(D-x)^2 + (f(x) - H_B)^2}} = 0 \qquad (4.12)$$

where prime denotes a derivative with respect to x. For arbitrarily shaped ground, this equation will need to be solved numerically (by Newton's method, for example). For $f(x) = 0$ (i.e. flat ground) we obtain

$$\frac{x}{\sqrt{H_A^2 + x^2}} + \frac{x - D}{\sqrt{(D-x)^2 + H_B^2}} = 0 \qquad (4.13)$$

which, as expected, yields the result that $x = DH_A/(H_A + H_B)$.

In most propagation problems the horizontal scales are much greater than vertical scales and so we can make the following parabolic approximation

$$s(x) \approx D + \frac{(H_A - f(x))^2}{2x} + \frac{(f(x) - H_B)^2}{2(D - x)} \qquad (4.14)$$

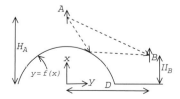

Figure 4.4 Reflection by an undulating ground.

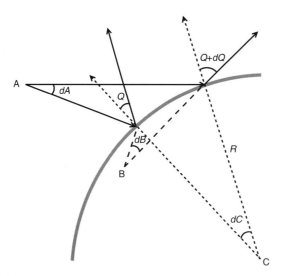

Figure 4.5 Effect of reflection at a curved surface on wavefront curvature.

From this, $s'(x) = 0$ implies that

$$2x(D - x)f'(x) + (D - x)(H_A - f(x)) - x(H_B - f(x)) = 0 \quad (4.15)$$

Then, for the simple linear ground profile $f(x) = H_0 + \alpha x$, we obtain the reflection point $x = D(H_A - H_0)/(H_A + H_B - 2H_0 - \alpha D)$.

Up to now we have assumed that, in free space, the field falls away as the inverse of the distance from the source. More correctly, the field falls away as the inverse of the radius of curvature of the wavefront (this is the distance from the source for an unimpeded spherical wave). In general, the wavefront will be described by two *principal radii of curvature* (ρ_1 and ρ_2) and the field will fall away as the inverse of their geometric average (i.e. as $1/\sqrt{\rho_1 \rho_2}$). When the wave encounters a curved surface, however, there can be focusing (or defocusing) and considerable modification to the curvature of the wavefront as a result of reflection. Consider Figure 4.5 in which point A is the source of a wave that is incident upon a cylindrical surface in a direction that is normal to the cylinder axis and at angle Q to the cylinder normal at the point of incidence. (At the point of incidence, the cylinder has radius of curvature R and center of curvature C.) The reflected wave can be considered to be radiating from an image source B. Let the source radiate within angle dA and the image source within angle dB, then the center of curvature will subtend the angle dC upon the same arc that is illuminated by the sources A and B. We assume that the wavefront radius of curvature is the principal radius ρ_1 at the cylinder, then

$$\frac{dA\rho_1}{\cos Q} = RdC = \frac{dB\hat{\rho}_1}{\cos Q} \quad (4.16)$$

where $\hat{\rho}_1$ is the radius of curvature of the reflected wave. We note that the angle of incidence changes over the arc length from Q to $Q + dQ$, then $dA = dQ - dC$ and

$dB = dQ + dC$. As a consequence, from Equation 4.16, we obtain that

$$\frac{(dQ - dC)\rho_1}{\cos Q} = RdC = \frac{(dQ + dC)\hat{\rho}_1}{\cos Q} \tag{4.17}$$

Equation 4.17 constitutes a pair of homogeneous equations that will only have a nonzero solution when

$$\frac{1}{\hat{\rho}_1} = \frac{2}{R \cos Q} + \frac{1}{\rho_1} \tag{4.18}$$

and from this it is clear that the radius of curvature of a wavefront will be modified by a curved reflecting surface. (Note that, for a surface defined by the equation $y = f(x)$, the radius of curvature will be given by $R = (1 + y'^2)^{\frac{3}{2}}/y''$.) If the surface is also curved in the transverse direction, the second principle radius of curvature ρ_2 is also affected. If the reflecting surface is spherical, the curvature of the surface in the lateral direction is also R and $1/\hat{\rho}_2 = 2/R \cos Q + 1/\rho_2$. Since the surface of the Earth is approximately spherical, one might think that such corrections are always required. In this case, however, $R = 6378$ km and it is clear that such corrections are not required for short-range terrestrial communications. (This is not the case for communications via the ionospheric duct since these can take place over many thousands of kilometers.)

4.3 Diffraction

If there are obstacles in the propagation path, Huygens' principle implies that radio waves can propagate over them by the mechanism of diffraction. We will investigate the propagation between two antennas that are separated by a screen with finite height (ground reflections will be ignored). Referring to Figure 4.6, the screen is located in the $x - y$ plane. We can perform these calculations using the integral equation 3.23, i.e.

$$I_A V_{AB} \approx \frac{2}{\eta_0} \int_S \mathbf{E}_A \cdot \mathbf{E}_B \, dS \tag{4.19}$$

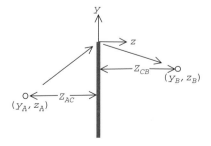

Figure 4.6 Diffraction over a screen.

where the field of antenna A will have the form

$$E_A = \frac{j\omega\mu_0 I_A}{4\pi} \mathbf{h}^A_{\text{eff}} \frac{\exp(-j\beta r_A)}{r_A} \tag{4.20}$$

and the field of antenna B the form

$$E_B = \frac{j\omega\mu_0 I_B}{4\pi} \mathbf{h}^B_{\text{eff}} \frac{\exp(-j\beta r_B)}{r_B} \tag{4.21}$$

where $\mathbf{h}^A_{\text{eff}}$ is the effective antenna length of antenna A and $\mathbf{h}^B_{\text{eff}}$ is the effective length of antenna B. The integration surface S in Equation 4.19 is taken to be the xy plane with the integrand set equal to zero on the screen (the power falling on the screen will be reflected, not transmitted). Due to the oscillatory nature of the integrals, the largest contribution will arise close to the origin of the $x - y$ plane. We will assume that horizontal distances are large in comparison with vertical distances. Consequently, if r_{AC} and r_{CB} are the distances from antennas A and B, respectively, to a point (x, y) above the screen, then $r_A \approx z_{AC} + (x^2 + (y - y_A)^2)/2z_{AC}$ and $r_B \approx z_{CB} + (x^2 + (y - y_B)^2)/2z_{CB}$. From Equation 4.19, the mutual impedance between antennas A and B is

$$Z_{AB} = \frac{-\omega^2 \mu_0^2}{8\pi^2 \eta_0} \mathbf{h}^A_{\text{eff}} \cdot \mathbf{h}^B_{\text{eff}} \exp(-j\beta D)$$
$$\times \int_0^\infty \int_{-\infty}^\infty \frac{\exp\left(-\frac{j\beta}{2}\left(\frac{x^2}{z_{AC}} + \frac{x^2}{z_{CB}} + \frac{(y-y_A)^2}{z_{AC}} + \frac{(y-y_B)^2}{z_{CB}}\right)\right)}{z_{AC} z_{CB}} dx dy \tag{4.22}$$

where $D = z_{AC} + z_{CB}$ is the horizontal distance between the antennas. We perform the x integral and then Equation 4.22 can be expressed as

$$Z_{AB} = \frac{-\beta^2 \eta_0}{8\pi^2} \sqrt{\frac{2\pi}{j\beta D z_{AC} z_{CB}}} \mathbf{h}^A_{\text{eff}} \cdot \mathbf{h}^B_{\text{eff}} \exp(-j\beta D)$$
$$\times \int_0^\infty \exp\left(-\frac{j\beta}{2}\left(\frac{y^2}{z_{AC}} + \frac{y^2}{z_{CB}} - \frac{2yy_A}{z_{AC}} - \frac{2yy_B}{z_{CB}} + \frac{y_A^2}{z_{AC}} + \frac{y_B^2}{z_{CB}}\right)\right) dy \tag{4.23}$$

and, after some algebra,

$$Z_{AB} = \frac{-\beta^2 \eta_0}{8\pi^2} \sqrt{\frac{2\pi}{j\beta D z_{AC} z_{CB}}} \mathbf{h}^A_{\text{eff}} \cdot \mathbf{h}^B_{\text{eff}} \exp(-j\beta D)$$
$$\times \exp\left(-\frac{j\beta}{2}\left(\frac{y_A^2}{z_{AC}} + \frac{y_B^2}{z_{CB}} - \frac{z_{AC} z_{CB}}{D}\left(\frac{y_A}{z_{AC}} + \frac{y_B}{z_{CB}}\right)^2\right)\right)$$
$$\times \int_0^\infty \exp\left(-\frac{jD\beta}{2z_{AC} z_{CB}}\left(y - \left(\frac{z_{CB} y_A}{D} + \frac{z_{AC} y_B}{D}\right)\right)^2\right) dy \tag{4.24}$$

We introduce the new variable $Y = \sqrt{D\beta/2 z_{AC} z_{BC}}(y - z_{CB} y_A/D - z_{AC} y_B/D)$ and, simplifying the exponential terms outside the integral,

4.3 Diffraction

$$Z_{AB} = \frac{j\omega\mu_0}{4\pi D}\sqrt{\frac{j}{\pi}} \mathbf{h}^A_{\text{eff}} \cdot \mathbf{h}^B_{\text{eff}}$$
$$\times \exp\left(-j\beta\left(D + \frac{(y_A - y_B)^2}{2D}\right)\right) \int_v^\infty \exp\left(-jY^2\right) dY \quad (4.25)$$

where $v = \sqrt{D\beta/2z_{AC}z_{CB}}\left(-z_{CB}y_A/D - z_{AC}y_B/D\right)$. We note from Equation 3.27 that, in the absence of the screen,

$$Z_{AB} = \frac{j\omega\mu_0 \mathbf{h}^A_{\text{eff}} \cdot \mathbf{h}^B_{\text{eff}}}{4\pi} \frac{\exp(-j\beta r_{AB})}{r_{AB}} \quad (4.26)$$

where r_{AB} is the distance between the antennas. Noting that $\exp(-j\beta r_{AB})/r_{AB} \approx \exp\left(-j\beta\left(D + (y_A - y_B)^2/2D\right)\right)/D$, it can be seen that the free space received power is modified by the *diffraction loss*

$$L_{\text{diff}} = \left|\sqrt{j/\pi}\int_v^\infty \exp(-jY^2)\,dY\right|^{-2} \quad (4.27)$$

where the above integral can be evaluated in terms of Fresnel integrals (see Appendix A). The height y_A above the screen for which $v = -0.6$ is known as the *Fresnel clearance* and is a height above which the loss due to the screen is less than about 1.5 dB. This clearance is often used to decide when the effect of an obstacle can be ignored.

Propagation over complex obstacles can be approximated by the use of an effective screen (Bullington, 1947). If we extend lines vertically from both receiver and transmitter antennas, we then lower both until they just touch the obstacles. The point where they intersect can now be taken as the top of an effective screen and the above techniques used. The results will usually be within a few dB of more exact calculations (Figure 4.7).

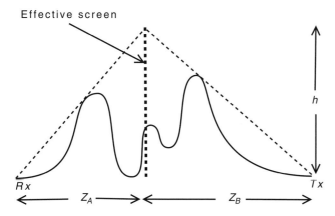

Figure 4.7 Approximating propagation over complex topography.

4.4 Surface Waves

In the generalized Friis equation, when either of the antennas comes close to the ground, the direct and reflected rays can cancel. This is only the case for the leading order (r_{AB}^{-1} terms), as higher-order terms (r_{AB}^{-2} terms) were neglected in the derivation. Consequently, there will still be power reaching the receiver, albeit weak. These higher-order terms are often referred to as *surface waves* and can be calculated using reciprocity techniques (Monteath, 1973). Consider the configuration shown in Figure 4.8 and consider the voltage V_{AB} in antenna A due to current I_B in antenna B. Then, according to Equation 3.23,

$$I_A V_{AB} \approx (2/\eta_0) \int_S \mathbf{E}_A \cdot \mathbf{E}_B \, dS \tag{4.28}$$

Let surface S be the $x - y$ plane at distance r_{AC} from A and r_{CB} from B. The field below the ground will be zero and so the integral will start at the level of the ground ($y = 0$). For points close to the ground, the path lengths of the direct and reflected field components will be approximately the same. Consequently, the fields in Equation 4.28 will consist of the direct field multiplied by a factor of the form $1 + R$ in order to take account of ground reflections. Due to the oscillatory nature of the integrals, the largest contribution will arise from around the origin of the $x - y$ plane (the z axis is the horizontal axis). Consequently, we can use the approximations

$$\mathbf{E}_A = \frac{j\omega\mu_0 I_A}{4\pi} \mathbf{h}_{\text{eff}}^A (1 + R_{AC}) \frac{\exp\left(-j\beta\left(r_{AC} + \frac{x^2+y^2}{2r_{AC}}\right)\right)}{r_{AC}} \tag{4.29}$$

and

$$\mathbf{E}_B = \frac{j\omega\mu_0 I_B}{4\pi} \mathbf{h}_{\text{eff}}^B (1 + R_{BC}) \frac{\exp\left(-j\beta\left(r_{CB} + \frac{x^2+y^2}{2r_{CB}}\right)\right)}{r_{CB}} \tag{4.30}$$

where R_{AC} and R_{CB} are the respective reflection coefficients for the paths to surface S. The angle of reflection at the ground (angle α) will be small and, consequently, in the case of vertical polarization,

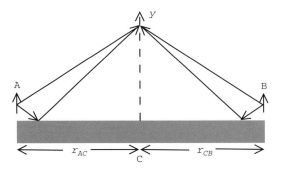

Figure 4.8 Effect of surface on propagation.

4.4 Surface Waves

$$R = \frac{\sin\alpha \, \eta_r^{-2} - \sqrt{\eta_r^{-2} - \cos^2\alpha}}{\sin\alpha \, \eta_r^{-2} + \sqrt{\eta_r^{-2} - \cos^2\alpha}}$$

$$\approx \frac{2\alpha}{\tilde{\eta}_r} - 1 \quad \text{for } \alpha \ll \eta_r \ll 1 \tag{4.31}$$

where $\tilde{\eta}_r = \eta_r\sqrt{1 - \eta_r^2}$ with η_r the relative impedance of the ground ($\tilde{\eta}_r \approx \eta_r$ when η_r is small). Providing the antennas are close to the ground, $\alpha \approx y/r_{AC}$ for R_{AC} and $\alpha \approx y/r_{CB}$ for R_{BC}. Consequently, $1 + R_{AC} \approx 2y/r_{AC}\tilde{\eta}_r$ and $1 + R_{CB} \approx 2y/r_{CB}\tilde{\eta}_r$. Replacing $\mathbf{E}_A$ and $\mathbf{E}_B$ in Equation 4.28, and using the above approximations, we obtain

$$Z_{AB} = \frac{2}{\eta_0}\left(\frac{-\omega^2\mu_0^2}{16\pi^2}\right)\mathbf{h}_{\text{eff}}^A \cdot \mathbf{h}_{\text{eff}}^B \frac{\exp(-j\beta r_{AB})}{r_{AC}r_{CB}}$$

$$\times \int_0^\infty \int_{-\infty}^\infty \frac{4y^2}{\tilde{\eta}_r^2 r_{AC}r_{CB}} \exp\left(\frac{-j\beta(x^2+y^2)r_{AB}}{2r_{AC}r_{CB}}\right) dxdy \tag{4.32}$$

Using $\int_0^\infty t^2 \exp(-jqt^2)\,dt = (1/2jq)\int_0^\infty \exp(-jqt^2)\,dt$ and $\int_0^\infty \exp(-jqt^2)\,dt = \frac{1}{2}\sqrt{\pi/jq}$, we obtain the following expression for the mutual impedance:

$$Z_{AB} \approx \frac{\eta_0 \mathbf{h}_{\text{eff}}^A \cdot \mathbf{h}_{\text{eff}}^B}{2\pi \tilde{\eta}_r^2 r_{AB}^2} \exp(-j\beta r_{AB}) \tag{4.33}$$

(Note that effective antenna lengths $\mathbf{h}_{\text{eff}}^A$ and $\mathbf{h}_{\text{eff}}^B$ are those appropriate to free space.) In a similar fashion to screen diffraction, we can regard the surface wave field as a modified free space field. In this case, the free space received power suffers the additional *ground loss*

$$L_{\text{gnd}} = \left|\frac{2c_0}{jr_{AB}\omega\tilde{\eta}_r^2}\right|^{-2} \tag{4.34}$$

The above arguments can be extended to the case where the ground slopes toward a peak (Monteath, 1973). Consider the geometry shown in Figure 4.9. Providing the antennas are close to the ground, $\alpha \approx (y - y_0)/r_{AC}$ for reflection coefficient R_{AC} and $\alpha \approx (y - y_0)/r_{CB}$ for R_{BC}. Consequently, $1 + R_{AC} \approx 2(y - y_0)/r_{AC}\tilde{\eta}_r$ and

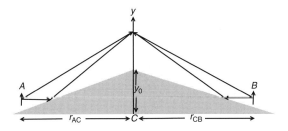

Figure 4.9 Effect of peaked ground on propagation.

$1 + R_{CB} \approx 2(y-y_0)/r_{CB}\tilde{\eta}_r$. Replacing $\mathbf{E}_A$ and $\mathbf{E}_B$ in Equation 4.28, and using the above approximations, we obtain

$$Z_{AB} = \frac{2}{\eta_0} \left(\frac{-\omega^2 \mu_0^2}{16\pi^2} \right) \mathbf{h}_{\text{eff}}^A \cdot \mathbf{h}_{\text{eff}}^B \frac{\exp(-j\beta r_{AB})}{r_{AC}r_{CB}}$$
$$\times \int_{y_0}^{\infty} \int_{-\infty}^{\infty} \frac{4(y-y_0)^2}{\tilde{\eta}_r^2 r_{AC}r_{CB}} \exp\left(\frac{-j\beta(x^2+y^2)r_{AB}}{2r_{AC}r_{CB}} \right) dxdy \quad (4.35)$$

Introducing the new coordinates $Y = (y-y_0)\sqrt{\beta r_{AB}/2r_{AC}r_{CB}}$ and $X = x\sqrt{\beta r_{AB}/2r_{AC}r_{CB}}$ we obtain the expression for the mutual impedance

$$Z_{AB} \approx -\frac{2\eta_0 \mathbf{h}_{\text{eff}}^A \cdot \mathbf{h}_{\text{eff}}^B}{\pi^2 \tilde{\eta}_r^2 r_{AB}^2} \exp(-j\beta r_{AB}) \int_0^{\infty} \int_{-\infty}^{\infty} Y^2 \exp\left(-j(X^2 + (Y+Y_0)^2)\right) dXdY \quad (4.36)$$

where $Y_0 = y_0\sqrt{\beta r_{AB}/2r_{AC}r_{CB}}$. For $Y_0 \ll 1$, $\exp\left(-j(X^2 + (Y+Y_0)^2)\right) \approx (1 - 2jYY_0)\exp\left(j(X^2 + Y^2)\right)$, and hence

$$Z_{AB} \approx \frac{\eta_0 \mathbf{h}_{\text{eff}}^A \cdot \mathbf{h}_{\text{eff}}^B}{2\pi \tilde{\eta}_r^2 r_{AB}^2} \exp(-j\beta r_{AB}) \left(1 + \frac{4}{\sqrt{\pi}} j^{-\frac{3}{2}} Y_0 \right) \quad (4.37)$$

using $\int_0^\infty t^3 \exp(-jt^2)\,dt = -\frac{1}{2}$, $\int_0^\infty t^2 \exp(-jt^2)\,dt = (1/4j)\sqrt{\pi/j}$ and $\int_{-\infty}^\infty \exp(-jt^2)\,dt = \sqrt{\pi/j}$. It can be seen that, provided $y_0^2 \ll r_{AC}r_{CB}/\beta r_{AB}$ (long ranges and/or low frequencies), obstructions (such as the horizon) do not significantly affect the surface wave propagation. At low frequencies, η_r can be quite small and so surface waves can provide relatively good over-the-horizon communication (a fact that is exploited by AM broadcasters).

Thus far our considerations have assumed a reflection coefficient that is appropriate to vertically polarized waves and hence to vertical antennas. For horizontal antennas, we need to use the reflection coefficient that is appropriate to horizontal polarization. In this case, however, it turns out that the surface wave is very much smaller than for the case of vertical polarization, a fact that emphasizes the importance of vertical polarization for surface wave communications.

4.5 The Geometric Theory of Diffraction

We have seen that the Friis equation requires considerable modification when obstacles and/or boundaries are present. The two major contributing phenomena are those of reflection and diffraction. The incorporation of these two phenomena can be difficult in complex environments, but the GTD provides a simplified approach (Keller, 1962). In general, energy will travel along those paths between receiver and transmitter for which the distance is a minimum (Fermat's principle), but these paths will be constrained by obstacles (a constrained minimization). In GTD, when an obstacle is encountered, the resulting diffraction or reflection is treated locally. Unimpeded, the wave field will fall

4.5 The Geometric Theory of Diffraction

away as $1/s$ as we move away from the source (s is the distance traveled). However, when the wave strikes an obstacle, it will either be reflected or diffracted. There will now be complex processes at work and there can be additional loss and/or a phase change. In the case of reflection, the effect will be expressed as a reflection coefficient R that depends on the properties of the reflecting boundary, the angle of incidence and the wave polarization. The *reflection loss* (a loss in addition to the spreading loss) will be $L_{rfl} = 1/|R|^2$. At frequencies above a few hundred MHz, however, $R = -1$ is often a good approximation and we simply have a phase advance of π.

Incorporating the effects of diffraction is not quite so easy. In GTD, at the points of diffraction, we can use *canonical* solutions (see Appendix J) to find the modification to the fields after they have negotiated the diffraction region. Figure 4.10 depicts some useful canonical solutions (a wedge and cylinder). For the case of the cylinder, it will be noted that the ray path follows the surface of the cylinder. In this case, the ray path obeys Fermat's principle, but with the variations constrained not to enter the obstacle. The surface wave that travels a round the object is sometimes known as a *creeping wave*. Solutions for wedge and cylinder diffraction problems exist (Jones, 1999), but are difficult to implement in general. Treating the wedge as a thin screen of the same height, however, can provide a useful approximation (see Figure 4.11). Sommerfeld (1954) has derived an expression for the diffraction over a screen (sometimes known as *knife edge diffraction*) and expressions can also be found in Jones (1999). Consider a field of amplitude E_B^A that is incident upon the screen (Figure 4.11). The incident field

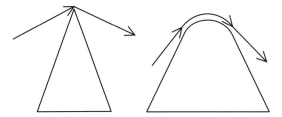

Figure 4.10 Useful canonical solutions.

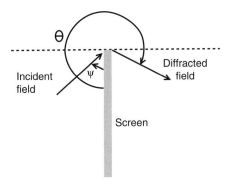

Figure 4.11 Diffraction by a thin screen.

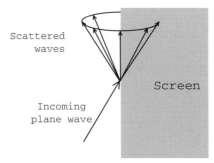

Figure 4.12 Oblique incidence at a screen edge.

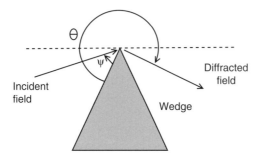

Figure 4.13 Diffraction by a wedge.

makes angle ψ with the screen (measured clockwise) and an angle δ with the screen edge ($\delta = \frac{\pi}{2}$ when orthogonal to the screen). The diffracted field has the form

$$E_B \approx D E_B^A \frac{\exp(-j\beta s)}{\sqrt{s}} \qquad (4.38)$$

where s is the distance from the screen edge to the observation point,

$$D = \frac{-\exp(-\frac{j\pi}{4})}{\sqrt{8\pi\beta}\sin\delta}\left(\frac{1}{\cos\frac{\theta-\psi}{2}} \mp \frac{1}{\cos\frac{\theta+\psi}{2}}\right) \qquad (4.39)$$

and θ is the angle that the transmitted wave makes with the screen (measured clockwise). In Equation 4.39, the plus sign applies to a PEC screen and a minus sign applies to a PMC screen. Oblique incidence with the screen edge causes diffraction into a cone with semi-angle δ, the cone reducing to a disc as $\delta \to \frac{\pi}{2}$ (see Figure 4.12).

There are situations where a thin screen is not a good approximation and a wedge is more appropriate. Luebbers (1984) has given an approximate expression for diffraction over a wedge that includes the thin screen as a special case. Let the wedge have exterior angle $n\pi$ ($n = 2$ when the wedge collapses to a screen) and the incoming and outgoing rays make angles ψ and θ, respectively, with the wedge face of incidence (see Figure 4.13). In this case, D will be given by

4.5 The Geometric Theory of Diffraction

$$D = \frac{-\exp(-\frac{j\pi}{4})}{n\sqrt{8\pi\beta}\sin\delta}\left(\cot\left(\frac{\pi+(\theta-\psi)}{2n}\right)F\left(\beta r a^+(\theta-\psi)\right)\right.$$

$$+ \cot\left(\frac{\pi-(\theta-\psi)}{2n}\right)F\left(\beta r a^-(\theta-\psi)\right)$$

$$+ R_i \cot\left(\frac{\pi-(\theta+\psi)}{2n}\right)F\left(\beta r a^-(\theta+\psi)\right)$$

$$\left. + R_d \cot\left(\frac{\pi+(\theta+\psi)}{2n}\right)F\left(\beta r a^+(\theta+\psi)\right)\right) \quad (4.40)$$

where $F(x) = 2j\sqrt{x}\exp(jx)\int_{\sqrt{x}}^{\infty}\exp(-j\tau^2)d\tau$ (≈ 1 for large x), δ is defined as for the screen and $a^{\pm}(\phi) = 2\cos^2\left(n\pi N^{\pm} - \phi/2\right)$. N^+ is the integer that best satisfies $2\pi n N^+ = \theta \pm \psi + \pi$ and N^- is the integer that best satisfies $2\pi n N^- = \theta \pm \psi - \pi$ (for most propagation problems $N^+ = 1$ and $N^- = 0$). R_i and R_d are the reflection coefficients for the incident and diffracted wave sides of the wedge, calculated for the appropriate polarization. The above expressions apply to a plane wave incident upon the wedge but, in the case of a spherical or cylindrical wave, s should be replaced by $s(s' + s)/s'$ where s' is the distance of the wedge tip from the source. The diffraction loss will now be given by $L_{diff} = s(s' + s)/s'|D|^2$. It should be noted that, after a diffraction on a GTD path, the source must now be regarded as the point of diffraction until the next diffraction is reached.

Diffraction over a cylinder is another problem for which there exist analytic results (Jones, 1999). In Figure 4.14 we see the top of a rounded hill modeled as part of the surface of a cylinder of radius a. The ray from the source to the observation point will follow the shortest path between them, constrained by the hill. This path will follow the ground for a distance L and the observation point will be distance s from where it leaves the ground. From Jones (1999), the field at the observation point will be reduced by factor F times its value where the ray first touches the ground where

$$F = \frac{4\exp\left(\frac{j\pi}{4} - \frac{jv_s}{a}L\right)}{\sqrt{2\pi\beta N_s^2}}\frac{\exp(-j\beta s)}{\sqrt{s}} \quad (4.41)$$

with $v_s = \beta a + a_1(\beta a/2)^{\frac{1}{3}}\exp(2j\pi/3)$ and $a_1 = -2.338$ (the first zero of the Airy function). N_s is defined in terms of Airy functions with $N_s^2 = -(2^7/\beta a)^{\frac{1}{3}}\exp(-2j\pi/3)\times$

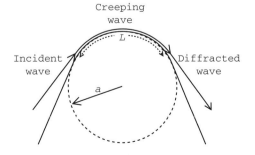

Figure 4.14 Diffraction over a rounded hill approximated as a cylinder.

$\{Ai'(a_1)\}^2$ (Ai is the Airy function and Ai' is its first derivative with $Ai'(a_s) \approx 0.699$). As v_s has a large negative imaginary part, it will be noted that there will be strong attenuation of the creeping waves as they travel over the hill. This loss can be much greater than what would have been predicted on the basis of a simple screen model via the Bullington method. Nevertheless, an appreciable amount of power can still be transmitted over a curved surface by the diffraction process. Jones (1999) derived the above factor F to be applied to a TE wave that is incident upon a PEC cylinder (electric field in the direction of the cylinder axis). The factor, however, can also be applied to the magnetic field of a TM wave that is incident on a PMC cylinder (the corresponding electric fields can be derived directly from the time harmonic Maxwell equations).

4.6 Propagation in Urban Environments

The ideas of GTD are most useful when studying propagation in an urban environment where there is little possibility of a full electromagnetic solution. Figure 4.15 shows some propagation mechanisms in a typical city environment (buildings shown as gray). For line-of-sight communication between receiver and transmitter, both direct and reflected propagation can be important (paths A, B and C in the figure) and their interference can be significant (the paths can be calculated using Fermat's principle). For non line of sight between receiver and transmitter, propagation can take place by both reflection and diffraction. For short ranges (paths D and E in the figure) both mechanisms can be significant and the diffraction around corners can be calculated

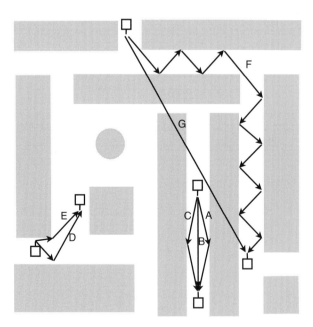

Figure 4.15 Propagation mechanisms in an urban environment.

using the wedge diffraction theory of the last section. If the change in direction is large, both of these mechanisms can be weak. For long paths, diffraction around corners is an insignificant mechanism due to the large number of diffractions that are needed over a path and the fact that each diffraction can be quite weak. Reflections (path F) can also be weak due to the significant loss that can occur on reflection from building material and the large number of reflections required. For long ranges, the most significant mechanism in the urban environment can often be diffraction over the tops of buildings (path G in the figure).

Diffraction over rooftops (Figure 4.16) can be quite complex but, as a first approximation, we could represent all the rooftops by a single screen through the Bullington method. This, however, can be quite inaccurate and it is far better to treat the buildings as a series of screens. Assume a single polarization, then Equation 4.19 can be reinterpreted as the integral equation

$$E(\mathbf{r}_0) = \int_S K_0(\mathbf{r}_0, \mathbf{r}) E(\mathbf{r}) dS \tag{4.42}$$

where

$$K_0(\mathbf{r}_0, \mathbf{r}) = \frac{j\beta_0}{2\pi} \frac{\exp\left(-j\beta_0 \left(z_0 - z + \frac{(x_0-x)^2+(y_0-y)^2}{2|z_0-z|}\right)\right)}{|z_0 - z|} \tag{4.43}$$

This equation provides the electric field to the fore of a surface on which the field is known and can be used to propagate the electric field from screen to screen. We assume the field of source A can be approximated as

$$E_A(\mathbf{r}) = \frac{j\omega\mu_0 I_A}{4\pi |z - z_A|} h^A_{\text{eff}} \exp\left(-j\beta \left(|z - z_A| + \frac{x^2 + (y - y_A)^2}{2|z - z_A|}\right)\right) \tag{4.44}$$

and the repeated application of Equation 4.42 then provides the field at B caused by source A

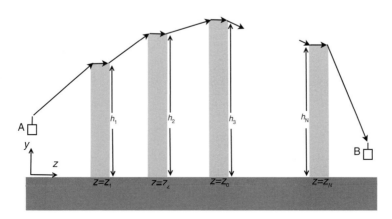

Figure 4.16 Propagation over rooftops.

$$E(\mathbf{r}_B) = \int_{S_N} \cdots \int_{S_3} \int_{S_2} \int_{S_1} K_0(\mathbf{r}_B, \mathbf{r}_N) K_0(\mathbf{r}_N, \mathbf{r}_{N-1}) \ldots K_0(\mathbf{r}_3, \mathbf{r}_2) K_0(\mathbf{r}_2, \mathbf{r}_1)$$
$$\times E_A(\mathbf{r_1}) dS_1 dS_2 dS_3 \ldots dS_N \tag{4.45}$$

where S_1 to S_N are the vertical half planes above the respective screens (we assume that the power that lands on a screen is absorbed by it). The x integrals can then be performed to yield

$$E(\mathbf{r}_B) = \left(\frac{j\beta_0}{2\pi}\right)^{\frac{N}{2}} \frac{\exp(-j\beta_0 |z_B - z_A|)}{\sqrt{|z_B - z_A|}}$$
$$\times \int_{h_N}^{\infty} \cdots \int_{h_3}^{\infty} \int_{h_2}^{\infty} \int_{h_1}^{\infty} \frac{\exp\left(-j\beta_0 \frac{(y_B - y_N)^2}{2|z_B - z_N|}\right)}{\sqrt{|z_B - z_N|}} \frac{\exp\left(-j\beta_0 \frac{(y_N - y_{N-1})^2}{2|z_N - z_{N-1}|}\right)}{\sqrt{|z_N - z_{N-1}|}} \times \cdots$$
$$\times \frac{\exp\left(-j\beta_0 \frac{(y_3 - y_2)^2}{2|z_3 - z_2|}\right)}{\sqrt{|z_3 - z_2|}} \frac{\exp\left(-j\beta_0 \frac{(y_2 - y_1)^2}{2|z_2 - z_1|}\right)}{\sqrt{|z_2 - z_1|}} \hat{E}_A(y_1) dy_1 dy_2 dy_3 \ldots dy_N \tag{4.46}$$

where

$$\hat{E}_A(y) = \frac{j\omega\mu_0 I_A}{4\pi\sqrt{|z_1 - z_A|}} h_{\text{eff}}^A \exp\left(-j\beta \frac{(y - y_A)^2}{2|z_1 - z_A|}\right) \tag{4.47}$$

The above expression can provide a means of estimating rooftop diffraction; a formula of this form was considered by Vogler (1982). Unfortunately, except for a small number of screens, it is difficult to make further progress analytically. Fast numerical techniques are available to evaluate the integrals (Saunders, 1994), especially the lattice methods developed by Korobov (1959) (see Appendix B). Further, in Chapter 7, we will see that the above Kirchhoff integral procedure can be generalized to provide an effective numerical technique for calculating propagation over complex topography within a complex propagation medium.

We will consider the simplified case where the distance between the screens is constant Δz and all screens have the same height h. We introduce new vertical coordinates such that $t_n = (y_n - h)\sqrt{\beta_0/2\Delta z}$, then Equation 4.46 reduces to

$$E(\mathbf{r}_B) = \frac{j\beta_0}{2\pi} \frac{\exp(-j\beta_0 |z_B - z_A|)}{\sqrt{|z_B - z_A|}} \int_h^{\infty} \int_h^{\infty} \frac{\exp\left(-j\beta_0 \frac{(y_B - y_N)^2}{2|z_B - z_N|}\right)}{\sqrt{|z_B - z_N|}}$$
$$\times K\left(\sqrt{\frac{\beta_0}{2\Delta z}}(y_N - h), \sqrt{\frac{\beta_0}{2\Delta z}}(y_1 - h)\right) \frac{\hat{E}_A(y_1)}{\sqrt{\Delta z}} dy_1 dy_N \tag{4.48}$$

where

$$K(t_1, t_N) = \left(\frac{j}{\pi}\right)^{\frac{N-2}{2}} \int_0^{\infty} \cdots \int_0^{\infty} \exp\left(-j\left((t_N - t_{N-1})^2 + \cdots + (t_2 - t_1)^2\right)\right) dt_2 \ldots dt_{N-1} \tag{4.49}$$

The above integral for K can be expanded in powers of t_1 and t_N (Xia and Bertoni, 1992) and the coefficients evaluated using results due to Boersma (1978). From Boersma (1978), we have that $K(0, 0) = 1/(N-1)^{\frac{3}{2}}$ and this suggests $\exp(-j(t_N - t_1)^2)/(N-1)^{\frac{3}{2}}$

4.6 Propagation in Urban Environments

as a first approximation to K when the arguments t_1 and t_N are small (this is exact if $N=2$). (The reader should consult Xia and Bertoni (1992) for higher-order approximations.) Introducing new variables $Y_B = (y_N - y_B)\sqrt{\beta_0/2|z_B - z_N|}$ and $Y_A = (y_1 - y_A)\sqrt{\beta_0/2|z_1 - z_A|}$ we obtain

$$E(\mathbf{r}_B) \approx E_0(\mathbf{r}_B) \frac{j}{\pi(N-1)^{\frac{3}{2}}} \sqrt{\frac{|z_B - z_A|}{\Delta z}}$$
$$\times \int_{H_A}^{\infty} \exp\left(-jY_A^2\right) dY_A \int_{H_B}^{\infty} \exp\left(-jY_B^2\right) dY_B \qquad (4.50)$$

where E_0 is the free space electric field, $H_B = (h-y_B)\sqrt{\beta_0/2|z_B - z_N|}$ and $H_A = (h-y_A)\sqrt{\beta_0/2|z_1 - z_A|}$. (Note that the expression is valid when the receiver and transmitter are close to their nearest building.) Obviously, we will rarely have a situation where the spacing and height of the screens is fixed, but the above expressions can be effective if we use an average height and an average spacing. For large N we have that $|z_B - z_A| \propto N$ and so we see that the field will fall away as $1/|z_B - z_A|^2$. We have already seen that, for open ground, the field close to the ground will also fall away as $1/|z_B - z_A|^2$ and, as a consequence, the inverse square law for fields is a general rule of thumb for mobile communications calculations (this is often found to be close to reality).

The above considerations have assumed that there is only a direct path from the edge of the nearest building to a station. For a station below the level of the rooftops, however, there will also be the possibility of reflections from the vertical sides of buildings and the ground. Figure 4.17 shows the case where the station receives signals by direct and reflected paths. The total field at the station will be the sum of the field for the direct path and those for the image paths (indicated by dotted lines in the figure). For an in-depth account of urban radio wave propagation, the reader should consult the books by Bertoni (1999) and Blaunstein (2000).

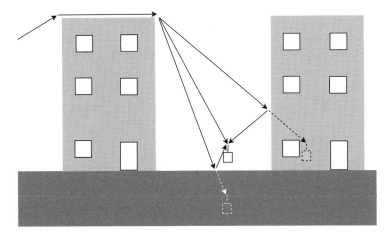

Figure 4.17 Propagation to a station near ground level.

4.7 The Channel Impulse Response Function

From the considerations of this chapter, it is obvious that there can often be many communication paths between two stations and so the received signal can be a complex function of the transmitted signal. This complexity is usually described in terms of the *channel impulse response function* $h(\tau)$, which relates the received signal $s_{Rx}(t)$ to the transmitted signal $s_{Tx}(t)$ through

$$s_{Rx}(t) = \int_{-\infty}^{\infty} h(\tau) s_{Tx}(t - \tau) d\tau \qquad (4.51)$$

(we have assumed the channel to be static or at least slowly varying). The channel function allows us to take into account the multiplicity of propagation modes that are caused by the multiple propagation paths in a complex environment. For a discrete number of modes in free space, we will have

$$h(\tau) = \sum_i \frac{\delta(\tau - \tau_p^i)}{\sqrt{L_i}} \qquad (4.52)$$

where, for the ith mode, τ_p^i is the phase delay and L_i the loss. (Note that L_i is the *total loss* on mode i and might be the product of losses such as spreading, diffraction and reflection.) The impulse response function is a major tool for studying the effect of propagation on a modulated signal and is an important element in the study of signal processing techniques to counteract the degrading effects of propagation. In particular, the existence of multiple modes can cause *delay spread*. This can result in *intersymbol interference* for the symbols of the modulation and can lead to errors in the decoding of a modulated signal.

Quite often, the channel conditions can change over time (the stations and/or scatterers can be moving) and this will lead to an impulse response function that evolves with time (τ_p^i and L_i will now vary with time). The time scale is very much longer than that of the signal and so the time variable in this case is often known as *slow time*. In this case, $h(\tau)$ is replaced by $h(\tau, t)$ in Equation 4.51 where t refers to changes with respect to slow time. In the case of multimode propagation, the modes will have the potential to interfere with each other and so changes to the channel over time can lead to changes in signal amplitude over time. This phenomena is known as *fading* and can be quite rapid when the stations are moving, leading to what is known as *flutter*. Change in the channel over time will also induce a Doppler shift in the signal frequency and, in a multipath environment, this will induce a *Doppler spread* of the signal.

4.8 References

L. Barclay (editor), *Propagation of Radio Waves*, 2nd edition, Institute of Electrical Engineers, London, 2003.

H.L. Bertoni, *Radio Propagation for Modern Wireless Systems*, Prentice Hall, Englewood Cliffs, NJ, 1999.

4.8 References

N. Blaunstein, *Radio Propagation in Cellular Networks*, Artech House, Boston, 2000.

J. Boersma, On certain multiple integrals occurring in a waveguide scattering problem, SIAM J. Math. Anal., vol. 9, pp. 377–393, 1978.

K. Bullington, Radio propagation at frequencies above 30 Mc, Proc. IRE, vol. 35, no. 10, pp. 1122–1136, 1947.

L.B. Felsen and N. Marcuvitz, *Radiation and Scattering of Waves*, IEEE Press and Oxford University Press, New York and Oxford, 1994.

D.S. Jones, *Methods in Electromagnetic Wave Propagation*, 2nd edition, IEEE/OUP series in Electromagnetic Wave Theory, Oxford University Press, New York and Oxford, 1999.

J.B. Keller, Geometrical theory of diffraction, J. Opt. Soc. Amer., vol. 52, pp. 116–130, 1962.

N.M. Korobov, The approximate computation of multiple integrals, Doklady Akademii Nauk SSSR, vol. 124, pp. 1207–1210, 1959.

R.L. Luebbers, Finite conductivity uniform GTD versus knife edge diffraction in prediction of propagation path loss, IEEE Trans. Antennas Propagat., vol. 32, pp. 70–76, 1983.

G.D. Monteath, *Applications of the Electro-magnetic Reciprocity Principle*, Pergamon Press, Oxford, 1973.

S.R. Saunders, *Antennas and Propagation for Wireless Communication Systems*, John Wiley, Chichester, 1999.

S.R. Saunders and F.R. Bonar, Prediction of mobile radio wave propagation over buildings of irregular heights and spacing, IEEE Trans. Antennas Propagat., vol. 42, 137–144, 1994.

A.J.W. Sommerfeld, *Optics*, Academic Press, Inc., New York, 1954.

L.E. Vogler, An attenuation function for multiple knife-edge diffraction, Radio Sci., vol. 17, pp. 1541–1546, 1982.

H.H. Xia and H.L. Bertoni, Diffraction of cylindrical and plane waves by an array of absorbing half-screens, IEEE Trans. Antennas Propagat., vol. 40, pp. 170–177, 1992.

5 Geometric Optics

In the current chapter we discuss the *geometric optics* (GO) approximation to Maxwell's equations, one of the more important techniques for obtaining approximate solutions to these equations. While it strictly only applies in the high-frequency limit, the GO approximation can nevertheless provide invaluable insights into many propagation problems outside this limit. We derive ray tracing equations of GO for both isotropic and anisotropic media and then consider their solution by both analytic and numerical solution techniques. Finally, the variational formulation of the ray tracing equations is described and its solution by direct methods, including the finite element technique, is discussed.

5.1 The Basic Equations

Geometric optics assumes that the wavelength is much smaller than the scale L of variations in the medium ($\beta \gg 1/L$). In GO we assume that the time harmonic field can be described by an ansatz of the form

$$\mathbf{E}(\mathbf{r}) = \mathbf{E}_0(\mathbf{r}) \exp(-j\beta_0 \phi(\mathbf{r})) \tag{5.1}$$

where $\mathbf{E}_0$ and ϕ are slowly varying on the scale of a wavelength. We first eliminate $\mathbf{H}$ between the time harmonic Maxwell equations and obtain the equation

$$\nabla \times \nabla \times \mathbf{E} - \omega^2 \mu \epsilon \mathbf{E} = 0 \tag{5.2}$$

where we have assumed μ to be constant (this is a good approximation for nearly all propagation media that we will study). Introducing the refractive index $N = \sqrt{\mu\epsilon}/\sqrt{\mu_0 \epsilon_0}$ ($c = c_0/N$ is the speed of light), we have

$$\nabla^2 \mathbf{E} + \nabla((\nabla \ln N^2) \cdot \mathbf{E}) + \beta_0^2 N^2 \mathbf{E} = 0 \tag{5.3}$$

on noting that $\nabla \times \nabla \times \mathbf{E} = \nabla(\nabla \cdot \mathbf{E}) - \nabla^2 \mathbf{E}$ and $\nabla \cdot (\epsilon \mathbf{E}) = 0$. Substituting the ansatz 5.1 into Equation 5.3,

$$-j\beta_0 \nabla^2 \phi \mathbf{E}_0 - 2j\beta_0 \nabla\phi \cdot \nabla \mathbf{E}_0 - \beta_0^2 \nabla\phi \cdot \nabla\phi \mathbf{E}_0 - j\beta_0 \nabla \ln(N^2) \cdot \mathbf{E}_0 \nabla\phi + \beta_0^2 N^2 \mathbf{E}_0 = 0 \tag{5.4}$$

if we only retain the two leading order terms in β_0. For the β_0^2 term we have

$$\nabla\phi \cdot \nabla\phi - N^2 = 0 \tag{5.5}$$

5.1 The Basic Equations

and for the β_0 term

$$-\nabla^2\phi E_0 - 2\nabla\phi \cdot \nabla E_0 - \nabla \ln(N^2) \cdot E_0\nabla\phi = 0 \qquad (5.6)$$

It will noted that the phase ϕ satisfies an equation of the form $F(\mathbf{x},\mathbf{p}) = 0$ where $\mathbf{p} = \nabla\phi$ is the wave vector. We can solve this equation in terms of a set of characteristic curves. The curves, and the values of ϕ on these curves, can be found by solving the generalized Charpit equations

$$\frac{dx_i}{\partial F/\partial p_i} = \frac{d\phi}{\sum_{j=1}^{3} p_j \partial F/\partial p_j} = \frac{dp_i}{-\partial F/\partial x_i} = dg \qquad (5.7)$$

where g is a parameter along the curve. Equation 5.5 suggests $F(\mathbf{x},\mathbf{p}) = \mathbf{p} \cdot \mathbf{p}/2 - N^2/2$ and, from the Charpit equations (Smith, 1967), we obtain

$$\frac{d\mathbf{r}}{dg} = \mathbf{p} \qquad (5.8)$$

$$\frac{d\mathbf{p}}{dg} = \frac{1}{2}\nabla N^2 \qquad (5.9)$$

$$\frac{d\phi}{dg} = N^2 \qquad (5.10)$$

where we have used the relation $\mathbf{p} \cdot \mathbf{p} = N^2$. Starting at a field source (position $\mathbf{r} = \mathbf{r}_S$), we can solve the above equations to trace out a set of curves $\mathbf{r} = \mathbf{r}(g)$, often known as *rays*, on which ϕ will be known. It will be noted that the rays are everywhere orthogonal to the surfaces of constant phase (see Figure 5.1). The parameter g can be related to the geometrical distance s along the ray path through $g = \int_{\mathbf{r}_s}^{\mathbf{r}} N^{-1} ds$ where the integral is taken along the path. In the case of an isotropic medium, g is the group distance along the path. Later in this chapter, we will see that we can solve the above equations analytically for some special cases of refractive media. In general, however, the equations will need to be solved by numerical techniques, the Runge–Kutta variety (and methods based around them) being particularly useful in ray tracing (see Appendix B).

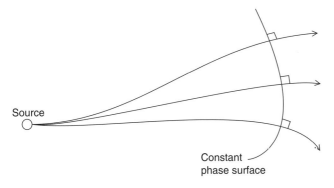

Figure 5.1 Propagation through a spatially varying medium.

The ray path equations can be reduced to a system of second-order equations in **r** alone

$$\frac{d^2\mathbf{r}}{dg^2} = \frac{1}{2}\nabla N^2 \tag{5.11}$$

which provides an alternative *ray tracing* formulation (see Cervey (2001) for a variety of ray tracing formulations in the related area of seismology). Further, these equations are the Euler–Lagrange equations for the variational principle

$$\delta \int_A^B N \left(\frac{d\mathbf{r}}{dg} \cdot \frac{d\mathbf{r}}{dg} \right)^{\frac{1}{2}} dg = 0 \tag{5.12}$$

where A and B are points at which the ray path is fixed. The variational equation can be rewritten as

$$\delta \int_A^B \frac{ds}{c} = 0 \tag{5.13}$$

which is the more general mathematical form of Fermat's principle. We note that the integral in Equation 5.13 is the phase delay (i.e. $\tau_p = \int_A^B c^{-1} ds$) and so the above variational equation implies the ray path between two points will be that for which the variation in delay is stationary. As originally stated, the principle implied that the ray path between two points is the path for which the variation in propagation delay is minimum. In the case of propagation through the ionospheric medium, however, we will see that a ray can occur at a stationary value other than a minimum.

A particularly important aspect of the variational approach is that, rather than solving the Euler–Lagrange equations to find the ray path, it is also possible to solve Equation 5.13 directly. In a direct approach to ray tracing, the ray is found by searching over the possible ray paths that join the end points (A and B) to find the one that makes the phase distance stationary. This approach can be implemented using Rayleigh–Ritz techniques, the finite element method being a particularly important example of such an approach (see the section on Fermat's principle later in this chapter and Appendix C). Such an approach has been considered by Simlauer (1970) for an isotropic ionosphere and by Thurber and Thurber (1987) for the related area of seismic ray tracing.

An important application of ray tracing is to be found in the modeling of ionospheric radio wave propagation. As mentioned in Chapter 2, solar radiation can generate ionized layers above the Earth, known as the *ionosphere*, in which the electrons are free to move and cause refraction. There are several layers with the D layer lying at a height of about 80 km, the E layer lying at a height of about 110 km and the F layer lying at a height of about 300 km. (There are various complexities to these ionospheric layers, which we will discuss further in the chapter on the ionospheric duct.) In the lower regions of each layer, the refractive index has a negative vertical gradient with the potential to refract energy back to the ground. The ionosphere is a complex plasma but, to a first approximation, its refractive can be expressed as $N = \sqrt{1 - f_p^2/f^2}$ where f is the wave frequency and f_p is the plasma frequency. (The plasma frequency f_p in Hz is related to the electron density N_e in electrons per cubic meter through the relation $f_p^2 = 80.61 N_e$.) Arguably the simplest realistic model of the ionospheric is a single layer (the F layer) for which f_p^2 has a parabolic profile

$$\frac{f_p^2}{f_m^2} = 1 - \frac{(z - h_m)^2}{y_m^2} \quad \text{for} \quad |z - h_m| < y_m$$
$$= 0 \quad \text{otherwise} \tag{5.14}$$

where h_m is the layer peak height, y_m is its thickness and f_m is the peak plasma frequency. On replacing dg by ds/N in Equation 5.11

$$\frac{d}{ds}\left(N\frac{dx}{ds}\right) = 0 \tag{5.15}$$

and integrating

$$N(z)\frac{dx}{ds} = C \tag{5.16}$$

where C is a constant of integration. If we define θ to be the angle between the z direction and the ray, Equation 5.16 becomes

$$N(z)\sin\theta = N(0)\sin\theta_0 \tag{5.17}$$

where θ_0 is the value of θ at the start of the ray ($x = y = z = 0$). The above expression is *Snell's law*, which, for a parabolic layer, now becomes

$$\frac{\sin^2\theta - \sin^2\theta_0}{\sin^2\theta} = \frac{f_m^2}{f^2}\left(1 - \frac{(z-h_m)^2}{y_m^2}\right) \quad \text{for} \quad |z - h_m| < y_m$$
$$= 0 \quad \text{otherwise} \tag{5.18}$$

If the propagation is to return to the ground, there will need to be a height h at which the angle θ becomes $\pi/2$ and, from Equation 5.18, this will be given by

$$h = h_m - y_m\sqrt{1 - \frac{f^2}{f_m^2}\cos^2\theta_0} \tag{5.19}$$

It can be seen that there will be no real solution for θ_0 less than $\theta_m = \cos^{-1}(f_m/f)$ and so rays cannot return to the ground for $\theta < \theta_m$ (see Figure 5.2). We can now rearrange Equation 5.11 into the form

$$N\frac{d}{ds}\left(N\frac{dz}{ds}\right) = \frac{1}{2}\frac{d(N^2)}{dz} \tag{5.20}$$

and combining this with Equation 5.16,

$$\frac{d^2z}{dx^2} = \frac{1}{2C^2}\frac{d(N^2)}{dz} \tag{5.21}$$

Multiplying by dz/dx and integrating

$$\left(C\frac{dz}{dx}\right)^2 = N^2(z) + B \tag{5.22}$$

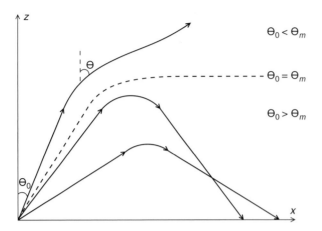

Figure 5.2 Ray paths in a single parabolic ionospheric layer.

where B is a constant of integration. If we assume $N(0) = 1$, we find from Equation 5.16 that $C = \sin \theta_0$ and from Equation 5.22 that $B = -\sin^2 \theta_0$. Then, integrating Equation 5.22,

$$x(z) = \int_0^z \sqrt{\frac{\sin^2 \theta_0}{N^2(z) - \sin^2 \theta_0}} \, dz \qquad (5.23)$$

If the ray returns to the ground, then $x(h)$ (h given by Equation 5.19) will be half the distance along the ground (the ray will be symmetric about its midpoint). As a consequence, the total ground range D will be (Davies, 1990)

$$D = 2(h_m - y_m) \tan \theta_0 + 2 \int_{h_m - y_m}^{h} \frac{\sin \theta_0}{\sqrt{\cos^2 \theta_0 - \frac{f_m^2}{f^2}(1 - \frac{(z-h_m)^2}{y_m^2})}} \, dz$$

$$= 2(h_m - y_m) \tan \theta_0 + \frac{y_m f \sin \theta_0}{f_m} \ln \left(\frac{1 + \frac{f}{f_m} \cos \theta_0}{1 - \frac{f}{f_m} \cos \theta_0} \right) \qquad (5.24)$$

For $f < f_m$, D will be a monotonic function of θ_0 and there will be one ray for each range. If $f > f_m$, however, things become far more complex. In this case, the minimum of D is greater than zero; i.e. there will be a region around the source (known as the *skip zone*) that rays cannot reach. This is best illustrated by considering ray tracing over a range of elevations. Figure 5.3 shows the rays from a source that propagate via a parabolic ionospheric layer ($f_m = 10$ MHz, $y_m = 100$ km and $h_m = 300$ km). From this it will be noted that the edge of the skip zone signifies a distinct change in ray behavior. The rays decrease in range with increasing elevation until they reach the edge of the skip zone and then increase in range with further increases in elevation. It will be further noted that those rays that increase in range with elevation reach much higher altitudes than those that decrease their range with elevation. As a consequence, these rays are known as *high rays*. For obvious reasons, the rays that decrease in range with

5.1 The Basic Equations

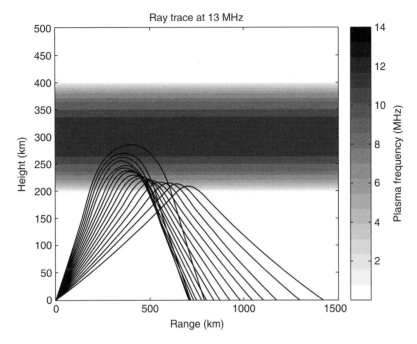

Figure 5.3 Example of ray paths through an parabolic ionospheric layer.

increasing elevation are known as *low rays*. It will be noted that the low rays cross over themselves and therefore focus the power at the crossover point (the edge of the skip zone is the point on the ground where such focusing occurs). The envelope of the low rays is a surface on which the focusing occurs and is known as a caustic surface. In theory, the GO approach breaks down at a caustic surface, but it can be shown (Jones, 1994) that GO can still be used providing that, when the rays cross in the direction of propagation, the phase is advanced by $\pi/2$ on passage through the crossover. The high rays do not cross each other and so the problem does not arise in this case. The above difference between the behaviors of high and low rays turns out to be of crucial importance when we come to further consider Fermat's principle later.

The ray equations 5.11 provide us with a means of calculating propagation paths, but they do not provide us with the fields themselves. To the leading order of β_0, the $\nabla \times \mathbf{E}$ Maxwell's equation will imply

$$\mathbf{H} \approx \frac{1}{\eta} \frac{d\mathbf{r}}{ds} \times \mathbf{E} \tag{5.25}$$

and the $\nabla \cdot (\epsilon \mathbf{E})$ equation

$$\mathbf{E} \cdot \frac{d\mathbf{r}}{ds} \approx 0 \tag{5.26}$$

where $d\mathbf{r}/ds$ is the unit vector in the propagation direction. From these expressions, the Poynting vector $\mathbf{S}$ will be

$$\mathbf{S} = \frac{1}{2\eta} |\mathbf{E}|^2 \frac{d\mathbf{r}}{ds} \tag{5.27}$$

and from which it can be seen that energy will flow along the ray paths to leading order. We consider a bundle of rays that intersect an area A_1 on a constant phase surface $\phi = \phi_1$ and then an area A_2 on a later surface $\phi = \phi_2$. No energy will flow across the rays of the bundle and so the power flowing through area A_1 will be the same as the power flowing through area A_2, i.e.

$$A_1 \frac{|\mathbf{E}^1|^2}{\eta_1} = A_2 \frac{|\mathbf{E}^2|^2}{\eta_2} \tag{5.28}$$

where $\mathbf{E}^1$ is the electric field on $\phi = \phi_1$, $\mathbf{E}^2$ is the electric field on $\phi = \phi_2$, η_1 is the impedance on $\phi = \phi_1$ and η_2 is the impedance on $\phi = \phi_2$. We can use Equation 5.28 to calculate the field magnitude along a ray tube given the power that enters the tube from the radiation source. Let total power P_{Tx} be transmitted by a source, assumed to be isotropic, then a power $P_{Rx} = P_{Tx}/L_{\text{sprd}}$ will arrive at the receiver (L_{sprd} is the spreading loss). Consider the ray that joins transmitter and receiver to be part of a bundle of rays that leave the transmitter with solid angle $d\Omega$, then $L_{\text{sprd}} = (dA/d\Omega)(4\pi/\lambda)^2$ if the bundle has cross sectional area dA at the receiver.

We now have a means of calculating the field magnitude, but not its direction. If we take the dot product of $\mathbf{E}_0^*$ with Equation 5.6, and noting that $\nabla\phi \cdot \mathbf{E}_0 = 0$ in the leading order,

$$-\nabla^2\phi \mathbf{E}_0 \cdot \mathbf{E}_0 - \nabla\phi \cdot \nabla(\mathbf{E}_0 \cdot \mathbf{E}_0) = 0 \tag{5.29}$$

We can now use Equation 5.29 to eliminate $\nabla^2\phi$ from Equation 5.6 and obtain

$$\nabla\phi \cdot \frac{\nabla(|\mathbf{E}_0|^2)}{|\mathbf{E}_0|^2} \mathbf{E}_0 - 2\nabla\phi \cdot \nabla\mathbf{E}_0 - 2\frac{\nabla N}{N} \cdot \mathbf{E}_0 \nabla\phi = 0 \tag{5.30}$$

Noting that $d\mathbf{r}/dg = \nabla\phi$,

$$\frac{1}{|\mathbf{E}_0|}\frac{d}{dg}(|\mathbf{E}_0|)\mathbf{E}_0 - \frac{d\mathbf{E}_0}{dg} - \frac{\nabla N}{N} \cdot \mathbf{E}_0 \frac{d\mathbf{r}}{dg} = 0 \tag{5.31}$$

If we define the *polarization* vector by $\mathbf{P} = \mathbf{E}_0/|\mathbf{E}_0|$, then

$$\frac{d\mathbf{P}}{dg} + \frac{\nabla N}{N} \cdot \mathbf{P} \frac{d\mathbf{r}}{dg} = 0 \tag{5.32}$$

This equation allows us to track the change in field direction as we move along a ray.

We have thus far concentrated on solving the ray tracing equations in terms of Cartesian coordinates, but other coordinate systems can be better suited to the problem at hand. In the case of propagation via the ionosphere, polar coordinates are certainly of use. If a ray path remains in a plane, we can use a 2D polar coordinate system (r, θ) and Fermat's principle

$$\delta \int_T^R N(\mathbf{r})\, ds = 0 \tag{5.33}$$

can be rewritten as

$$\delta \int_T^R N(r,\theta) \left[\left(\frac{dr}{d\theta}\right)^2 + r^2\right]^{1/2} d\theta = 0 \qquad (5.34)$$

In the case of ionospheric ray tracing, 2D polar coordinates are particularly useful when deviations from the great circle path are negligible, and this is certainly the case for a spherically stratified medium. In this case, coordinate r will be the distance from the center of Earth and θ the angular coordinate along the great circle path. Further, the rays will be confined to planes that pass through the center of Earth and the source of radiation. From the Euler–Lagrange equations for Equation 5.34, we obtain (Coleman, 1998)

$$\frac{dQ}{d\theta} = \frac{1}{2}\frac{r}{\sqrt{N^2 - Q^2}}\frac{\partial N^2}{\partial r} + \sqrt{N^2 - Q^2} \qquad (5.35)$$

and

$$\frac{dr}{d\theta} = \frac{rQ}{\sqrt{N^2 - Q^2}} \qquad (5.36)$$

where $Q = Ndr/\sqrt{dr^2 + r^2 d\theta^2}$. In this form of ray tracing, we only need to solve two first order ODEs to find a ray. In particular, these equations can provide an effective means of studying propagation in an ionosphere for which the horizontal variations are much weaker than those in altitude (a reasonable assumption in many circumstances).

The geometry of a bundle of rays emanating from a source allows us to calculate how the power of a wave varies as it propagates. Since power flows within the confines of the rays emanating from a source, we need to know how the differences between two nearby rays vary during propagation. Between nearby rays, the deviations (δr and δQ) in quantities r and Q can be calculated from

$$\frac{d(\delta Q)}{d\theta} = \frac{r}{\sqrt{N^2 - Q^2}}$$
$$\times \left[\frac{\delta r}{2}\left(\frac{1}{r}\frac{\partial N^2}{\partial r} + \frac{\partial^2 N^2}{\partial r^2}\right) - \left(\frac{1}{2(N^2 - Q^2)}\frac{\partial N^2}{\partial r} - \frac{1}{r}\right)\left(\frac{\delta r}{2}\frac{\partial N^2}{\partial r} - Q\delta Q\right)\right] \qquad (5.37)$$

and

$$\frac{d(\delta r)}{d\theta} = \frac{r}{\sqrt{N^2 - Q^2}}\left[\frac{\delta r Q}{r} + \delta Q - \frac{Q}{N^2 - Q^2}\left(\frac{\delta r}{2}\frac{\partial N^2}{\partial r} - Q\delta Q\right)\right] \qquad (5.38)$$

These equations follow from the equations for ray path deviations (the Jacobi accessory equation) that can be found in Appendix C. We will use the above results later when we come to consider propagation in the ionospheric duct in more detail.

One of the problems that arises when using the coordinate θ to parameterize points on the ray is that it does not allow for complex ionospheres in which the ray can pass the same range more than once. By introducing a parameter g that is the group distance along the ray ($dg = ds/N$) we can rearrange Equations 5.35 and 5.36 into a form that allows for this possibility (Coleman, 1998). First, we note that

$$\frac{dg}{d\theta} = \frac{\sqrt{\frac{dr}{d\theta}^2 + r^2}}{N} \tag{5.39}$$

which, combined with Equations 5.35 and 5.36, yields

$$\frac{dQ}{dg} = \frac{1}{2}\frac{\partial N^2}{\partial r} + \frac{N^2 - Q^2}{r} \tag{5.40}$$

$$\frac{dr}{dg} = Q \tag{5.41}$$

and

$$\frac{d\theta}{dg} = \frac{\sqrt{N^2 - Q^2}}{r} \tag{5.42}$$

that describe the rays in terms of the new parameter g. We can also produce equations that describe the ray deviations by using Equation 5.42 to replace the θ derivatives in Equations 5.35 and 5.36 by g derivatives.

5.2 Analytic Integration

Thus far, we have only studied propagation through media with very simple variations in refractive index and for which the ray tracing equations have been amenable to analytic solution. In general, the ray tracing equations will require numerical solution techniques (see Appendix B and the section on Fermat's principle). We can, however, derive further analytic results for a large class of more complex refractive indices and these can be useful, especially in testing numerical procedures.

We first consider the variational principle for a 2D ray path that is described in terms of Cartesian coordinates (x, y) and assume the refractive index only depends on the y coordinate (the medium is *horizontally stratified*). The variational principle can be expressed as

$$\delta \int_T^R N(y) \sqrt{\left(\frac{dy}{dx}\right)^2 + 1}\, dx = 0 \tag{5.43}$$

and for which the Euler–Lagrange equations yield the first integral (see Appendix C)

$$\frac{N(y)}{\sqrt{\left(\frac{dy}{dx}\right)^2 + 1}} = C \tag{5.44}$$

where C is a constant of integration. This is essentially Snell's law for a horizontally stratified ionosphere. For a parabolic ionospheric layer, we have previously seen that Snell's law leads to a full analytic integration of the ray tracing equations.

Now consider radial coordinates (r, θ) and a refractive medium that depends on r alone (the medium is *spherically stratified*). The Euler–Lagrange equations for the appropriate variational principle 5.34 yield the first integral

$$N(r) r^2 \frac{d\theta}{ds} = C \tag{5.45}$$

which is known as Bouger's law. This result is essentially a Snell's law for a spherically stratified refractive media (Kelso, 1968). Croft and Hoogasian (1968) have shown that this law can be further integrated in the case of an ionospheric medium with

$$N = \sqrt{\alpha + \frac{\beta}{r} + \frac{\gamma}{r^2}} \quad \text{for } r_b < r < r_m r_b / (r_b - y_m)$$
$$= 1 \quad \text{otherwise} \tag{5.46}$$

where

$$\alpha = 1 - \left(\frac{\omega_c}{\omega}\right)^2 + \frac{r_b^2}{y_m^2} \left(\frac{\omega_c}{\omega}\right)^2 \tag{5.47}$$

$$\beta = -\frac{2 r_b^2 r_m}{y_m^2} \left(\frac{\omega_c}{\omega}\right)^2 \tag{5.48}$$

and

$$\gamma = \frac{r_m^2 r_b^2}{y_m^2} \left(\frac{\omega_c}{\omega}\right)^2 \tag{5.49}$$

This refractive index represents what is known as a *quasi-parabolic layer*, an ionosphere with peak plasma frequency ω_c at radial distance r_m from the center of Earth (the base of the layer is at radial distance r_b and $r_m = r_b + y_m$). It was shown by Croft and Hoogasian that, for a ray that starts and ends on the ground, the ground range will be

$$D = 2 r_E \left[(\phi - \phi_0) - \frac{r_E \cos \phi_0}{2 \sqrt{\gamma_0}} \ln \frac{\beta^2 - 4 \alpha \gamma_0}{4 \gamma_0 \left(\sin \phi + \frac{1}{r_b} \sqrt{\gamma_0} + \frac{1}{2\sqrt{\gamma_0}} \beta \right)^2} \right] \tag{5.50}$$

where ϕ_0 is the initial elevation of the ray, $\cos \phi = (r_E / r_b) \cos \phi_0$, r_E is the radius of the Earth and $\gamma_0 = \gamma - r_E^2 \cos^2 \phi_0$.

Snell's and Bouger's laws are just two of a much larger class of first integrals. Consider the two-dimensional refractive index

$$N(x, y) = R(\Im \{g(z)\}) |g'(z)| \tag{5.51}$$

where $z = x + jy$ is a complex variable; it was shown in Coleman (2004) that the Euler–Lagrange equations for the corresponding Fermat's principle have the first integral

$$\frac{R(\Im \{g(z)\})}{|g'(z)|} \frac{d\Re \{g(z)\}}{ds} = C \tag{5.52}$$

The above result can be justified in the following manner. Consider Fermat's principle with the refractive index given by Equation 5.51, i.e.

$$\delta \int_A^B R(\Im\{g(z)\}) |g'(z)| \, ds = 0 \tag{5.53}$$

If we apply the conformal transformation $Z = g(z)$ where $Z = X + jY$, then Fermat's principle will take the form

$$\delta \int_A^B R(Y) \, dS = 0 \tag{5.54}$$

where $dS = |g'(z)| \, ds$ is the distance element in the new coordinates. This variational equation will have the first integral (see Appendix C)

$$R(Y) \frac{dX}{dS} = C \tag{5.55}$$

in the new (X, Y) coordinates. Transforming back to the original coordinates Equation 5.55 yields Equation 5.52. Equation 5.55 can be formally integrated (Coleman, 2004) to yield

$$X + C_0 = \int \frac{C}{\sqrt{R^2(Y) - C^2}} \, dY \tag{5.56}$$

Consequently, we can obtain the propagation through a medium with refractive index defined by Equation 5.51 from propagation through a horizontally stratified ionosphere.

Bouger's law follows from Equation 5.52 with $g(z) = j \ln z$ since, in terms of polar coordinates, $\ln z = \log r + j\theta$. To obtain the ray path in polar coordinates, we substitute $X = -\theta$ and $Y = \ln r$ into Equation 5.56. The case of a quasi-parabolic layer follows when $R(Y) = \sqrt{\gamma + \beta \exp(Y) + \alpha \exp(2Y)}$ and then the integral in Equation 5.56 can be performed analytically. Expression 5.56 then implies

$$-\theta + C_0 = \frac{-C}{\sqrt{\gamma - C^2}} \left(\ln \left(2\sqrt{\gamma - C^2}\sqrt{\gamma - C^2 + \beta r + \alpha r^2} + \beta r + 2\left(\gamma - C^2\right) \right) - \ln r \right) \tag{5.57}$$

It should be noted, however, that propagation below the ionosphere (two linear sections) will need to be treated separately. Taking this into account, the expression for ground range given by 5.50 can be derived from 5.57.

Many other generalizations of Snell's law are possible but, of particular interest, are ionospheres with horizontal gradients. If $g(z) = j \ln(z - z_0)$ we obtain the eccentric ionospheres discussed by Folkestad (1968) and, if $g(z) = (\cos(\alpha) + j\sin(\alpha))z$, we have a stratified ionosphere that has been tilted through an angle α.

5.3   Geometric Optics in an Anisotropic Medium

In this section we consider propagation in an ionosphere with the effects of Earth's magnetic field included (*magneto-ionic effects*). The medium is now anisotropic and the GO approximation is far more complicated. In this section, we will derive the

5.3 Geometric Optics in an Anisotropic Medium

ray tracing equations directly from the Maxwell equations using techniques based on Coleman (2008). We start with the time harmonic equation for the electric field in free space

$$\nabla \times \nabla \times \mathbf{E} - \omega^2 \mu_0 \epsilon_0 \mathbf{E} = -j\omega\mu_0 \mathbf{J} \tag{5.58}$$

and, from Chapter 2, the relationship

$$-j\omega X \epsilon_0 \mathbf{E} = U\mathbf{J} + j\mathbf{Y} \times \mathbf{J} \tag{5.59}$$

between the plasma current and electric field (X, U and $\mathbf{Y}$ defined as in Chapter 2). Assume the same ansatz as for the isotropic case, i.e.

$$\mathbf{E}(\mathbf{r}) = \mathbf{E}_0(\mathbf{r}) \exp(-j\beta_0 \phi(\mathbf{r})) \tag{5.60}$$

for the electric field and

$$\mathbf{J}(\mathbf{r}) = \mathbf{J}_0(\mathbf{r}) \exp(-j\beta_0 \phi(\mathbf{r})) \tag{5.61}$$

for the plasma current. We then obtain, to the leading order in β,

$$\nabla\phi \cdot \nabla\phi \mathbf{E}_0 - \nabla\phi \nabla\phi \cdot \mathbf{E}_0 - \mathbf{E}_0 = -j\frac{\omega\mu_0}{\beta_0^2} \mathbf{J}_0 \tag{5.62}$$

and from which

$$\nabla\phi \cdot \mathbf{E}_0 = j\frac{\omega\mu_0}{\beta_0^2} \nabla\phi \cdot \mathbf{J}_0 \tag{5.63}$$

From Equation 2.60 we have

$$-j\omega X \epsilon_0 \mathbf{E}_0 = U\mathbf{J}_0 + j\mathbf{Y} \times \mathbf{J}_0 \tag{5.64}$$

and this, together with Equation 5.63, reduces Equation 5.62 to

$$(\nabla\phi \cdot \nabla\phi - 1)(U\mathbf{J}_0 + j\mathbf{Y} \times \mathbf{J}_0) = -X(\mathbf{J}_0 - \nabla\phi \nabla\phi \cdot \mathbf{J}_0) \tag{5.65}$$

The dot product of $\nabla\phi$ with Equation 5.65 yields

$$U\nabla\phi \cdot \mathbf{J}_0 + j\nabla\phi \cdot (\mathbf{Y} \times \mathbf{J}_0) = X\nabla\phi \cdot \mathbf{J}_0 \tag{5.66}$$

from which

$$j\nabla\phi \cdot (\mathbf{Y} \times \mathbf{J}_0) = (X - U)\nabla\phi \cdot \mathbf{J}_0 \tag{5.67}$$

We can now use Equation 5.67 to rearrange Equation 5.65 into the form

$$(U\nabla\phi \cdot \nabla\phi - U + X)\mathbf{J}_0 = \frac{jX}{X - U}\nabla\phi\nabla\phi \cdot (\mathbf{Y} \times \mathbf{J}_0) - j(\nabla\phi \cdot \nabla\phi - 1)\mathbf{Y} \times \mathbf{J}_0 \tag{5.68}$$

Once $\mathbf{J}_0$ is known, we can calculate $\mathbf{E}_0$ from Equation 5.64 and $\mathbf{H}_0$ from

$$\mathbf{H}_0 = \frac{1}{\eta_0} \nabla\phi \times \mathbf{E}_0 \tag{5.69}$$

Without loss of generality, we now choose Cartesian coordinates with the x_1 axis in the direction of the magnetic field (i.e. $Y^2 = Y^3 = 0$) then $\mathbf{Y} \times \mathbf{J}_0 = -YJ_0^3 \hat{\mathbf{y}} + YJ_0^2 \hat{\mathbf{z}}$ where

$\hat{\mathbf{x}}$, $\hat{\mathbf{y}}$ and $\hat{\mathbf{z}}$ are the unit vectors along the x_1, x_2 and x_3 axes, respectively. Equation 5.68 can now be recast in the matrix form

$$\begin{pmatrix} Uq+1 & -\frac{jY}{X-U}p_1p_3 & \frac{jY}{X-U}p_1p_2 \\ 0 & Uq+1-\frac{jY}{X-U}p_2p_3 & \frac{jY}{X-U}p_2^2 - jqY \\ 0 & -\frac{jY}{X-U}p_3^2 + jqY & Uq+1+\frac{jY}{X-U}p_3p_2 \end{pmatrix} \begin{pmatrix} J_0^1 \\ J_0^2 \\ J_0^3 \end{pmatrix} = 0 \qquad (5.70)$$

where $q = (\mathbf{p} \cdot \mathbf{p} - 1)/X$ and $\mathbf{p} = \nabla\phi$. (Note that, through Equation 5.64, this is essentially an equation for the electric field.) For Equation 5.70 to have a non-zero solution, the determinant of the matrix will need to be zero, i.e.

$$(Uq+1)\left((Uq+1)^2 + \frac{qY^2}{X-U}(p_2^2 + p_3^2) - q^2Y^2\right) = 0 \qquad (5.71)$$

The set of eigenvectors corresponding to $Uq + 1 = 0$ will consist of electric fields that are parallel to the direction of the magnetic field, but those corresponding to $(Uq+1)^2 + (p_2^2 + p_3^2)qY^2/(X - U) - q^2Y^2 = 0$ are more complex. We now note that $p_1 Y = \mathbf{p} \cdot \mathbf{Y}$ and so, in a general coordinate system, the second set of eigenvalues will satisfy

$$(Uq+1)^2 + \frac{q}{X-U}(p^2Y^2 - (\mathbf{p} \cdot \mathbf{Y})^2) - q^2Y^2 = 0 \qquad (5.72)$$

where $p^2 = \mathbf{p} \cdot \mathbf{p}$. In the case of a plane wave, we will have $\phi = \beta_r \hat{\mathbf{p}} \cdot \mathbf{r}$ where $\hat{\mathbf{p}}$ is the unit vector in the propagation direction, and then Equation 5.72 will become an equation for β_r since we will now have $q = (\beta_r^2 - 1)/X$. We define the refractive index N in an anisotropic medium to be β_r and then, rearranging Equation 5.72, we obtain the Appleton–Hartree formula for the refractive index

$$N^2 = 1 - \frac{2X(U-X)}{2U(U-X) - Y_T^2 \pm \left(Y_T^4 + 4(U-X)^2 Y_L^2\right)^{\frac{1}{2}}} \qquad (5.73)$$

where Y_L is the component of $\mathbf{Y}$ that is parallel to the wave normal and Y_T is the component that is orthogonal (note that $Y_T^2 = Y^2 - Y_L^2$). This reduces to $N^2 = 1 - X$ when the Earth's magnetic field can be ignored.

Equation 5.72 is essentially a first-order differential equation for ϕ that we can solve, as in the homogeneous medium case, through the generalized Charpit equations 5.7. We first rearrange Equation 5.72 in the form $F(\mathbf{x}, \mathbf{p}) = 0$ where

$$\begin{aligned} F(\mathbf{x}, \mathbf{p}) = {} & (U-X)(Up^2 + X - U)^2 \\ & - (p^2 - 1)X(p^2Y^2 - (\mathbf{p}.\mathbf{Y})^2) - (U-X)(p^2-1)^2 Y^2 \end{aligned} \qquad (5.74)$$

Then the Charpit equations allow us to calculate a family of characteristics (or rays) on which the solution is defined. To simplify matters we assume the plasma to be collision free ($U = 1$), then

$$\begin{aligned} F(\mathbf{x}, \mathbf{p}) = {} & (p^2 - 1)^2(1 - X - Y^2) \\ & + (p^2 - 1)(2X(1-X) - XY^2) + (1-X)X^2 + (p^2 - 1)X(\mathbf{p}.\mathbf{Y})^2 \end{aligned} \qquad (5.75)$$

5.3 Geometric Optics in an Anisotropic Medium

From this

$$\frac{\partial F}{\partial p_i} = 4p_i(p^2 - 1)(1 - X - Y^2)$$
$$+ 2p_i(2X(1-X) - XY^2) + 2p_iX(\mathbf{p}.\mathbf{Y})^2 + 2Y_i(p^2 - 1)X\mathbf{p}.\mathbf{Y} \quad (5.76)$$

and, from the Charpit equations, we find

$$\frac{dx_i}{dt} = 4p_i(q+1)(1 - X - Y^2) + 2p_iY^2 + 2p_i(\mathbf{p}.\mathbf{Y})^2 + 2Y_iXq\mathbf{p}.\mathbf{Y} \quad (5.77)$$

on defining parameter t by $dt = Xd\phi / \sum_{j=1}^{3} p_j \partial F/\partial p_j$. We also have

$$\frac{\partial F}{\partial x_i} = -(p^2 - 1)^2 \left(\frac{\partial X}{\partial x_i} + 2\mathbf{Y} \cdot \frac{\partial \mathbf{Y}}{\partial x_i} \right)$$
$$+ (p^2 - 1)\left(\frac{\partial X}{\partial x_i}(2 - 4X - Y^2) - 2X\mathbf{Y} \cdot \frac{\partial \mathbf{Y}}{\partial x_i} \right)$$
$$+ X\frac{\partial X}{\partial x_i}(2 - 3X) + \frac{\partial X}{\partial x_i}(p^2 - 1)(\mathbf{p} \cdot \mathbf{Y})^2 + 2\mathbf{p} \cdot \frac{\partial \mathbf{Y}}{\partial x_i}X(p^2 - 1)\mathbf{p} \cdot \mathbf{Y} \quad (5.78)$$

and, from the Charpit equations, we find

$$\frac{dp_i}{dt} = \left(Xq^2 - (2 - 4X - Y^2 + (\mathbf{p} \cdot \mathbf{Y})^2)q - 2 + 3X \right) \frac{\partial X}{\partial x_i}$$
$$+ 2q(q+1)X\mathbf{Y} \cdot \frac{\partial \mathbf{Y}}{\partial x_i} - 2qX(\mathbf{p} \cdot \mathbf{Y})\mathbf{p} \cdot \frac{\partial \mathbf{Y}}{\partial x_i} \quad (5.79)$$

From Equation 5.72, q will satisfy

$$q^2(1 - X - Y^2) + q((\mathbf{p} \cdot \mathbf{Y})^2 + 2 - 2X - Y^2) + 1 - X = 0 \quad (5.80)$$

with the different roots corresponding to different propagation modes (O and X rays). It is possible to further reduce Equation 5.79 using Equation 5.80 to obtain

$$\frac{dp_i}{dt} = \left((1 - Y^2)(q+1)^2 - 2(1 - X - Y^2)(q+1) - Y^2 \right) \frac{\partial X}{\partial x_i}$$
$$+ 2q(q+1)X\mathbf{Y} \cdot \frac{\partial \mathbf{Y}}{\partial x_i} - 2qX(\mathbf{p} \cdot \mathbf{Y})\mathbf{p} \cdot \frac{\partial \mathbf{Y}}{\partial x_i} \quad (5.81)$$

which is one form of the Haselgrove ray tracing equations (Haselgrove, 1960). (It should be noted that these equations break down when $Y = 0$ and in this case the equations of section 5.1 should be used.) The parameter t that is used to parameterize a point on the ray path is not unique and Haselgrove showed that a suitable rescaling could produce other forms of her equations that were more convenient for some numerical techniques. In addition, she recast the equations in tensor form and hence derived a polar form of the equations that is perhaps more convenient for some ionospheric calculations (this form is used by Jones and Stephenson (1975) in their numerical implementation of Haselgrove's equations). Numerical experiments, however, have shown that the Cartesian form can lead to a more efficient numerical algorithm.

Once we have calculated a ray trajectory, and the associated vector **p** along it, we may calculate the associated current vector $\mathbf{J}_0$ along the ray as solutions to Equation 5.70. There will be a solution corresponding to each root of Equation 5.80 (the O and X modes), but they will only be determined up to a multiplicative constant. The associated $\mathbf{E}_0$ and $\mathbf{H}_0$ may then be found from Equations 5.64 and 5.69. Although we now have the ray paths and direction of the fields along the ray, we still do not have the actual field magnitudes. These will depend on the energy flowing from the source of the field and, as we have seen earlier, this flow is represented by the time harmonic Poynting vector

$$\mathbf{P} = \frac{1}{2}\Re\{\mathbf{E} \times \mathbf{H}^*\} \tag{5.82}$$

which is the power flow across a unit area that is orthogonal to **P**. For homogeneous media, the Poynting vector is parallel to the propagation direction and both the electric and magnetic fields are orthogonal to this direction (and each other). In Chapter 2, however, we saw that things are not quite so simple for anisotropic media. As in Chapter 2, we choose axes such that the propagation vector **p** is orientated along the x_3 axis ($p_1 = p_2 = 0$) and the magnetic field lies in the $x_2 x_3$ plane ($Y_1 = 0$). Then from the considerations of Chapter 2, we have that

$$\mathbf{E}_0 = E_0 \hat{\mathbf{x}} - \frac{jY_3 q}{1+q} E_0 \hat{\mathbf{y}} - \frac{jqY_2 X}{1-X} E_0 \hat{\mathbf{z}} \tag{5.83}$$

from which it will be noted that the electric field is not orthogonal to the propagation direction. It also turns out that the Poynting vector is not in the direction of propagation, but in the ray direction. To see this, consider Equation 5.77 for the above choice of axes, i.e.

$$\frac{d\mathbf{r}}{dt} \propto -XY_2 Y_3 \hat{\mathbf{y}} - Q\left(2(1-X) - Y^2 + Y_3^2 + \frac{2}{Q}(1 - X - Y^2 + XY_3^2)\right)\hat{\mathbf{z}} \tag{5.84}$$

where $Q = 1/q$. From Equation 2.94, we have

$$Q + 1 = \frac{Y_2^2 \mp \sqrt{Y_2^4 + 4(1-X)^2 Y_3^2}}{2(1-X)} \tag{5.85}$$

from which

$$\frac{1}{Q} = \frac{-2(1-X) + Y_2^2 \pm \sqrt{Y_2^4 + 4(1-X)^2 Y_3^2}}{2(1 - X - Y^2 + XY_3^2)} \tag{5.86}$$

Then, after some algebra, Equation 5.84 yields

$$\frac{d\mathbf{r}}{dt} \propto -XY_2 Y_3 \hat{\mathbf{y}} \mp Q\sqrt{Y_2^4 + 4(1-X)^2 Y_3^2}\,\hat{\mathbf{z}} \tag{5.87}$$

From expression 2.95 for the Poynting vector, it will now be seen that the Poynting and ray directions are both the same. Consequently, we can use the spread of rays from the radio source to calculate the spread of energy. In the case that the medium has collisions (i.e. $U \neq 1$) this is not necessarily the case, due to loss mechanisms other than spreading. However, in much of the ionosphere, the case that $U = 1$ is a useful approximation.

In applications of ray tracing to communications problems, two of the most important quantities to be calculated are the phase delay τ_p and the group delay τ_g. The phase delay is given by

$$\tau_p = \frac{1}{c_0} \int_A^B \frac{N}{p} \mathbf{p} \cdot d\mathbf{r} \qquad (5.88)$$

and the group delay by

$$\tau_g = \frac{1}{c_0} \int_A^B \frac{N_g}{p} \mathbf{p} \cdot d\mathbf{r} \qquad (5.89)$$

where A and B are the beginning and end of the communication path. Quantity N is the normal refractive index (sometimes known as the phase refractive index) and $N_g = N + f \partial N / \partial f$ is the group refractive index. We can obtain the phase and group refractive indices from equations given in Budden (1985). The phase refractive index N is obtained by solving

$$AN^4 + 2BN^2 + C = 0 \qquad (5.90)$$

where

$$A = 1 - X - Y^2 + XY_L^2 \qquad (5.91)$$

$$2B = -2(1-X)(1-X-Y^2) + XY^2 - XY_L^2 \qquad (5.92)$$

and

$$C = (1-X)((1-X)^2 - Y^2) \qquad (5.93)$$

Then, the group refractive index can be obtained from

$$N_g = \frac{N^4 A' - N^2 B' + C'}{N(N^2 A - B)} \qquad (5.94)$$

where

$$A' = 1 - \frac{3}{2}(X + Y^2) + 2XY_L^2 \qquad (5.95)$$

$$B' = (1-X)(1-3X) - 2Y^2 + \frac{3}{2}XY^2 + \frac{3}{2}XY_L^2 \qquad (5.96)$$

and

$$C' = -\frac{3}{2}X(1-X)^2 - Y^2 \left(\frac{1}{2} - X\right) \qquad (5.97)$$

5.4 Weakly Anisotropic Medium

The considerations of the previous section are required when the wave frequency ω is of the same order as the gyro frequency $\omega_H = |e\underline{B}_0|/m$. When the wave frequency is much greater than the gyro frequency (typically for frequencies greater than 5 MHz in ionospheric propagation), the analysis becomes much simpler. In this case, we can treat the magneto-ionic effects as a rotation of the field vectors around the propagation path. Based on Coleman (2008), we will derive a set of equations for this limit. Equation 2.60 can be inverted to yield

$$\mathbf{J} = \frac{-j\omega\varepsilon_0 X}{U^2 - Y^2}\left(U\mathbf{E} - j\mathbf{Y}\times\mathbf{E} - \frac{1}{U}\mathbf{Y}\cdot\mathbf{E}\mathbf{Y}\right) \tag{5.98}$$

We neglect collisions ($U = 1$) and assume that $\omega \gg \omega_p$ (i.e. we neglect terms of order Y^2), then Equation 5.58, together with Equation 5.98, implies

$$\nabla\times\nabla\times\mathbf{E} - \omega^2\mu_0\varepsilon_0(1-X)\mathbf{E} = j\omega^2\mu_0\varepsilon_0 X\mathbf{Y}\times\mathbf{E} \tag{5.99}$$

If we substitute the usual ansatz $\mathbf{E} = \mathbf{E}_0\exp(-j\beta_0\varphi)$ into this equation, the two leading orders in β_0 imply

$$-\nabla\varphi\nabla\varphi\cdot\mathbf{E}_0 + (\nabla\varphi\cdot\nabla\varphi)\mathbf{E}_0 - (1-X)\mathbf{E}_0 = 0 \tag{5.100}$$

and

$$-\nabla\varphi\nabla\cdot\mathbf{E}_0 - \nabla(\nabla\varphi\cdot\mathbf{E}_0) + 2\nabla\varphi\cdot\nabla\mathbf{E}_0 + \nabla^2\varphi\mathbf{E}_0 = \beta_0 X\mathbf{Y}\times\mathbf{E}_0 \tag{5.101}$$

on noting that $Y = O\left(\beta_0^{-1}\right)$. Taking the divergence of Equation 5.99, we obtain

$$\nabla\cdot((1-X)\mathbf{E} + jX\mathbf{Y}\times\mathbf{E}) = 0 \tag{5.102}$$

The two leading orders in β_0 imply

$$\nabla\varphi\cdot\mathbf{E}_0 = 0 \tag{5.103}$$

and

$$(1-X)\nabla\cdot\mathbf{E}_0 = \nabla X\cdot\mathbf{E}_0 - \beta_0 X\nabla\varphi\cdot(\mathbf{Y}\times\mathbf{E}_0) \tag{5.104}$$

Together with Equations 5.100 and 5.101, Equations 5.103 and 5.104 imply

$$(\nabla\varphi\cdot\nabla\varphi - 1 + X)\mathbf{E}_0 = 0 \tag{5.105}$$

and

$$\frac{-\nabla\varphi}{1-X}\nabla X\cdot\mathbf{E}_0 + \frac{\beta X}{1-X}\nabla\varphi\nabla\varphi\cdot(\mathbf{Y}\times\mathbf{E}_0) + 2\nabla\varphi\cdot\nabla\mathbf{E}_0 + \nabla^2\varphi\mathbf{E}_0 = \beta_0 X\mathbf{Y}\times\mathbf{E}_0 \tag{5.106}$$

Equation 5.105 simply yields the ray tracing equations for an isotropic medium (i.e. $d\mathbf{p}/dg = \nabla N^2/2$ and $d\mathbf{r}/dg = \mathbf{p}$). The effects of anisotropy, however, are evident in Equation 5.106. Noting that $d\mathbf{r}/dg = \nabla\varphi$, Equation 5.106 implies

$$2\frac{d\mathbf{E}_0}{dg} + \frac{d\mathbf{x}}{dg}\nabla\ln N^2 \cdot \mathbf{E}_0 + \nabla^2\varphi\mathbf{E}_0 = \beta_0 X\left(\mathbf{Y}\times\mathbf{E}_0 - \frac{1}{1-X}\frac{d\mathbf{x}}{dg}\frac{d\mathbf{x}}{dg}\cdot(\mathbf{Y}\times\mathbf{E}_0)\right) \tag{5.107}$$

which, after a dot product with $\mathbf{E}_0$, yields

$$\frac{d|\mathbf{E}_0|^2}{dg} + \nabla^2\varphi|\mathbf{E}_0|^2 = 0 \tag{5.108}$$

Combining Equation 5.107 with 5.108 we obtain

$$2\frac{d\mathbf{P}}{dg} + \frac{d\mathbf{r}}{dg}\nabla\ln N^2 \cdot \mathbf{P} = \beta_0 X\left(\mathbf{Y}\times\mathbf{P} - \frac{1}{1-X}\frac{d\mathbf{r}}{dg}\frac{d\mathbf{r}}{dg}\cdot(\mathbf{Y}\times\mathbf{P})\right) \tag{5.109}$$

which is an equation that describes how the polarization vector $\mathbf{P} = \mathbf{E}_0/|\mathbf{E}_0|$ changes as we move along the ray. It is obvious that the background magnetic field manifests itself as a rotation of the polarization vector about the ray direction, sometimes known as *Faraday rotation*. For the special case of propagation that lies in a plane, it is possible to obtain an explicit expression for this rotation. Consider a vector $\mathbf{T}$ that is orthogonal to the plane of propagation. Taking the dot product of $\mathbf{T}$ with Equation 5.109 we obtain

$$2\frac{d\cos\theta}{dg} = -\beta_0 X\sin\theta\mathbf{Y}\cdot\frac{d\mathbf{r}}{dg} \tag{5.110}$$

where θ is the angle between $\mathbf{T}$ and the polarization vector. The above equation can be integrated to yield

$$\theta = \frac{\beta_0}{2}\int_A^B X\mathbf{Y}\cdot d\mathbf{r} \tag{5.111}$$

where A and B are the start and end points of the ray path.

5.5 Fermat's Principle for Anisotropic Media

So far we have only studied Fermat's principle in the case of an isotropic medium. In this case, the principle has been found to be very useful for deriving analytic results concerning ray tracing. Usually, the variational principle is used to derive ordinary differential equations (the Haselgrove equations in particular) that are the starting point for most approaches to ray tracing. Unfortunately, ray tracing that starts with the differential equations is best suited to problems that only depend on *initial values*. In many circumstances, however, we need to find the propagation between two given points and such *two point boundary problems* are not directly amenable to the standard initial value techniques for solving ordinary differential equations (see Appendix B for such techniques). Instead, we need to guess initial values that will get the solution close to the required end point and then improve the guess through several iterations (this technique is known as homing). The direct solution of variational principles (such as that of Fermat) allows us to have a more natural approach to two-point boundary problems through Rayleigh–Ritz techniques, the finite element (FE) method being a particularly important example of such techniques. This approach is fairly straightforward in the case of isotropic media, but leads to complexities in the case of anisotropic media. In particular, we are interested in the anisotropic medium that is formed by the ionosphere when the effect of Earth's magnetic field is included. For such an anisotropic medium, Fermat's principle will be

$$\delta \int_A^B \frac{N}{p} \mathbf{p} \cdot d\mathbf{r} = 0 \qquad (5.112)$$

where $\mathbf{r}$ is a position vector on the ray path and $\mathbf{p}$ is a vector that is normal to the wavefront (note that $p = |\mathbf{p}| = N$). Haselgrove's equations for magneto-ionic media (Haselgrove, 1956) are the Hamiltonian equations that can be derived from this variational principle. It will be noted that the functional of the variational principle

$$P = \int_A^B \frac{N}{p} \mathbf{p} \cdot d\mathbf{r} \qquad (5.113)$$

is the phase distance between the points A and B. The square of refractive index N^2 is, from Equation 5.90,

$$N^2 = 1 - \frac{2X(1-X)}{2(1-X) - (Y^2 - Y_L^2) \pm \sqrt{(Y^2 - Y_L^2)^2 + 4(1-X)^2 Y_L^2}} \qquad (5.114)$$

where the $+$ sign corresponds to an ordinary ray and the $-$ sign corresponds to the extraordinary ray (Budden, 1985). Note that $Y_L = \mathbf{Y} \cdot \mathbf{w}$ where $\mathbf{w} = \mathbf{p}/p$.

The variational principle expressed by Equation 5.113 introduces difficulties for a direct implementation in that we must consider variations of the vector $\mathbf{p}$ as well as the geometry of the ray path. It would be preferable to eliminate $\mathbf{p}$ and we can do this through an approach introduced by Coleman (2011). In this approach we restrict the variations to those that partially satisfy the Haselgrove equations and in this manner are able to eliminate the variations in $\mathbf{p}$ and hence produce a variational principle in terms of ray geometry alone. The variational principle 5.112 can be written as

$$\delta \int_A^B N(\mathbf{r}, Y_L) \dot{r}_w \, dt = 0 \qquad (5.115)$$

where $\dot{r}_w = \dot{\mathbf{r}} \cdot \mathbf{w}$, $\dot{\mathbf{r}} = d\mathbf{r}/dt$ and t is a variable that parameterizes the path. The refractive index N only depends on $\mathbf{p}$ through the quantity Y_L and so N is a function of $\mathbf{r}$ and Y_L alone (i.e. $N(\mathbf{r}, \mathbf{p}) = N(\mathbf{r}, Y_L)$). The Euler–Lagrange equations for variations in $\mathbf{p}$ will now yield

5.5 Fermat's Principle for Anisotropic Media

$$\dot{r}_i = \dot{\mathbf{r}} \cdot \mathbf{p} \left(\frac{p_i}{pN} - \frac{p}{2N} \frac{\partial \ln(N^2)}{\partial Y_L} \frac{\partial Y_L}{\partial p_i} \right) \tag{5.116}$$

and noting that

$$p \frac{\partial Y_L}{\partial p_i} = Y_i - Y_L w_i \tag{5.117}$$

this can be expressed as

$$\dot{r}_i = \dot{r}_w \left(w_i - \frac{1}{2}(Y_i - Y_L w_i) \frac{\partial \ln(N^2)}{\partial Y_L} \right) \tag{5.118}$$

The dot product of Equation 5.118 with $\dot{\mathbf{r}}$ and $\mathbf{Y}$ will yield, respectively,

$$\dot{\mathbf{r}} \cdot \dot{\mathbf{r}} = \dot{r}_w \left(\dot{r}_w - \frac{1}{2}(\mathbf{Y} \cdot \dot{\mathbf{r}} - Y_L \dot{r}_w) \frac{\partial \ln(N^2)}{\partial Y_L} \right) \tag{5.119}$$

and

$$\dot{\mathbf{r}} \cdot \mathbf{Y} = \dot{r}_w \left(Y_L - \frac{1}{2}(Y^2 - Y_L^2) \frac{\partial \ln(N^2)}{\partial Y_L} \right) \tag{5.120}$$

Introducing the unit tangent $\mathbf{t} = \dot{\mathbf{r}}/|\dot{\mathbf{r}}|$ we can now rewrite Equations 5.119 and 5.120 as

$$1 = w_t \left(w_t - \frac{1}{2}(Y_t - Y_L p_t) \frac{\partial \ln(N^2)}{\partial Y_L} \right) \tag{5.121}$$

and

$$Y_t = w_t \left(Y_L - \frac{1}{2}(Y^2 - Y_L^2) \frac{\partial \ln(N^2)}{\partial Y_L} \right) \tag{5.122}$$

where $Y_t = \mathbf{Y} \cdot \mathbf{t}$ and $w_t = \mathbf{w} \cdot \mathbf{t}$. These last two equations provide a pair of simultaneous equations for w_t and Y_L from which we can obtain these quantities as functions of $\mathbf{r}$ and $\mathbf{t}$. We can use Equation 5.122 to eliminate w_t from Equation 5.121 and hence obtain an implicit equation for Y_L in terms of $\mathbf{r}$ and $\mathbf{t}$ (assuming Y and X are known functions of $\mathbf{r}$). Once this equation is solved, w_t can then be derived from Equation 5.121 in terms of $\mathbf{r}$ and $\mathbf{t}$. The values of Y_L and w_t thus derived will then be compatible with the Euler–Lagrange equations for 5.115. Returning to the variational principle, we now see that the functional

$$P = \int_A^B N(\mathbf{r}, Y_L) w_t \, ds \tag{5.123}$$

only depends on $\mathbf{p}$ through w_t and Y_L. Consequently, in the manner of complementary variational principles (Arthurs, 1980), we can use the above expressions for w_t and Y_L in terms of path geometry to eliminate $\mathbf{p}$ from the functional and hence obtain a functional in terms of ray geometry alone (i.e. in terms of $\mathbf{r}$ and $\mathbf{t}$ alone).

We now turn our attention to the implementation of the above procedure. As mentioned previously, we can solve the variational principle by means of a direct approach such as the Rayleigh–Ritz method. In this approach, the ray paths are restricted to a set of trial rays that are sufficiently general to cover the type of ray behavior that is to be expected for the anisotropic medium under consideration. These trial rays are characterized by a set of N parameters α_1 to α_N and, when substituted into the functional, they turn it into a function of α_1 to α_N (i.e. $P = P(\alpha_1, \ldots, \alpha_N)$). Our problem now reduces to that of finding the stationary value of P with respect to the parameters α_1 to α_N. The parameter values that make P stationary are found as the solutions to

$$\frac{\partial P}{\partial \alpha_i} = 0 \quad i = 1, 2, \ldots N \tag{5.124}$$

This is usually a highly nonlinear system of equations and will require an iterative solution such as that provided by the Newton–Raphson algorithm. This algorithm provides a linear system of equations for the increment in α between the Ith and $(I+1)$th stages of iteration

$$\frac{\partial P}{\partial \alpha_i} + \sum_{j=1}^{N} \left(\alpha_j^{I+1} - \alpha_j^I \right) \frac{\partial^2 P}{\partial \alpha_i \partial \alpha_j} = 0 \quad i = 1, 2, \ldots N \tag{5.125}$$

where the derivatives are evaluated at the Ith stage. We now have a linear system that can be solved by any of a number of standard techniques. To start the iteration, we will need a good initial estimate α^0. For a well-behaved ionosphere, this can often be provided by a simple analytic solution for an approximate ionosphere (a parabolic approximation to the ionosphere at the midpoint between the ends of the ray is usually adequate). For more complex ionospheres (if traveling ionospheric disturbances are present, for example) this is insufficient, but a 2D homing solution usually provides a good starting point. That is, the variational approach provides a means of upgrading a 2D solution to the full 3D case.

The above procedure is fairly straightforward, except for the fact that we are dealing with a variational problem for which the Lagrangian is homogeneous, a situation for which the Euler–Lagrange equations are known to be interdependent and for which the functional is parameter invariant (see Appendix C). One way to fix this problem is to choose the parameter t to be one of the ray coordinates or to be related to them in some way (the distance along the ray, for example). This will then remove the property of parameter invariance. In Coleman (2011) the parameter was chosen to be the distance along the great circle path in the plane through the ray end points and Earth's center.

We now need to choose the functional form of the approximate ray paths. One possibility is that of a linear combination of basis functions (such as a truncated Fourier series) where the basis coefficients are the parameters that characterize the candidate ray paths. Alternatively, as in the case of the finite element approach, they can be piecewise polynomial functions. Perhaps the simplest approximation is that of the finite element method with piecewise linear elements. We first note that, since the functional is parameter invariant, we can arbitrarily change the parameter t that characterizes a point on the ray path. In the case of an $M+1$ segment piecewise linear approximation, we will take this parameter to have the integer values 0 to $M+1$ at the interpolation

5.5 Fermat's Principle for Anisotropic Media

points of the ray path. In Cartesian coordinates, the piecewise linear approximation will take the form

$$\mathbf{r}(t) = \mathbf{r}_i + (t - i)(\mathbf{r}_{i+1} - \mathbf{r}_i) \quad \text{for } i < t < i+1 \tag{5.126}$$

The segment end points $\mathbf{r}_1$ to $\mathbf{r}_M$ will constitute the parameters that characterize the elements of the set of curves over which will make the variation (note that the end points $\mathbf{r}_0$ and $\mathbf{r}_{M+1}$ are fixed). Substituting the above form of the ray into the functional P, we obtain a function of the parameters $\mathbf{r}_1$ to $\mathbf{r}_M$ (i.e. $P = P(\mathbf{r}_1, ..., \mathbf{r}_M)$). In theory we can now find the values of $\mathbf{r}_1$ to $\mathbf{r}_M$ for which P is stationary from Equation 5.124. Unfortunately, due to parameter invariance, this procedure will not yield a unique solution unless we introduce some form of constraint. For approximations of the form 5.126 one of the simplest constraints is to require that the segments be of constant length. There is, however, the possibility that we could constrain the lengths of the segments so that they bunch around the points of maximum curvature. Figure 5.4 shows the results of a point-to-point ray trace using a direct approach to Fermat's principle for both O and X rays (X rays solid and O rays dashed). It will be noted that there are low and high rays for both of these modes.

The original formulation of Fermat's principle implied that, of all the possible paths, the ray path was that which made phase delay (or, equivalently, the phase distance) a minimum. A minimum, however, is not the only type of stationary behavior and the more general form of Fermat's principle includes other stationary paths as potential

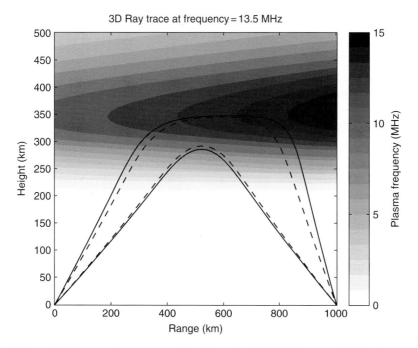

Figure 5.4 Point-to-point ray trace using Fermat's principle (X rays, solid line; O rays, dashed line).

rays. The variational principle for an ionosphere that is free from a background magnetic field is the simplest case with variational principle of the form

$$\delta \int_T^R N(x,y,z) \sqrt{\left(\frac{dx}{dt}\right)^2 + \left(\frac{dy}{dt}\right)^2 + \left(\frac{dz}{dt}\right)^2} \, dt = 0 \qquad (5.127)$$

This functional certainly satisfies the Legendre property of minima (see Appendix C) for frequencies above the plasma frequency and so it seems reasonable to use one of the vast range of effective minimization techniques (Press et al., 1992) that are available. This is certainly a very effective approach for transionospheric propagation and high rays at high frequencies (HF). Unfortunately, however, the approach breaks down for low rays. The reason for this is that the Legendre property is a necessary condition for a minimum, but it is not sufficient. A necessary and sufficient condition is provided by the Jacobi test for minima (see Appendix C). According to the Jacobi test for a minimum, it is necessary for there to be no focal points on the ray for the Legendre condition to predict a minimum (i.e. that the rays do not cross over). Although this is satisfied by transionospheric and high rays, it can be seen from Figure 5.3 that the low rays cross over and hence will have focal points. The low rays result from non-minimum stationary phase and such rays need to be found by an approach such as the Newton–Raphson procedure outlined earlier.

5.6 References

A.M. Arthurs, *Complementary Variational Principles*, Clarendon Press, Oxford University, Oxford, 1980.

J.A. Bennett, J. Chen and P.L. Dyson, Analytic ray tracing for the study of HF magneto-ionic radio propagation in the ionosphere, Appl. Comput. Electromag. Soc. J., vol. 6, pp. 192–210, 1991.

D. Bilitza, International reference ionosphere 1990, Rep. 90-22, Natl. Space Sci. Data Cent., Greenbelt, Md., 1990.

K.G. Budden, *The Propagation of Radio Waves*, Cambridge University Press, 1985.

V. Cervey, *Seismic Ray Tracing*, Cambridge University Press, 2001.

C.J. Coleman, A ray tracing formulation and its application to some problems in OTHR, Radio Sci., vol. 33, pp. 1187–1197, 1998.

C.J. Coleman, A general purpose ionospheric ray tracing procedure, Defence Science and Technology Organisation Australia, Technical Report SRL-0131-TR, 1993.

C.J. Coleman, *An Introduction to Radio Frequency Engineering*, Cambridge University Press, Cambridge, 2004.

C.J. Coleman, Ionospheric ray tracing equations and their solution, Radio Sci. Bull., vol. 325, pp. 17–23, 2008.

C.J. Coleman, On the generalisation of Snell's law, Radio Sci., vol. 39, 2004.

C.J. Coleman, Point to point ionospheric ray tracing by a direct variational method, Radio Science, vol. 46, RS5016,7PP., doi:10.1029/2011RS004748, 2011.

T.A. Croft and H. Hoogasian, Exact ray calculations in a quasi-parabolic ionosphere with no magnetic field, Radio Sci., vol. 3, pp. 69–74, 1968.

K. Folkestad, Exact ray computations in a tilted ionosphere with no magnetic field, Radio Sci., vol. 3, pp. 81–84, 1968.

C. Fox, *An Introduction to Calculus of Variations*, Dover Publ., New York, 2010.

C.B. Haselgrove and J. Haselgrove, Twisted ray paths in the ionosphere, Proc. Phys. Soc., vol. 75, pp. 357–363, 1960.

J. Haselgrove, Ray theory and a new method for ray tracing, Proc. Phys. Soc., pp. 355–364, 1954.

J. Haselgrove, The Hamilton ray path equations, J. Atmos. Terr. Phys., vol. 2, pp. 397–399, 1963.

I.M. Gelfand, and S.V. Fomin, *Calculus of Variations*, Dover Publ., New York, 2000.

D.S. Jones, *Methods in Electromagnetic Wave Propagation*, IEEE Series on Electromagnetic Wave Theory, New York and Oxford, 1994.

R.M. Jones and J.J. Stephenson, A versatile three-dimensional ray tracing computer program for radio waves in the ionosphere, OT Report 75-76. US Govt. Printing Office, Washington, DC, 20402, 1975.

J.M. Kelso, Ray tracing in the ionosphere, Radio Sci., vol. 3, pp. 1–12, 1968.

J.F. Mathews, *Numerical Methods for Computer Science, Engineering and Mathematics*, Prentice Hall, Englewood Cliffs, NJ, 1987.

W.H. Press, S. Teukolsky, W. Vetterling and B. Flannery, *Numerical Recipes in FORTRAN: The Art of Scientific Computing*, 2nd edition, Cambridge University Press, Cambridge, 1992.

M.H. Reilly, Upgrades for efficient three-dimensional ionospheric ray-tracing: Investigation of HF near vertical incidence sky wave effects, Radio Sci., vol. 26, pp. 981–980, 1991.

H. Sadeghi, S. Suzuki and H. Takenaka, A two-point, three-dimensional seismic ray tracing using genetic algorithms, Phys. Earth Planet. Inter., vol. 113, pp. 355–365, 1999.

J. Smilauer, The variational method ray path calculation in an isotropic, generally inhomogeneous ionosphere, J. Atmos. Terr. Phys., vol. 32, pp. 83–96, 1970.

J.R. Smith, *Introduction to the Theory of Partial Differential Equations*, Van Nostrand, London, 1967.

H.J. Strangeways, Effects of horizontal gradients on ionospherically reflected or transionospheric paths using a precise homing-in method, J. Atmos. Solar-Terr. Phys., vol. 62, pp. 1361–1376, 2000.

J. Um and C.H. Thurber, A fast algorithm for two-point seismic ray tracing, Bull. Seismol. Soc. Am., vol. 77, pp. 972–986, 1987.

A. Vasterberg, Investigations of the ionospheric HF channel by ray-tracing, IRF Scientific Technical Report 241, Swedish Institute of Space Physics, Uppsala, Sweden, 1997.

6 Propagation through Irregular Media

Propagation media can often contain irregular structures as a result of turbulence in the environment. Except for resonant structures, however, such scattering will normally be relatively weak. Both the atmosphere and ionosphere suffer from turbulence that can generate small-scale structure and this can result in significant back and forward scatter. The forward-scatter mechanism can have an impact on line-of-sight communication by causing a multiplicity of signal paths, and this can result in a spread of transmission delay. These multiple paths can cause fluctuations in signal level when there is relative motion between the irregularity and the line of sight. This phenomenon, known as *scintillation*, is most evident in satellite communications and radio astronomical observations (the twinkle of stars being its manifestation at optical frequencies). We will derive expressions for the mutual coherence function (MCF), a function that relates the statistics of propagation to the statistics of the propagation medium. Further, we will show how the MCF can be used to create simulations of a propagation channel.

Rough surfaces can also cause significant scatter (both backward and forward), and this can be quite strong, especially in the case of the sea. Sea scatter can cause significant problems for radar since it can result in radar returns (*clutter*) that can mask targets. We consider various techniques for modeling scatter by surface roughness and derive expressions that relate the scatter to the statistics of the roughness.

6.1 Scattering by Permittivity Anomalies

In our original derivation of the reciprocity theorem we assumed that, for the separate electromagnetic fields $(\mathbf{H}_A, \mathbf{E}_A)$ and $(\mathbf{H}_B, \mathbf{E}_B)$, their associated media had the same permittivity ($\epsilon_A = \epsilon_B = \epsilon$). We will now assume, however, that there is a region V^+ over which the permittivity of field B differs from that of A ($\epsilon_B = \epsilon + \delta\epsilon$) and then the reciprocity result will exhibit the modified form

$$\int_V \mathbf{J}_A \cdot \mathbf{E}_B \, dV - \int_V \mathbf{J}_B \cdot \mathbf{E}_A \, dV = -j\omega \int_{V^+} \delta\epsilon \mathbf{E}_B \cdot \mathbf{E}_A \, dV \qquad (6.1)$$

This is a specialization of result 3.38 for an inhomogeneous anisotropic medium. Consider the field $(\mathbf{H}_A, \mathbf{E}_A)$ caused by an antenna A when driven by current I_A and field $(\mathbf{H}_B, \mathbf{E}_B)$ caused by antenna B when driven by current I_B (see Figure 6.1). Equation 6.1 will reduce to

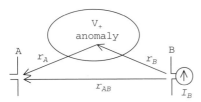

Figure 6.1 Scattering by a permittivity anomaly.

$$I_A \delta V = -j\omega \int_{V^+} \delta\epsilon \, \mathbf{E}_B \cdot \mathbf{E}_A \, dV \tag{6.2}$$

where δV is the additional voltage in the antenna A caused by scatter from the permittivity anomaly that disturbs field B. This formula will allow us to extend formula 3.27 for the mutual impedance Z_{AB} between two antennas to the case where there is a permittivity anomaly between the antennas (Monteath, 1973) and the mutual impedance is increased by δZ_{AB}. If we assume the anomaly to be a small perturbation, we can use the unperturbed $\mathbf{E}_B$ when evaluating the integral in Equation 6.2. As a consequence

$$I_A \delta V = j \frac{\omega^3 \mu^2 \epsilon I_A I_B \mathbf{h}^A_{\text{eff}} \cdot \mathbf{h}^B_{\text{eff}}}{16\pi^2} \int_{V^+} \frac{\delta\epsilon}{\epsilon} \frac{\exp(-j\beta(r_A + r_B))}{r_A r_B} dV \tag{6.3}$$

where $\mathbf{h}^A_{\text{eff}}$ and $\mathbf{h}^B_{\text{eff}}$ are the effective antenna lengths of antennas and r_A and r_B are their distances from the integration point. As a consequence, in terms of the mutual impedance Z_{AB} between the antennas in a uniform medium (Equation 3.27), the additional mutual impedance caused by the anomaly is

$$\frac{\delta Z_{AB}}{Z_{AB}} = \frac{\beta^2 r_{AB}}{4\pi} \exp(j\beta r_{AB}) \int_{V^+} \frac{\delta\epsilon}{\epsilon} \frac{\exp(-j\beta(r_A + r_B))}{r_A r_B} dV \tag{6.4}$$

where r_{AB} is the distance between the antennas. If the dimensions of the anomaly are much smaller than a wavelength,

$$\frac{\delta Z_{AB}}{Z_{AB}} = \frac{\beta^2 r_{AB} V^+}{4\pi r_A r_B} \frac{\delta\epsilon}{\epsilon} \exp(j\beta(r_{AB} - r_A - r_B)) \tag{6.5}$$

where V^+ is the volume of the scatterer. Such scatter is known as *Rayleigh scatter* and it is possible to extend this formula to the case where $\delta\epsilon$ is of arbitrary magnitude (Ishimaru, 1997). Assuming that the background is free space ($\epsilon_r = 1$), then $\delta\epsilon_r = \epsilon_r - 1$ where ϵ_r is the relative permittivity of the scatterer (assumed uniform). Since the scatterer is small, we can use the quasistatic approximation (i.e. we treat the electric fields as if static), then the field inside the scatterer will be $3(\epsilon_r + 2)$ times the field in the absence of the scatterer. For arbitrary ϵ_r, Equation 6.5 will then become

$$\frac{\delta Z_{AB}}{Z_{AB}} = \frac{\beta^2 r_{AB} V^+}{4\pi r_A r_B} \frac{3\epsilon_r - 3}{\epsilon_r + 2} \exp(j\beta(r_{AB} - r_A - r_B)) \tag{6.6}$$

We now consider the situation where the anomaly is irregular in nature. Irregularity is often described in a statistical sense and, in this case, is best described in terms of stochastic averages over the possible irregularity realizations. It is assumed that

$\langle \delta\epsilon \rangle = 0$ and hence, from Equation 6.4, we will find that $\langle \delta Z_{AB} \rangle = 0$ ($\langle \ldots \rangle$ denotes a stochastic average). Importantly, however, it is possible to relate the variance of the mutual impedance $\langle |\delta Z_{AB}|^2 \rangle$ to the autocorrelation of the permittivity $\langle \delta\epsilon' \delta\epsilon^* \rangle$ (note that a prime denotes a quantity evaluated at a primed coordinate and that $*$ denotes a complex conjugate). Monteath (1975) has used Equation 6.4 to derive the stochastic scattering formula of the variety described by Booker and Gordon (1950). The following stochastic relation follows from Equation 6.4

$$\langle |\delta Z_{AB}|^2 \rangle = Z_{AB}^2 \frac{\beta^4 r_{AB}^2}{16\pi^2} \int_{V^+} \int_{V^+} \frac{\langle \delta\epsilon' \delta\epsilon^* \rangle \exp(-j\beta(r'_A + r'_B - r_A - r_B))}{|\epsilon|^2 \; r'_A r'_B r_A r_B} \, dV dV' \quad (6.7)$$

We assume the region of irregularity V^+ to be relatively compact and base our integration coordinates on a representative point inside V^+ that has vector positions $\mathbf{R}_A$ and $\mathbf{R}_B$ with respect to antennas A and B, respectively. If $\mathbf{R} = (X, Y, Z)$ is a position vector in the integration coordinates, $r_A \approx R_A + \mathbf{R} \cdot \hat{\mathbf{R}}_A$ and $r_B \approx R_B + \mathbf{R} \cdot \hat{\mathbf{R}}_B$ where $\hat{\mathbf{R}}_A = \mathbf{R}_A / |\mathbf{R}_A|$ and $\hat{\mathbf{R}}_B = \mathbf{R}_B / |\mathbf{R}_B|$. As a consequence,

$$\frac{\langle |\delta Z_{AB}|^2 \rangle}{Z_{AB}^2} \approx \frac{\beta^4 r_{AB}^2}{16\pi^2 R_A^2 R_B^2} \int_{V^+} \int_{V^+} \frac{\langle \delta\epsilon' \delta\epsilon^* \rangle}{|\epsilon|^2} \exp(-j\beta(\mathbf{R}' - \mathbf{R}) \cdot (\hat{\mathbf{R}}_A + \hat{\mathbf{R}}_B)) \, dV dV' \quad (6.8)$$

The autocorrelation of the irregularity is normally assumed to have the form $\langle \delta\epsilon' \delta\epsilon^* \rangle = \rho(\mathbf{r}' - \mathbf{r})$ and we define an additional function by $W(\mathbf{k}) = (1/8\pi^3) \int_{R^3} \rho(\mathbf{r}') \exp(-j\mathbf{k} \cdot \mathbf{r}') \, dV'$. Function W is the *spectrum of the irregularity* and describes its distribution in terms of irregularity wavelength $\mathbf{k}$. We assume the autocorrelation to be well localized in comparison with the size of V^+ and so it is possible to replace one of the integrals in Equation 6.8 with W (note that $V^+ \approx R^3$ as far as this integral is concerned). Furthermore, we assume the irregularity spectrum peaks around $\mathbf{k} = 0$ and its width in $|\mathbf{k}|$ is small in comparison with β. As a consequence, W is also well localized and Equation 6.8 will reduce to

$$\frac{\langle |\delta Z_{AB}|^2 \rangle}{Z_{AB}^2} \approx \frac{\pi \beta^4 r_{AB}^2}{2 R_A^2 R_B^2} W(\beta(\hat{\mathbf{R}}_A + \hat{\mathbf{R}}_B)) \int_{V^+} \frac{\langle |\delta\epsilon|^2 \rangle}{|\epsilon|^2} \, dV \quad (6.9)$$

which is the Booker scattering formula. It will be noted that the mutual impedance will peak in the case of back scatter ($\hat{\mathbf{R}}_B = -\hat{\mathbf{R}}_A$). In general, however, there can be quite strong communication between antennas A and B whatever their positions. In the case of patches of atmospheric turbulence, this is the origin of what is known as *tropospheric scatter* communication (a form of propagation that can be important when the line of sight has been blocked). A fairly common model of isotropic irregularity is the Gaussian autocorrelation given by $\rho(\mathbf{r}) = \langle |\delta\epsilon|^2 \rangle \exp(-\mathbf{r} \cdot \mathbf{r}/L^2)$ where L is a typical length scale of the irregularity. For this autocorrelation function, W is given by the Gaussian distribution $W(\mathbf{k}) = \langle |\delta\epsilon|^2 \rangle (L^2/4\pi)^{\frac{3}{2}} \exp(-\mathbf{k} \cdot \mathbf{k} L^2/4)$.

In the ionosphere, turbulence in the ionization is strongly affected by the magnetic field of Earth, the irregularity being stretched along the field lines. As a consequence, the correlation length along a field line will be much greater than in a direction orthogonal to it. A simple spectrum that reflects this has the form $W(\mathbf{k}) = W(K)$ where

$K^2 = L_\perp^2(k_x^2 + k_y^2) + L_\parallel^2 k_z^2$ in a field aligned coordinate system (the z axis is in the direction of Earth's magnetic field). For situations where $L_\parallel \gg L_\perp$, the dominant contribution to back scatter will arise when the incident wave is orthogonal to the field lines. This fact can be exploited for communications at higher latitudes. In the auroral regions, a large amount of ionospheric turbulence exists and this is stretched along the field lines. Providing that suitable irregularity is present, a signal that becomes orthogonal to a field line will be scattered in all directions orthogonal to that field line. Consequently, two stations with radiation that becomes orthogonal to the field lines at the same point can communicate via scatter (Figure 6.2).

In forward scatter, the propagation indicated by Equation 6.9 can still be relatively strong. As illustrated in Figure 6.3, the scatter will cause a large number of additional paths and hence there will be a spread of delay times. In addition, if the irregularity moves across the propagation path, there will also be a spread of Doppler shifts and fluctuations in signal level. We can use Equation 6.4 to study these effects from a deterministic viewpoint, but the irregularity is usually only known in a statistical sense (i.e. the spectrum of irregularity). In this case, we can use Equation 6.9 to find some

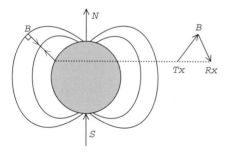

Figure 6.2 Communication via back scatter from field-aligned irregularity.

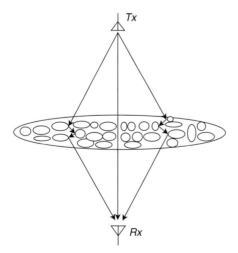

Figure 6.3 Forward scatter causing scintillation.

overall statistics. If actual values are required, however, an alternative approach is to create a realization that exhibits the correct statistics. We will consider this approach later in this chapter.

6.2 The Rytov Approximation

In our earlier studies on reciprocity, we obtained expression 3.39 for the perturbation to a field caused by an perturbation in permittivity. Assuming the medium to be isotropic, this relation reduces to

$$\delta \mathbf{E}_B(\mathbf{r}_A) \cdot \hat{\mathbf{s}} = j\omega \int_V \delta\varepsilon \mathbf{E}_A \cdot \mathbf{E}_B dV \qquad (6.10)$$

where $\delta \mathbf{E}_B$ is the deviation in electric field caused by a perturbation $\delta\epsilon$ in permittivity. In this expression, $\mathbf{E}_A$ is a field caused by current element $\mathbf{J}_A = \delta(\mathbf{r}-\mathbf{r}_A)\hat{\mathbf{s}}$, where $\hat{\mathbf{s}}$ is an arbitrary unit vector. In the previous section, we have assumed that we can replace $\mathbf{E}_B$ by its value for the unperturbed medium. This gives a useful solution for the situation where $\delta\varepsilon$ is small and is known as the Born approximation (Ishimaru, 1997). The Born approximation can be extended to a perturbation series

$$\mathbf{E}_B = \mathbf{E}_B^0 + \mathbf{E}_B^1 + \mathbf{E}_B^2 + \cdots \qquad (6.11)$$

where $\mathbf{E}_B^0$ is the field for a medium without perturbations. Expression 6.10 then yields

$$\mathbf{E}_B^{i+1}(\mathbf{r}_A) \cdot \hat{\mathbf{s}} = j\omega \int_V \mathbf{E}_A \cdot \delta\varepsilon \cdot \mathbf{E}_B^i dV \qquad (6.12)$$

for generating the terms of the series.

Although, in theory, the above series approach can provide solutions to any accuracy, the convergence tends to be slow. Another type of series that can be faster converging is that provided by the Rytov approach. We start from the equation for the electric field $\mathbf{E}$ for a medium with the permeability of free space, but a permittivity that can vary, i.e.

$$\nabla \times \nabla \times \mathbf{E} - \omega^2 \mu_0 \epsilon \mathbf{E} = 0 \qquad (6.13)$$

We have that $\nabla \times \nabla \times \mathbf{E} = -\nabla^2 \mathbf{E} + \nabla(\nabla \cdot \mathbf{E})$ and so, on noting that $\nabla \cdot (\epsilon \mathbf{E}) = 0$,

$$\nabla^2 \mathbf{E} + \omega^2 \mu_0 \epsilon_0 \epsilon_r \mathbf{E} - \nabla\left(\frac{\nabla\epsilon \cdot \mathbf{E}}{\epsilon}\right) = 0 \qquad (6.14)$$

The last term in the equation can be neglected provided that the length scales of the medium are much larger than a wavelength, i.e.

$$\nabla^2 \mathbf{E} + \omega^2 \mu_0 \epsilon \mathbf{E} = 0 \qquad (6.15)$$

We will study the generic Helmholtz equation

$$\nabla^2 E + \omega^2 \mu_0 (\epsilon + \delta\epsilon) E = 0 \qquad (6.16)$$

where $\delta\epsilon$ represents a perturbation to background permittivity ϵ. For the Born approximation, the zeroth order term satisfies

$$\nabla^2 E_B^0 + \omega^2 \mu_0 \epsilon E_B^0 = 0 \tag{6.17}$$

and the first-order term satisfies

$$\nabla^2 E_B^1 + \omega^2 \mu_0 \epsilon E_B^1 + \omega^2 \mu_0 \delta\epsilon E_B^0 = 0 \tag{6.18}$$

The Rytov approach (Ishimaru, 1997) introduces a new field ϕ such that

$$E = \exp\phi \tag{6.19}$$

Then, substituting Equation 6.19 into 6.16, we obtain

$$\nabla^2 \phi + \nabla\phi \cdot \nabla\phi + \omega^2 \mu_0 (\epsilon + \delta\epsilon) = 0 \tag{6.20}$$

We expand this new field in a perturbation series

$$\phi = \phi_0 + \phi_1 + \phi_2 + \cdots \tag{6.21}$$

then, in the leading order, we obtain from Equation 6.20 that

$$\nabla^2 \phi_0 + \nabla\phi_0 \cdot \nabla\phi_0 + \omega^2 \mu_0 \epsilon = 0 \tag{6.22}$$

Consequently, $E_0 = \exp\phi_0$ satisfies

$$\nabla^2 E_0 + \omega^2 \mu_0 \epsilon E_0 = 0 \tag{6.23}$$

i.e. E_0 is the electric field in the unperturbed case. In the next order, Equation 6.20 yields

$$\nabla^2 \phi_1 + 2\nabla\phi_0 \cdot \nabla\phi_1 + \omega^2 \mu_0 \delta\epsilon = 0 \tag{6.24}$$

and this can be rearranged into

$$\left(\nabla^2 + \omega^2 \mu_0 \epsilon\right)(E_0 \phi_1) + \omega^2 \mu_0 \delta\epsilon E_0 = 0 \tag{6.25}$$

It will be noted that the equation for $\phi_1 E_0$ will be same as the equation for E_1 (the Born approximation to the field perturbation) and so we have $\phi_1 = E_1/E_0$. As a consequence, the two-term Rytov approximation can be written as

$$E = E_0 \exp\frac{E_1}{E_0} \tag{6.26}$$

where E_0 is the unperturbed field and E_1 is the Born approximation to the field perturbation. In general, the Rytov solution is found to have a far greater range of validity than the Born approximation and so Equation 6.25 can be regarded as a means of extending the range of applicability of the Born approximation.

We will consider the special case of a plane wave that illuminates a volume V of irregularity in a homogeneous background. Without loss of generality, we take the

Propagation through Irregular Media

propagation direction to be the z axis so that $E_0 = \exp(-j\beta z)$. From Equation 6.10, we have that

$$E_1(\mathbf{r}_A) = j\omega \int_V \delta\varepsilon K(\mathbf{r}_A, \mathbf{r}) \exp(-j\beta z) dV \qquad (6.27)$$

where

$$K(\mathbf{r}_A, \mathbf{r}) = -\frac{j\omega\mu_0}{4\pi} \frac{\exp(-j\beta|\mathbf{r}_A - \mathbf{r}|)}{|\mathbf{r}_A - \mathbf{r}|} \qquad (6.28)$$

For the integral in Equation 6.27, the greatest contribution will come from around the axis through the observation point and parallel to the direction of the incident wave. Consequently, we can use the approximation

$$K(\mathbf{r}_A, \mathbf{r}) \approx -\frac{j\omega\mu_0}{4\pi} \frac{\exp\left(-j\beta\left(|z_A - z| + \frac{(x_A-x)^2+(y_A-y)^2}{2(z_A-z)}\right)\right)}{|z_A - z|} \qquad (6.29)$$

so that

$$E_1(\mathbf{r}_A) = \frac{\omega^2\mu_0}{4\pi} \int_V \delta\varepsilon \exp(-j\beta z) \frac{\exp\left(-j\beta\left(|z_A - z| + \frac{(x_A-x)^2+(y_A-y)^2}{2|z_A-z|}\right)\right)}{|z_A - z|} dV \qquad (6.30)$$

For a given point of observation $\mathbf{r}_A$, consider the transverse (x and y) integrals. As x moves away from x_A, and y moves away from y_A, the exponential factor increases in its frequency of oscillation and eventually reaches a point where the wavelength of oscillation is less than the scale of variations l_0 in the irregularity. At this point, the oscillations start to cancel out the contributions from the irregularity. Essentially, for a given observation point, the major contribution to the forward scatter will come from a cone that has a half angle λ/l_0. Since we will normally have that $l_0 \gg \lambda$, this cone will have a very small angle (see Figure 6.4). The Rytov approximation will now yield

$$E(\mathbf{r}_A) \approx \exp(-j\beta z_A) \exp(\chi_1 + jS_1) \qquad (6.31)$$

where

$$\chi_1 + jS_1 = \frac{\omega^2\mu_0}{4\pi} \int_V \delta\varepsilon \frac{\exp\left(-j\beta\left(|z_A - z| - z_A + z + \frac{(x_A-x)^2+(y_A-y)^2}{2|z_A-z|}\right)\right)}{|z_A - z|} dV \qquad (6.32)$$

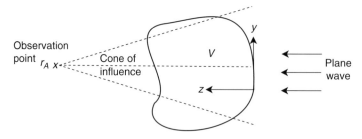

Figure 6.4 Influence of scatterers in the Rytov approximation.

The back-scatter contributions in Equation 6.32 (i.e. $z > z_A$) will include a highly oscillatory factor $(\exp(-2j\beta|z_A - z|))$ that will cancel out these contributions and so we can restrict contributions to $z \leq z_A$. Consequently,

$$\chi_1 + jS_1 = \frac{\omega^2 \mu_0}{4\pi} \int_0^{z_A} \int_S \delta\varepsilon \frac{\exp\left(-j\beta \frac{(x_A-x)^2+(y_A-y)^2}{2|z_A-z|}\right)}{|z_A - z|} dS dz \quad (6.33)$$

where $dS = dxdy$ and S is the cross section of the irregularity region at z.

Thus far, we have assumed that the wave that is incident upon the irregularity is the plane wave $E_0 = \exp(-j\beta z)$ and for which the Rytov approximation yields

$$E(\mathbf{r}_A) = \exp(-j\beta z_A) \exp\left(\frac{\omega^2 \mu_0}{4\pi} \int_0^{z_A} \int_S \delta\varepsilon \frac{\exp\left(-j\beta \frac{|\mathbf{w}_A-\mathbf{w}|^2}{2|z_A-z|}\right)}{|z_A - z|} dS dz\right) \quad (6.34)$$

where $\mathbf{w} = x\hat{\mathbf{x}} + y\hat{\mathbf{y}}$ and $\mathbf{w}_A = x_A\hat{\mathbf{x}} + y_A\hat{\mathbf{y}}$. We will now, however, investigate the case where the incident wave is spherical. Obviously, for a small patch of irregularity, we can treat the incident wave as locally plane. We would, however, like to go a little further than this approximation and consider the effect of a curved wavefront. For a source at point $\mathbf{r}_0$, the spherical wave is given by

$$E_0(\mathbf{r}_A) = \frac{\exp(j\beta|\mathbf{r}_A - \mathbf{r}_0|)}{4\pi|\mathbf{r}_A - \mathbf{r}_0|} \quad (6.35)$$

and, if we assume the source lies on the z axis, we can expand about this axis to obtain

$$E_0(\mathbf{r}_A) \approx \frac{\exp\left(j\beta|z_A - z_0| + \frac{|\mathbf{w}_A|^2}{2|z_A - z_0|}\right)}{4\pi|z_A - z_0|} \quad (6.36)$$

The Rytov approximation will now yield

$$E(\mathbf{r}_A) \approx \frac{\exp(j\beta|\mathbf{r}_A - \mathbf{r}_0|)}{4\pi|\mathbf{r}_A - \mathbf{r}_0|} \exp\phi \quad (6.37)$$

where

$$\phi(\mathbf{r}_A) = \frac{\omega^2 \mu_0}{4\pi} \int_0^{z_A} \int_S \delta\varepsilon |z_A - z_0| \frac{\exp\left(-j\beta \left(\frac{|\mathbf{w}_A-\mathbf{w}|^2}{2|z_A-z|} + \frac{|\mathbf{w}|^2}{2|z-z_0|} - \frac{|\mathbf{w}_A|^2}{2|z_A-z_0|}\right)\right)}{|z_A - z||z - z_0|} dS dz \quad (6.38)$$

Comparing the plane and spherical wave Rytov approximation, it turns out that the spherical wave approximation can be obtained from the plane wave approximation by replacing $\mathbf{w}_A$ by $\gamma \mathbf{w}_A$ and $|z_A - z|$ by $\gamma|z_A - z|$ where $\gamma = (z - z_0)/(z_A - z_0)$ (Ishimaru, 1997), i.e.

$$E(\mathbf{r}_A) = \frac{\exp(j\beta|\mathbf{r}_A - \mathbf{r}_0|)}{4\pi|\mathbf{r}_A - \mathbf{r}_0|} \exp\left(\frac{\omega^2 \mu_0}{4\pi} \int_0^{z_A} \int_S \delta\varepsilon \frac{\exp\left(-j\beta \frac{|\gamma \mathbf{w}_A-\mathbf{w}|^2}{2\gamma|z_A-z|}\right)}{\gamma|z_A - z|} dS dz\right) \quad (6.39)$$

For the case of perturbation that is fairly compact in the z direction, we could replace γ by a $\gamma_R = (z_R - z_0)/(z_A - z_0)$ where z_R is a representative point within the medium. Assuming that the irregularity straddles the origin, the spherical wave case is thus obtained from the plane wave case by replacing occurrences of $\mathbf{w}_A$ by $\gamma_R \mathbf{w}_A$ and z_A by $\gamma_R z_A$, where $\gamma_R = |z_0|/(z_A - z_0)$.

6.3 Mutual Coherence

Irregularity has the effect of causing multiple propagation paths that can result in the blurring of a radio signal. If a plane wave travels through a region of irregularity, the consequence of the multipath will be that observations of the wave across the receiver plane will vary from point to point, i.e. the observations at one point will not necessarily give a reliable prediction of the observations at another point (see Figure 6.5). The degree of reliability is essentially the coherence of the field at these two points. Knowledge of such coherence is essential for systems that use arrays of antennas and require them to be coherent across the array in order to achieve effective beamforming. A useful measure of this coherence is the MCF

$$\Gamma(\mathbf{r}', \mathbf{r}'') = \langle E' E''^* \rangle \qquad (6.40)$$

where E' represents the field evaluated at $\mathbf{r}'$ and E'' the field evaluated at $\mathbf{r}''$. It is often more convenient to work in terms of a normalized MCF that is given by $\Gamma_a = \langle E'_a E''^*_a \rangle$ where $E = E_0 E_a$ and E_0 is the field in the background irregularity free medium (note that $\Gamma = \Gamma_0 \Gamma_a$ where $\Gamma_0 = E'_0 E''^*_0$.) For Γ_a it will be noted that $\Gamma_a = 0$ when

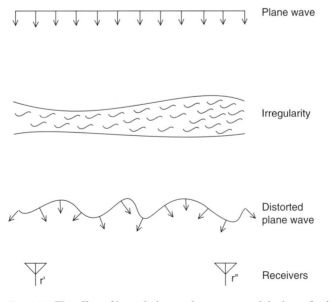

Figure 6.5 The effect of irregularity on plane waves and the loss of coherence.

the fields are completely uncorrelated and $\Gamma_a = 1$ when the fields are completely correlated (totally incoherent and totally coherent). Consequently, Γ_a is a measure of the correlation between the field values at points $\mathbf{r}'$ and $\mathbf{r}''$.

The significance of the MCF can be seen by considering the average intensity $I' = \langle E'E'^*\rangle$ of a field measured at point $\mathbf{r}'$ and the average intensity $I' = \langle E''E''^*\rangle$ measured at point $\mathbf{r}''$. If the fields are summed, i.e. $E = E' + E''$, the intensity $I = \langle EE^*\rangle$ will be

$$I = I' + I'' + 2\Re\Gamma\langle E'E''^*\rangle \tag{6.41}$$

If the measured fields are totally incoherent, $\Gamma_a = 0$ and $I = I' + I''$. The powers are said to add incoherently and the power of these summed fields is twice that for an individual field sample. If we now consider the fields to be totally coherent, i.e. $\Gamma_a = 1$, we will have $I = |E' + E''|^2$. The powers are said to add coherently and, for measurement points of the same wavefront, the power of the summed fields is four times that for an individual sample.

6.4 The Rytov Approximation and Irregular Media

We can use the Rytov approximation to calculate the MCF in the case that the irregularity can be regarded as a perturbation of the background medium. From the Rytov approximation

$$\Gamma_a(\mathbf{r}', \mathbf{r}'') = \langle \exp\left(\chi' + \chi'' + j(S' - S'')\right)\rangle \tag{6.42}$$

Since S and χ will both be the sum of a great number of random elements, they will both have Gaussian distributions. If ϕ is a Gaussian random variable, with mean $\mu = \langle\phi\rangle$ and variance $\sigma^2 = \langle(\phi - \mu)^2\rangle$, then

$$\langle\exp\phi\rangle = \exp\mu \exp\left(\frac{\sigma^2}{2}\right) \tag{6.43}$$

Further, if ϕ is the linear combination $\phi = aX + bY$ of random variables X and Y, then $\langle aX + bY\rangle = a\langle X\rangle + b\langle Y\rangle$ and $\langle(aX + bY)^2\rangle = a^2\langle X^2\rangle + 2ab\langle XY\rangle + b^2\langle Y^2\rangle$. Consequently, 6.31 will yield

$$\Gamma_a(\mathbf{r}', \mathbf{r}'') = \exp\left(\langle\phi' + \phi''^*\rangle - \frac{1}{2}\langle\phi' + \phi''^*\rangle^2\right)$$

$$\times \exp\left(\frac{1}{2}\langle(\chi' + \chi'')^2\rangle + j\langle(\chi' + \chi'')(S' - S'')\rangle - \frac{1}{2}\langle(S' - S'')^2\rangle\right) \tag{6.44}$$

To calculate the MCF we will need several second moments. It is sufficient, however, to calculate $\langle(\chi'_1 + jS'_1)(\chi''_1 + jS''_1)\rangle$ and $\langle(\chi'_1 + jS'_1)(\chi''_1 - jS''_1)\rangle$ as all the requisite second moments can be derived from these. From Equation 6.33

$$\langle(\chi_1' + jS_1')(\chi_1'' + jS_1'')\rangle$$

$$= \frac{\beta^4}{16\pi^2\epsilon^2} \int_0^{z'} \int_S \int_0^{z''} \int_S \langle\delta\varepsilon_a\delta\varepsilon_b\rangle \frac{\exp\left(-j\beta\frac{|\mathbf{w}'-\mathbf{w}_a|^2}{2|z'-z_a|}\right)\exp\left(-j\beta\frac{|\mathbf{w}''-\mathbf{w}_b|^2}{2|z''-z_b|}\right)}{|z'-z_a||z''-z_b|} dS_b dz_b dS_a dz_a$$

(6.45)

on observing that $\chi_1 + jS_1 = \phi_1$ (note that $\langle\phi_1\rangle = 0$). (Note that we have used subscripts a and b to denote quantities evaluated at $\mathbf{r}_a$ and $\mathbf{r}_b$, respectively. In addition, $\mathbf{w}' = x'\hat{\mathbf{x}} + y'\hat{\mathbf{y}}$, $\mathbf{w}'' = x''\hat{\mathbf{x}} + y''\hat{\mathbf{y}}$, $\mathbf{w}_a = x_a\hat{\mathbf{x}} + y_a\hat{\mathbf{y}}$ and $\mathbf{w}_b = x_b\hat{\mathbf{x}} + y_b\hat{\mathbf{y}}$.) It is assumed that the correlation length l_0 is much smaller than the overall size of the irregularity patch and therefore the autocorrelation has the character of a delta function on this scale. Further, it will be noted that the major contribution to Equation 6.45 comes from around two axes in the z direction, the axis passing through the point $\mathbf{r}'$ for the integral with respect to coordinates $\mathbf{r}_a$ and through $\mathbf{r}''$ for the integral with respect to coordinates $\mathbf{r}_b$. Along these axes, however, the autocorrelation can be treated as a delta function for integration purposes since $l_0 \gg \lambda$. Consequently, we assume that $\langle\delta\varepsilon_a\delta\varepsilon_b\rangle = \epsilon^2\delta(z_a - z_b)A(\mathbf{w}_a - \mathbf{w}_b)$ where $\mathbf{w} = x\hat{\mathbf{x}} + y\hat{\mathbf{y}}$ are the transverse coordinates and the *structure function* A is defined by $A(\mathbf{w}_a - \mathbf{w}_b) = \int_{-\infty}^{\infty}(\langle\delta\varepsilon_a\delta\varepsilon_b\rangle/\epsilon^2)dz$. In addition, we assume that the correlation lengths, and the wavelength, are small enough for the transverse limits of integration to be replaced by infinity. Equation 6.45 will then reduce to

$$\langle(\chi_1' + jS_1')(\chi_1'' + jS_1'')\rangle = \frac{\beta^4}{16\pi^2} \int_{z_S}^{z_E} \int_{R^2} \int_{R^2} A(\mathbf{w}_a - \mathbf{w}_b)$$

$$\times \frac{\exp\left(-j\beta\left(\frac{|\mathbf{w}'-\mathbf{w}_a|^2}{2|z'-z_a|} + \frac{|\mathbf{w}''-\mathbf{w}_b|^2}{2|z''-z_a|}\right)\right)}{|z'-z_a||z''-z_a|} dx_b dy_b dx_a dy_a dz_a$$

(6.46)

where z_S is the point on the z axis where the irregularity starts and $z_E = \min(z', z'', z_F)$ and z_F is the point on the z axis where the irregularity finishes (note that we assume the length scale l_0 of the irregularity is small enough for z_S and z_F to be approximately unchanged between the axes through the observation points $\mathbf{r}'$ and $\mathbf{r}''$). We now consider the MCF in the transverse direction alone (i.e. we take $z'' = z' = z$), then

$$\langle(\chi_1' + jS_1')(\chi_1'' + jS_1'')\rangle = \frac{\beta^4}{16\pi^2} \int_{z_S}^{z_E} \int_{R^2} \int_{R^2} A(\mathbf{w}_a - \mathbf{w}_b)$$

$$\times \frac{\exp\left(-j\beta\frac{|\mathbf{w}'-\mathbf{w}_a|^2+|\mathbf{w}''-\mathbf{w}_b|^2}{2|z-z_a|}\right)}{|z-z_a|^2} dx_b dy_b dx_a dy_a dz_a \quad (6.47)$$

Note that $|\mathbf{w}' - \mathbf{w}_a|^2 + |\mathbf{w}'' - \mathbf{w}_b|^2 = 2|\mathbf{f} - (\mathbf{w}' + \mathbf{w}'')/2|^2 + 2|\mathbf{g} - (\mathbf{w}' - \mathbf{w}'')/2|^2$ where $\mathbf{f} = (\mathbf{w}_a + \mathbf{w}_b)/2$ and $\mathbf{g} = (\mathbf{w}_a - \mathbf{w}_b)/2$ (equivalently, $\mathbf{w}_a = \mathbf{f} + \mathbf{g}$ and $\mathbf{w}_b = \mathbf{f} - \mathbf{g}$). Equation 6.47 will now become

6.4 The Rytov Approximation and Irregular Media

$$\langle(\chi_1' + jS_1')(\chi_1'' + jS_1'')\rangle = \frac{\beta^4}{16\pi^2} \int_{z_S}^{z_E} \int_{R^2} \int_{R^2} A(2\mathbf{g})$$

$$\times \frac{\exp\left(-j\beta \frac{\left|\mathbf{f} - \frac{\mathbf{w}'+\mathbf{w}''}{2}\right|^2 + \left|\mathbf{g} - \frac{\mathbf{w}'-\mathbf{w}''}{2}\right|^2}{|z-z_a|}\right)}{|z - z_a|^2} 4 d g_1 d g_2 d f_1 d f_2 d z_a \quad (6.48)$$

for which it is possible to perform the integrals with respect to f_1 and f_2 and obtain

$$\langle(\chi_1' + jS_1')(\chi_1'' + jS_1'')\rangle = \frac{\beta^3}{16j\pi} \int_{z_S}^{z_E} \int_{R^2} A(\mathbf{s}) \times \frac{\exp\left(-j\beta \frac{|\mathbf{s}-\mathbf{w}'+\mathbf{w}''|^2}{4|z-z_a|}\right)}{|z-z_a|} ds_1 ds_2 dz_a \quad (6.49)$$

where $s_1 = 2g_1$ and $s_2 = 2g_2$. We can represent the transverse cross correlation $A(\mathbf{s})$ in terms of the transverse spectrum of irregularity $W_T(\mathbf{k})$,

$$A(\mathbf{s}) = \int_{R^2} W_T(\mathbf{p}) \exp(j\mathbf{p} \cdot \mathbf{s}) dp_1 dp_2 \quad (6.50)$$

where $W_T(\mathbf{p}) = 2\pi W(p_1, p_2, 0)$. Further, we note that

$$W_T(\mathbf{p}) = \frac{1}{4\pi^2} \int_{R^2} A(\mathbf{s}) \exp(-j\mathbf{p} \cdot \mathbf{s}) ds_1 ds_2 \quad (6.51)$$

As a consequence,

$$\langle(\chi_1' + jS_1')(\chi_1'' + jS_1'')\rangle = \frac{\beta^3}{16j\pi} \int_{z_S}^{z_E} \int_{R^2} \int_{R^2} W_T(\mathbf{p})$$

$$\times \exp(j\mathbf{p} \cdot \mathbf{s}) \frac{\exp\left(-j\beta \frac{|\mathbf{s}-\mathbf{w}'+\mathbf{w}''|^2}{4|z-z_a|}\right)}{|z-z_a|} ds_1 ds_2 dp_1 dp_2 dz_a \quad (6.52)$$

and rearranging the exponentials

$$\langle(\chi_1' + jS_1')(\chi_1'' + jS_1'')\rangle = \frac{\beta^3}{16j\pi} \int_{z_S}^{z_E} \int_{R^2} \int_{R^2} W_T(\mathbf{p}) \exp\left(\frac{j|z-z_a|}{\beta}|\mathbf{p}|^2 + j\Delta\mathbf{w} \cdot \mathbf{p}\right)$$

$$\times \frac{\exp\left(-\frac{j\beta}{4|z-z_a|}\left|\mathbf{s} - \Delta\mathbf{w} - \frac{2|z-z_a|}{\beta}\mathbf{p}\right|^2\right)}{|z-z_a|} ds_1 ds_2 dp_1 dp_2 dz_a \quad (6.53)$$

where $\Delta\mathbf{w} = \mathbf{w}' - \mathbf{w}''$. We now perform the integrals with respect to s_1 and s_2 and obtain

$$\langle(\chi_1' + jS_1')(\chi_1'' + jS_1'')\rangle = -\frac{\beta^2}{4} \int_{z_S}^{z_E} \int_{R^2} W_T(\mathbf{p}) \exp\left(\frac{j|z-z_a|}{\beta}|\mathbf{p}|^2 + j\Delta\mathbf{w} \cdot \mathbf{p}\right) dp_1 dp_2 dz_a$$

$$(6.54)$$

and then, integrating with respect to z_a,

$$\langle (\chi_1' + jS_1')(\chi_1'' + jS_1'') \rangle = -\frac{j\beta^3}{4} \int_{R^2} \frac{W_T(\mathbf{p})}{|\mathbf{p}|^2} \left(\exp\left(-\frac{jz_E}{\beta}|\mathbf{p}|^2\right) \right.$$

$$\left. - \exp\left(-\frac{jz_S}{\beta}|\mathbf{p}|^2\right) \right) \exp\left(\frac{jz}{\beta}|\mathbf{p}|^2 + j\Delta\mathbf{w}\cdot\mathbf{p}\right) dp_1 dp_2 \tag{6.55}$$

If the irregularity is compact in the z direction, the above expression behaves as

$$\langle (\chi_1' + jS_1')(\chi_1'' + jS_1'') \rangle = -\frac{\beta^2}{4} W_T(0)(z_F - z_S)\frac{\pi\beta}{jz} \tag{6.56}$$

in the limit $z \to \infty$ (i.e. $\langle (\chi_1' + jS_1')(\chi_1'' + jS_1'') \rangle \to 0$). If, on the other hand, we restrict z so that $z \ll \beta l_0^2$ we find that Equation 6.55 reduces to

$$\langle (\chi_1' + jS_1')(\chi_1'' + jS_1'') \rangle = -\frac{\beta^2}{4}(z_E - z_S)A(\Delta\mathbf{w}) \tag{6.57}$$

which is the GO limit.

In order to find all the necessary second moments, we also need $\langle (\chi_1'+jS_1')(\chi_1''-jS_1'') \rangle$. In a similar fashion to Equation 6.47, we obtain

$$\langle (\chi_1' + jS_1')(\chi_1'' - jS_1'') \rangle = \frac{\beta^4}{16\pi^2} \int_{z_S}^{z_E} \int_{R^2} \int_{R^2} A(\mathbf{w}_a - \mathbf{w}_b)$$

$$\times \frac{\exp\left(-j\beta\frac{|\mathbf{w}'-\mathbf{w}_a|^2-|\mathbf{w}''-\mathbf{w}_b|^2}{2|z-z_a|}\right)}{|z-z_a|^2} dx_b dy_b dx_a dy_a dz_a \tag{6.58}$$

then, noting that $|\mathbf{w}' - \mathbf{w}_a|^2 - |\mathbf{w}'' - \mathbf{w}_b|^2 = |\mathbf{w}'|^2 - |\mathbf{w}''|^2 + 4\mathbf{f}\cdot(\mathbf{g}-(\mathbf{w}'-\mathbf{w}'')/2) - 4\mathbf{g}\cdot(\mathbf{w}'+\mathbf{w}'')/2$ where $\mathbf{w}_a = \mathbf{f}+\mathbf{g}$ and $\mathbf{w}_b = \mathbf{f}-\mathbf{g}$,

$$\langle (\chi_1' + jS_1')(\chi_1'' - jS_1'') \rangle = \frac{\beta^4}{16\pi^2} \int_{z_S}^{z_E} \int_{R^2} \int_{R^2} A(2\mathbf{g})$$

$$\times \frac{\exp\left(-j\beta\frac{|\mathbf{w}'|^2-|\mathbf{w}''|^2+4\mathbf{f}\cdot\left(\mathbf{g}-\frac{\mathbf{w}'-\mathbf{w}''}{2}\right)-4\mathbf{g}\cdot\frac{\mathbf{w}'+\mathbf{w}''}{2}}{2|z-z_a|}\right)}{|z-z_a|^2} 4 dg_1 dg_2 df_1 df_2 dz_a \tag{6.59}$$

On integrating with respect to f_1 and f_2,

$$\langle (\chi_1' + jS_1')(\chi_1'' - jS_1'') \rangle = \frac{\beta^2}{16} \int_{z_S}^{z_E} \int_{R^2} A(2\mathbf{g})$$

$$\times \exp\left(-j\beta\frac{|\mathbf{w}'|^2 - |\mathbf{w}''|^2 - 4\mathbf{g}\cdot\frac{\mathbf{w}'+\mathbf{w}''}{2}}{2|z-z_a|}\right) \delta\left(\mathbf{g} - \frac{\Delta\mathbf{w}}{2}\right) 4 dg_1 dg_2 dz_a \tag{6.60}$$

6.4 The Rytov Approximation and Irregular Media

Then, integrating with respect to g_1 and g_2,

$$\langle (\chi_1' + jS_1')(\chi_1'' - jS_1'') \rangle = \frac{\beta^2}{4} \int_{z_S}^{z_E} A(\Delta \mathbf{w}) dz_a \tag{6.61}$$

from which

$$\langle (\chi_1' + jS_1')(\chi_1'' - jS_1'') \rangle = \frac{\beta^2}{4}(z_E - z_S) A(\Delta \mathbf{w}) \tag{6.62}$$

To complete the MCF, we need the next order in the Rytov approximation (Manning, 2008). The next order (i.e. ϕ_2) will satisfy

$$\nabla^2 \phi_2 + \nabla \phi_1 \cdot \nabla \phi_1 + 2 \nabla \phi_0 \cdot \nabla \phi_2 = 0 \tag{6.63}$$

from which

$$\phi_2(\mathbf{r}_A) = \frac{1}{4\pi} \int_0^{z_A} \int_S \nabla \phi_1 \cdot \nabla \phi_1 \frac{\exp\left(-j\beta \frac{(x_A - x)^2 + (y_A - y)^2}{2|z_A - z|}\right)}{|z_A - z|} dSdz \tag{6.64}$$

where

$$\phi_1(\mathbf{r}) = \frac{\omega^2 \mu_0}{4\pi} \int_0^z \int_S \delta\varepsilon \frac{\exp\left(-j\beta \frac{(x - x^a)^2 + (y - y^a)^2}{2|z - z^a|}\right)}{|z - z^a|} dS_a dz_a \tag{6.65}$$

By calculating $\langle \phi \rangle$, we can find $\langle \chi \rangle$ and $\langle S \rangle$. We will assume $\langle \delta\varepsilon \rangle = 0$ and then $\langle \phi \rangle = \langle \phi_2 \rangle$. Noting that $z_E = \min(z, z_F)$, we have from Equation 6.64,

$$\langle \phi \rangle (\mathbf{r}_A) = \frac{1}{4\pi} \int_{z_S}^{z_A} \int_S \langle \nabla \phi_1 \cdot \nabla \phi_1 \rangle \frac{\exp\left(-j\beta \frac{(x_A - x)^2 + (y_A - y)^2}{2|z_A - z|}\right)}{|z_A - z|} dSdz \tag{6.66}$$

where

$$\langle \nabla \phi_1 \cdot \nabla \phi_1 \rangle (\mathbf{r}) = \frac{\beta^4}{16\pi^2} \int_{z_S}^{z_E} \int_S \int_{z_S}^{z_E} \int_S \langle \delta\varepsilon_a \delta\varepsilon_b \rangle$$

$$\times \nabla_a \cdot \nabla_b \frac{\exp\left(-j\beta \left(\frac{(x - x_a)^2 + (y - y_a)^2}{2|z - z_a|} + \frac{(x - x_b)^2 + (y - y_b)^2}{2|z - z_b|}\right)\right)}{|z - z_a||z - z_b|} dS_a dz_a dS_b dz_b \tag{6.67}$$

(Note that subscript a and subscript b denote quantities evaluated at $\mathbf{r}_a$ and $\mathbf{r}_b$, respectively.) As before, we assume that $\langle \delta\varepsilon_a \delta\varepsilon_b \rangle = \epsilon_0^2 \delta(z_a - z_b) A(\mathbf{w}_a - \mathbf{w}_b)$ and then Equation 6.67 will now reduce to

$$\langle \nabla \phi_1 \cdot \nabla \phi_1 \rangle (\mathbf{r}) = \frac{\beta^6}{16\pi^2} \int_{z_S}^{z_E} \int_{R^2} \int_{R^2} A(\mathbf{w}_a - \mathbf{w}_b)$$

$$\times (\mathbf{w} - \mathbf{w}_a) \cdot (\mathbf{w} - \mathbf{w}_b) \frac{\exp\left(-j\beta \left(\frac{|\mathbf{w} - \mathbf{w}_a|^2 + |\mathbf{w} - \mathbf{w}_b|^2}{2|z - z_A|}\right)\right)}{|z - z_a|^4} dx_a dy_a dx_b dy_b dz_a \tag{6.68}$$

where we have only retained terms consistent with the approximation 6.29. Once again, noting that $|\mathbf{w} - \mathbf{w}_a|^2 + |\mathbf{w} - \mathbf{w}_b|^2 = 2|\mathbf{w} - (\mathbf{w}_a + \mathbf{w}_b)|^2/2 + |\mathbf{w}_a - \mathbf{w}_b|^2/2$ and also that $(\mathbf{w} - \mathbf{w}_a) \cdot (\mathbf{w} - \mathbf{w}_b) = |\mathbf{w} - (\mathbf{w}_a + \mathbf{w}_b)|^2/2 - |\mathbf{w}_a - \mathbf{w}_b|^2/4$, we obtain

$$\langle \nabla \phi_1 \cdot \nabla \phi_1 \rangle (\mathbf{r}) = \frac{\beta^6}{16\pi^2} \int_{z_S}^{z} \int_{R^2} \int_{R^2} A(2\mathbf{g})$$

$$\times (|\mathbf{w} - \mathbf{f}|^2 - |\mathbf{g}|^2) \frac{\exp\left(-j\beta \left(\frac{|\mathbf{w}-\mathbf{f}|^2+|\mathbf{g}|^2}{|z-z_a|}\right)\right)}{|z - z_a|^4} 4 d\mathbf{g}_1 d\mathbf{g}_2 d\mathbf{f}_1 d\mathbf{f}_2 dz_a$$

(6.69)

Then, performing the integration with respect to f_1 and f_2,

$$\langle \nabla \phi_1 \cdot \nabla \phi_1 \rangle (\mathbf{r}) = \frac{\beta^6}{4\pi^2} \int_{z_S}^{z_E} \int_{R^2} A(2\mathbf{g})$$

$$\times \left(-\frac{\pi |z - z_a|^2}{\beta^2} - \frac{\pi |z - z_a|}{j\beta} |\mathbf{g}|^2 \right) \frac{\exp\left(-j\beta \left(\frac{|\mathbf{g}|^2}{|z-z_a|}\right)\right)}{|z - z_a|^4} d\mathbf{g}_1 d\mathbf{g}_2 dz_a$$

(6.70)

We introduce the new variable $\tau = -(\beta |\mathbf{g}|^2)/(z - z_a)$, then $d\tau = -dz_a \beta |\mathbf{g}|^2/(z - z_a)^2$ and the above integral reduces to

$$\langle \nabla \phi_1 \cdot \nabla \phi_1 \rangle (\mathbf{r}) = \frac{\beta^6}{4\pi^2} \int_{-\frac{\beta |\mathbf{g}|^2}{z-z_S}}^{-\frac{\beta |\mathbf{g}|^2}{z-z_E}} \int_{R^2} A(2\mathbf{g}) \left(-\frac{\pi}{\beta^2} + \frac{\pi}{j\beta^2} \tau \right) \frac{\exp(j\tau)}{-\beta |\mathbf{g}|^2} d\mathbf{g}_1 d\mathbf{g}_2 d\tau \quad (6.71)$$

On noting that $\int (1 + a\tau) \exp(a\tau) d\tau = \tau \exp(a\tau)$ we obtain that

$$\langle \nabla \phi_1 \cdot \nabla \phi_1 \rangle (\mathbf{r}) = -\frac{\beta^4}{4\pi} \int_{R^2} A(2\mathbf{g}) \left(\frac{\exp\left(-j\beta \left(\frac{|\mathbf{g}|^2}{|z-z_E|}\right)\right)}{|z - z_E|} - \frac{\exp\left(-j\beta \left(\frac{|\mathbf{g}|^2}{|z-z_S|}\right)\right)}{|z - z_S|} \right) d\mathbf{g}_1 d\mathbf{g}_2$$

(6.72)

Expression 6.72 must be treated with caution for points z within the irregularity since, in this case, we can have $z_E = z$. We can, however, deal with this by only taking the integral up to point $z - \delta$ and then taking the limit $\delta \to 0$ after performing the integrals with respect to g_1 and g_2. From Equation 6.70 it will be noted that $\langle \nabla \phi_1 \cdot \nabla \phi_1 \rangle$ depends only on the lateral coordinate through the geometry of the irregularity. Consequently, if we assume that the z extent only changes slowly over the lateral correlation distance, we can treat $\langle \nabla \phi_1 \cdot \nabla \phi_1 \rangle$ as constant for the purpose of the lateral integral in Equation 6.66 and hence obtain

$$\langle \phi \rangle (\mathbf{r}_A) \approx \frac{j}{2\beta} \int_{z_S}^{z_A} \langle \nabla \phi_1 \cdot \nabla \phi_1 \rangle dz \quad (6.73)$$

As a consequence, $\langle \chi' + \chi'' \rangle + j\langle S' - S'' \rangle \approx 2\langle \chi \rangle$ since we have already assumed that $z'' = z' = z$. If we now substitute Equation 6.72 into 6.73,

$$\langle \phi \rangle (\mathbf{r}_A) \approx -\frac{\beta^2}{8} A(0)(z_F - z_S)$$
$$- \frac{j\beta^3}{8\pi} \int_{R^2} A(2\mathbf{g}) \left(\int_{z_F}^{z_A} \frac{\exp\left(-j\beta\left(\frac{|\mathbf{g}|^2}{|z-z_F|}\right)\right)}{|z - z_F|} dz - \int_{z_S}^{z_A} \frac{\exp\left(-j\beta\left(\frac{|\mathbf{g}|^2}{|z-z_S|}\right)\right)}{|z - z_S|} dz \right) dg_1 dg_2$$
(6.74)

where the integral-free term represents that part of the first integral between z_S and z_F. We can now combine the integrals to yield

$$\langle \phi \rangle (\mathbf{r}_A) \approx -\frac{\beta^2}{8} A(0)(z_F - z_S) - \frac{j\beta^3}{8\pi} \int_{R^2} A(2\mathbf{g}) \int_{z_A - z_S}^{z_A - z_F} \frac{\exp\left(-j\beta\left(\frac{|\mathbf{g}|^2}{|z|}\right)\right)}{|z|} dz\, dg_1 dg_2 \quad (6.75)$$

In the limit that $z_A \ll \beta l_0^2$ (the GO limit), we can approximately evaluate the g_1 and g_2 integrals and obtain that $\langle \chi \rangle \approx 0$. Alternatively, in the limit $z_A \to \infty$ and for irregularity of bounded extent,

$$\langle \phi \rangle (\mathbf{r}_A) \approx -\frac{\beta^2}{8} A(0)(z_F - z_S) \quad (6.76)$$

Bringing the above results together, we have the MCF

$$\Gamma_a(\mathbf{r}', \mathbf{r}'') = \exp\left(-(z_E - z_S)\frac{\beta^2}{4}(A(0) - A(\Delta\mathbf{w}))\right) \quad (6.77)$$

where $\Delta\mathbf{w} = \mathbf{w}' - \mathbf{w}''$ and $z_E = \min(z, z_F)$. This is valid in the GO limit or, alternatively, at large distances following irregularity of bounded extent. It should be noted, however, that the contributions from the various moments are completely different in these different limits.

The above result refers to a plane wave that is incident upon the irregularity. In the case of irregularity of finite extent, however, we have seen that the Rytov approximation can be transformed into the case of a spherical wave arising from a source a finite distance from the irregularity. Assuming that the irregularity straddles the origin (i.e. $z_S < 0 < z_E$), and is a distance z_t from the source, we can obtain the spherical wave case from the plane wave case by replacing occurrences of $\mathbf{w}_A$ by $\gamma_R \mathbf{w}_A$ and z_A by $\gamma_R z_A$ where $\gamma_R = z_t/(z_A + z_t)$. Then, in the case of the MCF given by Equation 6.77, this replacement will imply

$$\Gamma(\mathbf{r}', \mathbf{r}'') = \Gamma_0 \exp\left(-(z_E - z_S)\frac{\beta^2}{4}(A(0) - A(\gamma_R \Delta\mathbf{w}))\right) \quad (6.78)$$

where $\Gamma_0 = E_0' E_0''^*$ and E_0 is the solution for an irregularity free medium.

The Rytov approximation provides a useful means of calculating the MCF, and it has been shown by Manning (2008) that the approximation can be applied to severe irregularity in limiting cases. Key to this, however, is the retention of the second-order terms in the Rytov approximation. Gherm et al. (2005) have used the Rytov approach to investigate propagation through ionospheric irregularity. In their case, they use a geometric optics approximation as the zeroth-order solution.

6.5 Parabolic Equations for the Average Field and MCF

Consider the generic perturbed Helmholtz equation

$$\nabla^2 E + \omega^2 \mu_0 (\epsilon + \delta\epsilon) E = 0 \qquad (6.79)$$

and assume the propagation to be predominantly in one direction. Without loss of generality, we assume this direction to be that of the z axis, then

$$E(\mathbf{r}) = \exp(-j\beta z) E_a(\mathbf{r}) \qquad (6.80)$$

where E_a is slowly varying on the scale of a wavelength. We substitute Equation 6.80 into 6.79, then

$$-\beta^2 E_a - 2j\beta \frac{\partial E_a}{\partial z} + \frac{\partial^2 E_a}{\partial z^2} + \nabla_T^2 E_a + \omega^2 \mu_0 (\epsilon + \delta\epsilon) E_a = 0 \qquad (6.81)$$

where $\nabla_T^2 = \partial^2/\partial x^2 + \partial^2/\partial y^2$ is the Laplace operator in the transverse direction. In the parabolic equation approximation, we assume that the $\exp(-j\beta z)$ factor embodies the dominant variations of E in the z direction and that E_a varies in this direction on a scale much greater than it varies in the transverse directions. Consequently, we ignore $\partial^2 E_a/\partial z^2$ in comparison to all other terms and Equation 6.81 reduces to

$$-2j\beta \frac{\partial E_a}{\partial z} + \nabla_T^2 E_a + \omega^2 \mu_0 \delta\epsilon E_a = 0 \qquad (6.82)$$

We will first look at the average of the field E_a, that is $\langle E_a \rangle$. From Equation 6.82,

$$-2j\beta \frac{\partial \langle E_a \rangle}{\partial z} + \nabla_T^2 \langle E_a \rangle + \omega^2 \mu_0 \langle \delta\epsilon E_a \rangle = 0 \qquad (6.83)$$

We first recall the Rytov approximation

$$E' \approx E_0' \exp \frac{E_1'}{E_0'} = E_0' \exp(\phi_1') \qquad (6.84)$$

where

$$E_1' = \frac{\omega^2 \mu_0}{4\pi} \int_V \delta\varepsilon E_0(\mathbf{r}) \frac{\exp\left(-j\beta\left(|z'-z| + \frac{(x'-x)^2 + (y'-y)^2}{2|z'-z|}\right)\right)}{|z'-z|} dV \qquad (6.85)$$

Quantity ϕ_1' is a weighted sum of $\delta\epsilon$ over the volume V (i.e. it is a linear combination of the random variables that describe the irregularity). We consider the volume V to be divided into slices of thickness Δz and consider the slice whose end contains the apex of the cone of influence for point $\mathbf{r}'$ (see Figure 6.4). Since the cone will have a small

6.5 Parabolic Equations for the Average Field and MCF

angle, we assume that $\delta\epsilon$ in the slice takes its values from those on the axis through $\mathbf{r}'$. The contribution from this slice is $\Delta\phi_1'$ where

$$\Delta\phi_1' \approx \frac{\omega^2\mu_0}{4\pi} \int_{z-\Delta z}^{z} \delta\varepsilon(x',y',Z) \int_{S} \frac{\exp\left(-j\beta\frac{(x'-x)^2+(y'-y)^2}{2|z-Z|}\right)}{|z-Z|} dS dZ$$

$$= -\frac{j\beta}{2\epsilon} \int_{z-\Delta z}^{z} \delta\varepsilon(x',y',Z) dZ \qquad (6.86)$$

for small Δz. As $\phi_1' - \Delta\phi_1'$ will be independent of $\delta\varepsilon'$

$$\langle \delta\epsilon(x',y',z)E_a'\rangle = \langle \delta\epsilon(x',y',z)\exp(\Delta\phi')\rangle \langle E_0\exp(\phi' - \Delta\phi')\rangle$$

$$\approx \langle \delta\epsilon(x',y',z)\rangle - \frac{j\beta}{2\epsilon}\int_{z-\Delta z}^{z}\delta\epsilon(x',y',z)\delta\varepsilon(x',y',Z)dZ\rangle \langle E_0\exp(\phi'-\Delta\phi')\rangle$$

(6.87)

Now $\langle\delta\epsilon\rangle = 0$ and, as before, we assume $\langle(\delta\epsilon(x',y',z')\delta\epsilon(x'',y'',z''))\rangle = \epsilon^2 A(x'-x'', y'-y'')\delta(z'-z'')$. Consequently,

$$\langle\delta\epsilon(x',y',z)E_a'\rangle \approx -\frac{j\beta\epsilon}{4}A(0)\langle E_0\exp(\phi'-\Delta\phi')\rangle \qquad (6.88)$$

and so, in the limit $\Delta z \to 0$,

$$\langle\delta\epsilon(x',y',z)E_a'\rangle \approx -\frac{j\beta\epsilon}{4}A(0)\langle E_a'\rangle \qquad (6.89)$$

As a consequence, Equation 6.83 reduces to

$$-2j\beta\frac{\partial\langle E_a\rangle}{\partial z} + \nabla_T^2\langle E_a\rangle - \frac{j\beta^3}{4}A(0)\langle E_a\rangle = 0 \qquad (6.90)$$

This equation has the solution

$$\langle E_a\rangle = \exp\left(-\frac{\beta^2}{8}A(0)z\right) \qquad (6.91)$$

and from which it can be seen that the effect of the irregularity is to cause an attenuation of the average field.

We will now derive an equation for the MCF by using similar arguments to those for the average field. Consider Equation 6.82 at the point $\mathbf{r}'$ multiplied by $E_a''^*$, i.e.

$$-2j\beta\frac{\partial E_a'}{\partial z'}E_a''^* + \nabla_T'^2 E_a' E_a''^* + \omega^2\mu_0\delta\epsilon' E_a' E_a''^* = 0 \qquad (6.92)$$

and the conjugate of Equation 6.82 at the point $\mathbf{r}''$ and multiplied by E_a', i.e.

$$+2j\beta E_a'\frac{\partial E_a''^*}{\partial z''} + E_a'\nabla_T''^2 E_a''^* + \omega^2\mu_0\delta\epsilon'' E_a' E_a''^* = 0 \qquad (6.93)$$

(note that E'_a denotes E_a evaluated at $\mathbf{r}'$ and E''_a denotes E_a evaluated at $\mathbf{r}''$). We will only consider transverse correlations and so set $z'' = z' = z$. Then, taking the difference between the stochastic averages of Equation 6.92 and 6.93, we obtain

$$-2j\beta \frac{\partial \Gamma_a}{\partial z'} + (\nabla_T'^2 - \nabla_T''^2)\Gamma_a + \omega^2 \mu_0 \langle (\delta\epsilon' - \delta\epsilon'') E'_a E''^*_a \rangle = 0 \qquad (6.94)$$

where $\Gamma_a = \langle E'_a E''^*_a \rangle$. We now need to consider the term $\langle (\delta\epsilon' - \delta\epsilon'') E'_a E''^*_a \rangle$ and, as above, proceed by recalling the Rytov approximation. It will be noted that $E'_a E''^*_a$ is of the form $|E_0|^2 \exp(\psi)$ where ψ is, as before, a weighted sum of $\delta\epsilon$ over the volume V. We consider the volume V to be divided into slices of thickness Δz and consider the slice whose end contains the apex of the cones of influence for points $\mathbf{r}'$ and $\mathbf{r}''$. For the cone with apex $\mathbf{r}'$ we assume that $\delta\epsilon$ in the slice takes its values from those on the axis through $\mathbf{r}'$ and for the cone with apex $\mathbf{r}''$ we take the values from those on the axis through $\mathbf{r}''$. Let $\Delta\psi$ be the contribution to ψ from the slice at the apex, then

$$\Delta\psi \approx \frac{\omega^2 \mu_0}{4\pi} \int_{z-\Delta z}^{z} \delta\epsilon(x', y', Z) \int_S \frac{\exp\left(-j\beta \frac{(x'-x)^2+(y'-y)^2}{2|z-Z|}\right)}{|z-Z|} dS dZ$$

$$+ \frac{\omega^2 \mu_0}{4\pi} \int_{z-\Delta z}^{z} \delta\epsilon(x'', y'', Z) \int_S \frac{\exp\left(+j\beta \frac{(x''-x)^2+(y''-y)^2}{2|z-Z|}\right)}{|z-Z|} dS dZ$$

$$= -\frac{j\beta}{2\epsilon} \int_{z-\Delta z}^{z} \left(\delta\epsilon(x', y', Z) - \delta\epsilon(x'', y'', Z)\right) dZ \qquad (6.95)$$

for small Δz. As $\psi - \Delta\psi$ is now independent of $\delta\epsilon'$ and $\delta\epsilon''$, we have that

$$\langle \left(\delta\epsilon(x', y', z) - \delta\epsilon(x'', y'', z)\right) E'_a E''^*_a \rangle \qquad (6.96)$$
$$= \langle \left(\delta\epsilon(x', y', z) - \delta\epsilon(x'', y'', z)\right) \exp(\Delta\psi) \rangle \langle |E_0|^2 \exp(\phi - \Delta\psi) \rangle \qquad (6.97)$$

Since $\Delta\psi$ will be small for small Δz,

$$\langle (\delta\epsilon(x', y', z) - \delta\epsilon(x'', y'', z)) \exp(\Delta\psi) \rangle \approx \langle (\delta\epsilon(x', y', z) - \delta\epsilon(x'', y'', z)) \rangle$$

$$- \frac{j\beta}{2\epsilon} \int_{z-\Delta z}^{z} (\langle \delta\epsilon(x', y', z)\delta\epsilon(x', y', Z) \rangle + \langle \delta\epsilon(x'', y'', z)\delta\epsilon(x'', y'', Z) \rangle) dZ$$

$$+ \frac{j\beta}{2\epsilon} \int_{z-\Delta z}^{z} (\langle (\delta\epsilon(x', y', z)\delta\epsilon(x'', y'', Z) \rangle + \langle \delta\epsilon(x'', y'', z)\delta\epsilon(x', y', Z) \rangle) dZ$$

As before, we assume that $\langle (\delta\epsilon(x', y', z')\delta\epsilon(x'', y'', z'') \rangle = \epsilon^2 A(x' - x'', y' - y'')\delta(z' - z'')$, and also note that $\langle \delta\epsilon \rangle = 0$. Consequently,

$$\langle (\delta\epsilon(x', y', z)\delta\epsilon(x'', y'', z)) \exp(\Delta\psi) \rangle \approx -\frac{j\beta\epsilon}{2}(A(0) - A(x' - x'', y' - y'')) \qquad (6.98)$$

6.5 Parabolic Equations for the Average Field and MCF

and so, taking the limit $\Delta z \to 0$, we obtain from Equation 6.96 that

$$\langle (\delta\epsilon(x',y',z) - \delta\epsilon(x'',y'',z))E'_a E''^{*}_a \rangle = -\frac{j\beta\epsilon}{2}(A(0) - A(x'-x'', y'-y''))\Gamma_a \quad (6.99)$$

As a result of Equation 6.99, Equation 6.94 for the transverse MCF can be rewritten as

$$-2j\beta\frac{\partial \Gamma_a}{\partial z} + (\nabla_T'^2 - \nabla_T''^2)\Gamma_a - \frac{j\beta^3}{2}(A(0) - A(x'-x'', y'-y''))\Gamma_a = 0 \quad (6.100)$$

(Ishimaru, 1999).

We consider the simple case of a plane wave that propagates through a disturbed medium in the z direction. Since this is a plane wave, its coherence behavior will be uniform across the wavefront and so Γ_a will only depend on transverse coordinates through $x'-x''$ and $y'-y''$ alone, i.e. $\Gamma_a = \Gamma_a(x'-x'', y'-y'', z)$. Substituting this form of Γ into Equation 6.100 we find that

$$-2j\beta\frac{\partial \Gamma_a}{\partial z} - \frac{j\beta^3}{2}(A(0) - A(x'-x'', y'-y''))\Gamma_a = 0 \quad (6.101)$$

which provides a simple ODE for Γ_a. If we stipulate that, at the start of propagation ($z = 0$), the wave will be perfectly coherent across its wavefront, then $\Gamma_a(z, x'-x'', y'-y'', 0) = 1$. Consequently, given this initial condition, Equation 6.101 can be integrated to yield

$$\Gamma_a(x'-x'', y'-y'', z) = \exp\left(-\frac{z\beta^2}{4}(A(0) - A(x'-x'', y'-y''))\right) \quad (6.102)$$

From this expression it is clear that the correlation across the wavefront is directly related to the transverse correlation of the irregularity.

In all our developments thus far, we have only considered the coherence of signals in the spatial domain. It is, however, also important to consider the coherence of signals in terms of their frequency and in terms of time at scales consistent with the dynamics of the propagation medium. It turns out that a loss of coherence in the frequency domain is related to the spread of propagation delay in a signal. Further, a loss of coherence in the time domain turns out to be related to a spread of a signal in the frequency domain. We will consider the irregularity to be time varying, but on a scale very much greater than that of the wave. Consequently, we can still analyze the problem in terms of a time harmonic signal. We consider the two-time, two-frequency MCF (Dana, 1986)

$$\Gamma_a(x', x'', y', y'', t', t'', \omega', \omega'', z) = \langle E_a(x', y', t', \omega', z)E^*_a(x'', y'', t'', \omega'', z)\rangle \quad (6.103)$$

and assume that

$$\langle(\delta\epsilon(x',y',z',t',\omega')\delta\epsilon(x'',y'',z'',t'',\omega''))\rangle \quad (6.104)$$
$$= \epsilon^2 A(x'-x'', y'-y'', t'-t'')\delta(z'-z'')B(\omega')B(\omega'')$$

Note that we have assumed that $\delta\epsilon$ has the form $\delta\epsilon = \xi(x,y,z,t)B(\omega)$ with $B(\omega) = 1$ for a nondispersive medium. This assumption holds good for most important cases, including for propagation in the ionospheric plasma for which $B(\omega) = \omega_p^2/(\omega^2 - \omega_p^2)$

and $\beta = \left(1 - \omega_p^2/\omega^2\right)^{\frac{1}{2}} \beta_0$ where ω_p is the plasma frequency. Under the above assumptions, the equation for the MCF will take the form (Dana, 1986)

$$\frac{\partial \Gamma_a}{\partial z} + \left(\frac{j}{2\beta'}\nabla_T'^2 - \frac{j}{2\beta''}\nabla_T''^2\right)\Gamma_a + \left(\frac{\beta'^2 B'^2}{8} + \frac{\beta''^2 B''^2}{8}\right)A(0)\Gamma_a$$

$$-\frac{\beta'\beta'' B' B''}{4} A(x'-x'', y'-y'', t'-t'')\Gamma_a = 0 \qquad (6.105)$$

where $B' = B(\omega')$ and $B'' = B(\omega'')$.

We can solve Equation 6.105 for a point source in free space to yield

$$\Gamma_a(\mathbf{w}', \mathbf{w}'', \omega', \omega'', z) = \frac{\beta'\beta''}{4\pi^2 z^2} \exp\left(-j\frac{\beta'|\mathbf{w}'|^2}{2z} + j\frac{\beta''|\mathbf{w}''|^2}{2z}\right) \qquad (6.106)$$

which, as $z \to 0$, becomes the product of delta functions $\delta(\mathbf{w}')\delta(\mathbf{w}'')$ (note that $\mathbf{w}' = x'\hat{\mathbf{x}} + y'\hat{\mathbf{y}}$ and $\mathbf{w}'' = x''\hat{\mathbf{x}} + y''\hat{\mathbf{y}}$). The delta function nature of the above solution turns out to be most useful. Assume Γ_a has the known behavior $\Gamma_S(\mathbf{w}', \mathbf{w}'', t', t'', \omega', \omega'')$ for $\mathbf{w}'$ and $\mathbf{w}''$ on the plane $z = z_S$, then

$$\Gamma_a(\mathbf{w}', \mathbf{w}'', t', t'', \omega', \omega'', z) = \int_{-\infty}^{\infty}\int_{-\infty}^{\infty}\int_{-\infty}^{\infty}\int_{-\infty}^{\infty} K(\mathbf{w}', \mathbf{w}_a, \mathbf{w}'', \mathbf{w}_b, \omega', \omega'', z, z_S)$$

$$\times \Gamma_S(\mathbf{w}_a, \mathbf{w}_b, t', t'', \omega', \omega'') dx_a dy_a dx_b dy_b \qquad (6.107)$$

where

$$K(\mathbf{w}', \mathbf{w}_a, \mathbf{w}'', \mathbf{w}_b, \omega', \omega'', z, z_S) = \frac{\beta'\beta''}{4\pi^2 |z-z_S|^2} \exp\left(-j\frac{\beta'|\mathbf{w}' - \mathbf{w}_a|^2}{2|z-z_S|} + j\frac{\beta''|\mathbf{w}'' - \mathbf{w}_b|^2}{2|z-z_S|}\right) \qquad (6.108)$$

will be solution to the free space version of the parabolic equation for the MCF. Furthermore, because of the delta function nature of the kernel in the limit that $z \to z_S$, $\Gamma_a(\mathbf{w}', \mathbf{w}'', t', t'', \omega', \omega'', z) \to \Gamma_S(\mathbf{w}', \mathbf{w}'', t', t'', \omega', \omega'')$. Consequently, we can use Equation 6.107 to calculate Γ away from a plane where it is known.

6.6 The Phase Screen Approximation

In many situations, the irregularity can be treated as a thin screen. This is particularly the case when studying the effect of ionospheric irregularity on radio astronomical observations or on satellite signals. We will consider the direct path between transmitter and receiver to be the z axis and adapt the parabolic equation approach of the previous section. We first consider the case of a plane wave $E_0 = \exp(-jz)$ that is incident upon a phase screen of thickness L_t (the screen starting at $z = 0$). We consider the case where $L_t \ll \beta l_0^2$, then we can ignore the Laplace operator terms in Equation 6.105 and integrate to find the MCF

6.6 The Phase Screen Approximation

$$\Gamma_a(\mathbf{w}', \mathbf{w}'', t', t'', \omega', \omega'', z) \tag{6.109}$$

$$= \exp\left(-z\left(\frac{\beta'^2 B'^2}{8} + \frac{\beta''^2 B''^2}{8}\right) A(0) + z\frac{\beta'\beta''B'B''}{4} A(\mathbf{w}' - \mathbf{w}'', t' - t'')\right)$$

for $0 \leq z \leq L_t$. (Note that $\Gamma_a = 1$ when $z = 0$, the condition that the plane wave be coherent when it is incident upon the layer of irregularity.) Since the irregularity ends at $z = L_t$, we can use Equation 6.107 to find Γ for $z > L_t$, i.e.

$$\Gamma_a(\mathbf{w}', \mathbf{w}'', t', t'', \omega', \omega'', z) = \int_{-\infty}^{\infty}\int_{-\infty}^{\infty}\int_{-\infty}^{\infty}\int_{-\infty}^{\infty} K(\mathbf{w}', \mathbf{w}_a, \mathbf{w}'', \mathbf{w}_b, \omega', \omega'', z, L_t)$$

$$\times \Gamma_a(\mathbf{w}_a, \mathbf{w}_b, t', t'', \omega', \omega'', L_t) dx_a dy_a dx_b dy_b \tag{6.110}$$

In the GO limit the kernel K behaves as the product of delta functions and so we find that

$$\Gamma_a(\mathbf{w}', \mathbf{w}'', t', t'', \omega', \omega'', z)$$

$$= \exp\left(-L_t\left(\frac{\beta'^2 B'^2}{8} + \frac{\beta''^2 B''^2}{8}\right) A(0) + L_t\frac{\beta'\beta''B'B''}{4} A(\mathbf{w}' - \mathbf{w}'', t' - t'')\right) \tag{6.111}$$

This is the value of Γ_a on exit from the screen and is the same as the result we obtain from the Rytov theory in the GO limit.

We can take the analysis a little further in the special case where $\omega' = \omega'' = \omega$. Then, from Equation 6.110

$$\Gamma_a(\mathbf{w}', \mathbf{w}'', t', t'', \omega, z)$$

$$= \int_{-\infty}^{\infty}\int_{-\infty}^{\infty}\int_{-\infty}^{\infty}\int_{-\infty}^{\infty} \frac{\beta^2}{4\pi^2|z - L_t|^2} \exp\left(-j\beta\frac{|\mathbf{w}' - \mathbf{w}_a|^2 - |\mathbf{w}'' - \mathbf{w}_b|^2}{2|z - L_t|}\right)$$

$$\times \exp\left(-L_t\frac{\beta^2 B^2}{4} A(0) + L_t\frac{\beta^2 B^2}{4} A(\mathbf{w}' - \mathbf{w}'', t' - t'')\right) dx_a dy_a dx_b dy_b \tag{6.112}$$

and, noting that $|\mathbf{w}' - \mathbf{w}_a|^2 - |\mathbf{w}'' - \mathbf{w}_b|^2 = |\mathbf{w}'|^2 - |\mathbf{w}''|^2 + 4\mathbf{f} \cdot (\mathbf{g} - (\mathbf{w}' - \mathbf{w}'')/2) - 4\mathbf{g} \cdot (\mathbf{w}' + \mathbf{w}'')/2$ where $\mathbf{w}_a = \mathbf{f} + \mathbf{g}$ and $\mathbf{w}_b = \mathbf{f} - \mathbf{g}$, we can transform to $\mathbf{g}$ and $\mathbf{f}$ coordinates (note that $dx_a dy_a dx_b dy_b = 4 dg_1 dg_2 df_1 df_2$)

$$\Gamma_a(\mathbf{w}', \mathbf{w}'', t', t'', \omega, z)$$

$$= \int_{-\infty}^{\infty}\int_{-\infty}^{\infty}\int_{-\infty}^{\infty}\int_{-\infty}^{\infty} \frac{\beta^2}{4\pi^2|z - L_t|^2} \exp\left(-j\beta\frac{|\mathbf{w}'|^2 - |\mathbf{w}''|^2 + 4\mathbf{f}\cdot\left(\mathbf{g} - \frac{\mathbf{w}'-\mathbf{w}''}{2}\right) - 4\mathbf{g}\cdot\frac{\mathbf{w}'+\mathbf{w}''}{2}}{2|z - L_t|}\right)$$

$$\times \exp\left(-L_t\frac{\beta^2 B^2}{4} A(0) + L_t\frac{\beta^2 B^2}{4} A(2\mathbf{g}, t' - t'')\right) 4 dg_1 dg_2 df_1 df_2 \tag{6.113}$$

Propagation through Irregular Media

Integrating with respect to f_1 and f_2,

$$\Gamma_a(\mathbf{w}', \mathbf{w}'', t', t'', z) = \int_{-\infty}^{\infty} \int_{-\infty}^{\infty} \exp\left(-j\beta \frac{|\mathbf{w}'|^2 - |\mathbf{w}''|^2 - 4\mathbf{g} \cdot \frac{\mathbf{w}' + \mathbf{w}''}{2}}{2|z - L_t|}\right) \delta\left(\mathbf{g} - \frac{\Delta \mathbf{w}}{2}\right)$$

$$\times \exp\left(-L_t \frac{\beta^2 B^2}{4} A(0) + L_t \frac{\beta^2 B^2}{4} A(2\mathbf{g}, t' - t'')\right) dg_1 dg_2 \quad (6.114)$$

and, integrating with respect to g_1 and g_2,

$$\Gamma_a(\mathbf{w}', \mathbf{w}'', t', t'', z) = \exp\left(-L_t \frac{\beta^2 B^2}{4} A(0) + L_t \frac{\beta^2 B^2}{4} A(\Delta \mathbf{w}, t' - t'')\right) \quad (6.115)$$

This is again the result we would have expected from the Rytov approach in the GO limit. It will be noted that the above MCF has the form $\Gamma(\Delta\mathbf{w}, \Delta t, \omega', \omega'', z)$ where $\Delta\mathbf{w} = \mathbf{w}' - \mathbf{w}''$ and $\Delta t = t' - t''$. As a consequence, we can convert the above solution for an incident plane wave into one for an incident spherical wave in the same manner as for the Rytov approximation. Referring to Figure 6.6, we can do this by replacing $\Delta\mathbf{w}$ by $\gamma \Delta\mathbf{w}$ and z by γz_r where $\gamma = z_t/(z_t + z_r)$. From Equation 6.111, we will now have

$$\Gamma_a(\Delta\mathbf{w}, \Delta t, z_r) = \exp\left(-L_t \frac{\beta^2 B^2}{4} A(0) + L_t \frac{\beta^2 B^2}{4} A(\gamma \Delta \mathbf{w}, t' - t'')\right) \quad (6.116)$$

In the case that $\omega' \neq \omega''$, things get a bit more complicated and so we shall restrict ourselves to the important special case in which $\mathbf{w}' = \mathbf{w}'' = \mathbf{0}$. This is the case of a single receiver that is fixed in position and so we are only interested in the coherence in time and frequency. Let $\beta_d = (\beta' - \beta'')/2$, $\beta_s = (\beta' + \beta'')/2$, then

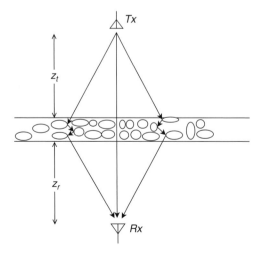

Figure 6.6 Scattering by a phase screen.

6.6 The Phase Screen Approximation

$$K(\mathbf{0}, \mathbf{w}_a, \mathbf{0}, \mathbf{w}_b, \omega', \omega'', z, L_t) = \frac{\beta_s^2 - \beta_d^2}{4\pi^2 |z - L_t|^2} \exp\left(-j\beta_s \frac{2\mathbf{f} \cdot \mathbf{g}}{|z - L_t|} - j\beta_d \frac{|\mathbf{f}|^2 + |\mathbf{g}|^2}{|z - L_t|}\right)$$

$$= \frac{\beta_s^2 - \beta_d^2}{4\pi^2 |z - L_t|^2} \exp\left(-j\beta_d \frac{|\mathbf{g}|^2}{|z - L_t|} - j\beta_d \frac{\left|\mathbf{f} + \frac{\beta_s}{\beta_d}\mathbf{g}\right|^2 - \frac{\beta_s^2}{\beta_d^2}|\mathbf{g}|^2}{|z - L_t|}\right)$$

(6.117)

We substitute Equations 6.117 and 6.111 into Equation 6.110 and change to $\mathbf{g}$ and $\mathbf{f}$ coordinates. Then, integrating with respect to f_1 and f_2,

$$\Gamma_a(\mathbf{0}, \mathbf{0}, t', t'', \omega', \omega'', z)$$

$$= \int_{-\infty}^{\infty} \int_{-\infty}^{\infty} \frac{\beta_s^2 - \beta_d^2}{j\beta_d \pi |z - L_t|} \exp\left(j\frac{\beta_s^2 - \beta_d^2}{\beta_d} \frac{|\mathbf{g}|^2}{|z - L_t|}\right)$$

$$\times \exp\left(-L_t \left(\frac{\beta'^2 B'^2}{8} + \frac{\beta''^2 B''^2}{8}\right) A(0) + L_t \frac{\beta' \beta'' B' B''}{4} A(2\mathbf{g}, t' - t'')\right) d\mathbf{g}_1 d\mathbf{g}_2$$

(6.118)

In the limit $\beta_d \to 0$, this once again yields the GO result. The above expression, however, can be integrated analytically in the special case when the structure function A is quadratic in its spatial arguments.

Because of the difficulties of solving the full equations for the MCF, the phase-screen approximation has proven a very useful idealization, especially for problems of transionospheric propagation where it provides a fairly realistic model. For a thin screen, Knepp (1983) has developed a full analytic solution for the spatial and frequency aspects when the structure function has the form $A = A_0 + |\Delta\mathbf{w}|^2 A_2$ and $B = \omega_p^2/\omega^2$ (the high frequency limit). The MCF now becomes

$$\Gamma_a(\Delta\mathbf{w}, \omega_d) = \exp\left(-\frac{\sigma_\phi^2 \omega_d^2}{2\omega^2}\right) \exp\left(-\frac{\omega_{\text{coh}}|\Delta\mathbf{w}|^2}{l_0^2(\omega_{\text{coh}} + j\omega_d)}\right) \frac{\omega_{\text{coh}}}{\omega_{\text{coh}} + j\omega_d} \quad (6.119)$$

where $\omega_d = \omega' - \omega''$, l_0 is the correlation length of the electric field in the transverse direction, ω_{coh} is the correlation bandwidth and σ_ϕ is the standard deviation of the phase fluctuations (note that we have suppressed the variable denoting the distance from the irregularity to the receiver plane since we assume this to be fixed). In terms of the structure function,

$$l_0^2 = -\frac{A_0(z_t + z_r)^2}{A_2 z_t^2 \sigma_\phi^2} \quad (6.120)$$

$$\omega_{\text{coh}} = -\frac{\pi \omega A_0(z_t + z_r)}{\lambda A_2 z_t z_r \sigma_\phi^2} = \frac{\pi \omega l_0^2 z_t}{\lambda z_r(z_t + z_r)} \quad (6.121)$$

and

$$\sigma_\phi^2 = \frac{\omega_p^4}{4\omega^2 c_0^2} L_t A_0 \quad (6.122)$$

where L_t is the screen thickness. For a irregularity with Gaussian autocorrelation $\langle \delta\epsilon_a \delta\epsilon_b \rangle = \langle |\delta\epsilon|^2 \rangle \exp(-|\mathbf{r}_a - \mathbf{r}_b|^2/L^2)$ we will have $A(\mathbf{w}) = (L\sqrt{\pi}\langle|\delta\epsilon|^2\rangle/\epsilon^2) \exp(-|\mathbf{w}|^2/L^2)$ and so $A_0 = L\sqrt{\pi}\langle|\delta\epsilon|^2\rangle/\epsilon^2$ and $A_2 = -(\sqrt{\pi}/L)\langle|\delta\epsilon|^2\rangle/\epsilon^2$.

For more extended irregularity, multiple phase screen methods have been developed (Knepp, 1983). Such multiple phase screen techniques have also been used to study the effect of ionospheric turbulence on propagation at HF frequencies (Nickisch, 1992). At these frequencies, propagation can pass through extended regions of irregularity and this makes a single phase screen an unrealistic model.

An important quantity, known as the *power impulse response function*, is defined by

$$G(z,t) = \frac{1}{2\pi} \int_{-\infty}^{\infty} \exp(j\omega_d t) \Gamma(\mathbf{0}, \omega_d, z) d\omega_d \qquad (6.123)$$

(Ishimaru, 1997; Knepp, 1983). This is the received power when a delta function pulse is transmitted. The effect of irregularity is to cause the pulse to spread in delay, and this behavior can be described in terms of the mean excess delay $\langle \tau \rangle$ and the delay spread (or *jitter*) σ_τ. These quantities are defined by

$$\langle \tau \rangle = \int_{-\infty}^{\infty} \tau G(z,\tau) d\tau \qquad (6.124)$$

and

$$\sigma_\tau^2 = \int_{-\infty}^{\infty} \tau^2 G(z,\tau) d\tau - \langle \tau \rangle^2 \qquad (6.125)$$

From these definitions, it can be shown (Knepp, 1983) that the delay, and delay spread, can be calculated from the MCF through

$$\langle \tau \rangle = j \frac{\partial \Gamma}{\partial \omega_d}\Big|_{\omega_d=0} \qquad (6.126)$$

and

$$\sigma_\tau^2 = -\frac{\partial^2 \Gamma}{\partial \omega_d^2}\Big|_{\omega_d=0} - \langle \tau \rangle^2 \qquad (6.127)$$

The above results are general and do not depend on the use of a phase screen. For a thin phase screen, $\langle \tau \rangle = 1/\omega_{coh}$ and $\sigma_\tau^2 = (1 + 1/\alpha^2)/\omega_{coh}^2$ where $\alpha = -A_2 \lambda z_t z_r / \pi A_0 (z_t + z_r)$.

6.7 Channel Simulation

Thus far, we have only considered the effect of irregularity on a monochromatic wave, i.e. a wave of the form $E_0 = \exp(j(\omega t - \beta z))$. In practice, however, we are usually concerned with the effect of the propagation medium on a modulated signal. We have already seen that the modulation and carrier can have different delays (the group and phase delays), even in a benign medium. In the case of a narrow band signal, if $s_{Tx}(t) = m(t)\exp(j\omega_c t)$ is transmitted (ω_c is the carrier frequency), then the signal $s_{Rx}(t) = m(t - \tau_g)\exp(j\omega_c(t - \tau_p))/\sqrt{L}$ will be received where τ_g is the group delay, τ_p is the phase delay and L is the propagation loss. The introduction of disturbances into the propagation medium adds considerable complexity to this simple picture. Further, if the system is wide band, the additional problem of dispersion is also added. We can, however, consider a Fourier representation of the modulation $m(t) = (1/2\pi) \int_{-\infty}^{\infty} M(\nu) \exp(j\nu t) d\nu$ and thus consider the signal as a sum of monochromatic waves, which can be treated separately. We normally consider the communications channel in terms of the *channel impulse response function* $h(\mathbf{w}, \tau)$. This function relates the signal $s_{Rx}(\mathbf{w}, t)$ that is received to the signal $s_{Tx}(t)$ transmitted through

$$s_{Rx}(\mathbf{w}, t) = \int_{-\infty}^{\infty} h(\mathbf{w}, \tau) s_{Tx}(t - \tau) d\tau \tag{6.128}$$

where $\mathbf{w}$ is a coordinate in the receiver plane (for the moment we have assumed the channel to be static or at least slowly varying). The channel function allows us to take into account the multiplicity of propagation modes that are caused by the multiple propagation paths in an irregular medium. For a discrete number of narrowband modes, we will have

$$h(\mathbf{w}, \tau) = \sum_i \frac{\delta(\tau - \tau_g^i) \exp\left(j\omega_c(\tau_g^i - \tau_p^i)\right)}{\sqrt{L_i}} \tag{6.129}$$

where, for the ith mode, τ_g^i is the group delay, τ_p^i is the phase delay and L_i is the loss. Impulse response functions are a major tool for studying the effect of propagation on a modulated signal and are important in the study of signal processing techniques to counteract the degrading effects of propagation. The impulse function can be studied in the frequency domain through its Fourier transform $\hat{h}$, i.e.

$$\hat{h}(\mathbf{w}, \omega) = \int_{-\infty}^{\infty} h(\mathbf{w}, \tau) \exp(-j\omega\tau) d\tau \tag{6.130}$$

For a single mode, we consider the factorization $\hat{h} = E_0 E_a$ where E_0 is $\hat{h}$ when no irregularity is present. Field E_a can now be studied using the methods of the previous sections. The major problem in calculating E_a, however, is that the behavior of the irregularity is often only known in a statistical sense. Knepp and Wittwer (1984) overcome this by calculating a realization of the channel impulse response that exhibits the correct statistics, specifically the correct MCF. To proceed with the approach of Knepp and Wittwer, we introduce the *generalized power spectrum*, which is obtained as a Fourier transform of the MCF

$$S(\mathbf{K}, \tau) = \frac{1}{8\pi^3} \int_{-\infty}^{\infty} \int_{-\infty}^{\infty} \int_{-\infty}^{\infty} \Gamma_a(\Delta \mathbf{w}, \omega_d) \exp(-j\mathbf{w} \cdot \mathbf{K} + j\omega_d \tau) d\Delta w_1 d\Delta w_2 d\omega_d \tag{6.131}$$

for which

$$\Gamma_a(\Delta \mathbf{w}, \omega_d) = \int_{-\infty}^{\infty} \int_{-\infty}^{\infty} \int_{-\infty}^{\infty} S(\mathbf{K}, \tau) \exp(j\Delta \mathbf{w} \cdot \mathbf{K} - j\omega_d \tau) dK_1 dK_2 d\tau \tag{6.132}$$

(Note that, for the present, we have assumed the propagation medium to be static.) We express $E_a(\mathbf{w}, \omega)$ in terms of its time-dependent transverse spatial spectrum $H(\mathbf{K}, \tau)$ through

$$E_a(\mathbf{w}, \omega) = \int_{-\infty}^{\infty}\int_{-\infty}^{\infty}\int_{-\infty}^{\infty} H(\mathbf{K}, \tau) \exp(jK_1 w_1) \exp(jK_2 w_2) \exp(-j\omega\tau) d\tau dK_1 dK_2$$

(6.133)

and assume that this can be adequately approximated by its discretized form

$$E_a(f\Delta w, g\Delta w, h\Delta \omega) = \sum_{l=0}^{L-1}\sum_{m=0}^{L-1}\sum_{n=0}^{N-1} H(l\Delta K, m\Delta K, n\Delta \tau)$$
$$\times \exp(jfl\Delta K\Delta w + jgm\Delta K\Delta w - jhn\Delta\omega\Delta\tau) \Delta K\Delta K\Delta\tau$$

(6.134)

for $f = 0, \ldots, L-1$, $g = 0, \ldots, L-1$ and $h = 0, \ldots, N-1$ where $\Delta w = D/L$ and $\Delta \tau = T/N$ (distance D encompasses the extent of coherence in the transverse direction and T is the maximum delay). In order to satisfy the Nyquist limit, we choose $\Delta K = 2\pi/D$ and $\Delta \omega = 2\pi/T$. We can now generate a realization of H through

$$H(l\Delta K, m\Delta K, n\Delta \tau) = \left(\frac{S(l\Delta K, m\Delta K, n\Delta \tau)}{\Delta K\Delta K\Delta\tau}\right)^{\frac{1}{2}} r_{lmn}$$

(6.135)

where r_{lmn} is the complex random variable

$$r_{lmn} = \sqrt{\frac{1}{2}(a_{lmn} + jb_{lmn})}$$

(6.136)

with a_{lmn} and b_{lmn} independent Gaussian random variables that have zero mean and unit variance (note that $\langle r_{lmn} r^*_{ijk}\rangle = \delta_{li}\delta_{li}\delta_{li}$ and $\langle r_{lmn}\rangle = 0$). From Equation 6.134, we obtain that

$$\langle E_a(f\Delta w, g\Delta w, h\Delta\omega) E_a^*(p\Delta w, q\Delta w, r\Delta\omega)\rangle$$
$$= \sum_{l=0}^{L-1}\sum_{m=0}^{L-1}\sum_{n=0}^{N-1} S(l\Delta K, m\Delta K, n\Delta\tau)$$
$$\times \exp(j(f-p)l\Delta K\Delta w + j(g-q)m\Delta K\Delta w - j(h-r)n\Delta\omega\Delta\tau) \Delta K\Delta K\Delta\tau$$

(6.137)

which is a discretized version of Equation 6.132. Consequently, through Equation 6.135 we can generate a discrete realization of the spatial spectrum of the impulse response function with the correct MCF. We can generate the discretized form of E_a through Equation 6.134, and the impulse response function through a discretized form of the inverse of Equation 6.130, i.e.

$$h(f\Delta w, g\Delta w, h\Delta\tau) = \frac{\Delta\omega}{2\pi}\sum_{n=0}^{N-1} \hat{h}(f\Delta w, g\Delta w, n\Delta\omega) \exp(jhn\Delta\omega\Delta\tau)$$

(6.138)

where $\hat{h} = E_0 E_a$. In Equation 6.138, the effect of the field E_a can be regarded as a modulation of the background field E_0. Consequently, for narrow band signals, its effect will be to replace the delta functions in Equation 6.129 by

$$h_a(f\Delta w, g\Delta w, h\Delta\tau) = \frac{\Delta\omega}{2\pi} \sum_{n=0}^{N-1} E_a(f\Delta w, g\Delta w, n\Delta\omega) \exp(jhn\Delta\omega\Delta\tau) \quad (6.139)$$

with possibly different h_a for each mode.

The above techniques have been used by Knepp and Wittwer (1984) to simulate the effects of ionospheric turbulence on wideband signals. The methods, however, have a wide range of applicability and have been used by Nickisch et al. (2012) to simulate the effect of ionospheric turbulence on an HF propagation channel.

6.8 Rough Surface Scattering

We have previously considered the effect of a plane interface on an incoming plane wave. In this case, there is a single direction of reflection and the wave remains plane (this is often referred to as *specular reflection* and the direction of reflection is known as the *specular direction*). If the reflecting surface is now rough, the reflected wave becomes far more complex and the wavefront is distorted (see Figure 6.7). If we consider reflection from a point height Δh above a reference plane, the wavefront will be advanced a distance $2h \cos\theta$ beyond that for reflection in the reference plane. Parameter $\gamma = 2\beta h_{rms} \cos\theta$ (h_{rms} is the rms height of the interface) is known as the Rayleigh roughness parameter and is a measure of the fluctuations in phase caused by the rough surface. The condition $\gamma > \pi/2$ is known as the Rayleigh criterion and is frequently used to decide whether a surface is to be considered as rough. If the criterion is satisfied, the destructive interference of the reflected waves has become significant and the specular component (that in the direction of reflection for the averaged surface) is much reduced. In this case, there will be significant scatter in nonspecular directions (backward and forward scatter).

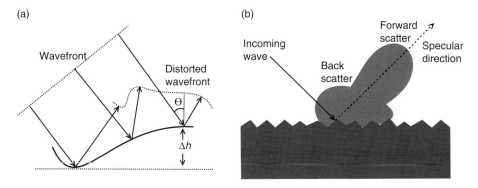

Figure 6.7 Effect of a rough surface showing (a) the distortion of wavefront and (b) forward and backward scatter.

The effect of scatter is particularly important in radar where back scatter from a rough surface can result in radar returns that can mask legitimate targets. This kind of unwanted signal is termed *clutter*, the strongest clutter arising from back scatter at sea surfaces (surfaces well approximated as PEC). There are several approaches to this problem (Ishimaru, 1997), but we will study this kind of scattering by means of the extinction theorem (Nieto-Vesperinas, 1982; Coleman, 1996), i.e.

$$\mathbf{E}^i(\mathbf{a}) + \frac{\eta_0}{j\beta} \nabla \times \nabla \times \int_S \frac{\exp(-j\beta|\mathbf{r}-\mathbf{a}|)}{4\pi|\mathbf{r}-\mathbf{a}|} \mathbf{J}(\mathbf{a})\, dS'$$
$$= C\mathbf{E}(\mathbf{a}) \tag{6.140}$$

where S is the rough surface (assumed to be approximately plane). The surface is illuminated from above by electric field $\mathbf{E}_i$ and constant C is 1 for points above the surface and 0 for points below the surface. Assuming the patch of roughness to be finite (Figure 6.8), we consider the scattered field at large distances. From the extinction theorem

$$\mathbf{E}^i(\mathbf{a}) + \frac{j\beta\eta_0 \exp(-j\beta a)}{4\pi a} \hat{\mathbf{a}} \times \hat{\mathbf{a}} \times \int_S \exp(j\beta \hat{\mathbf{a}} \cdot \mathbf{r}) \mathbf{J}(\mathbf{r})\, dS' = C\mathbf{E}(\mathbf{a}) \tag{6.141}$$

where $\hat{\mathbf{a}}$ is a unit vector in the direction of $\mathbf{a}$ and we have assumed a coordinate system that is centered on the patch. Let the surface have perturbations that are described by $z = h(x,y)$ where the unperturbed surface is the $z = 0$ plane. Let $\mathbf{J} = \mathbf{J}^0 + \mathbf{J}^1$ and $\mathbf{E} = \mathbf{E}^0 + \mathbf{E}^1$ where $\mathbf{J}^1$ and $\mathbf{E}^1$ are the perturbations to current and electric field, respectively. For small perturbations, we can now reduce the integral using the approximation

$$\int_S \exp(j\beta \hat{\mathbf{a}} \cdot \mathbf{r}) \mathbf{J}(\mathbf{r}) dS$$
$$\approx \int_{R^2} \exp(j\beta(\hat{a}_x x + \hat{a}_y y))(\mathbf{J}^0(x,y) + j\beta \hat{a}_z h(x,y)\mathbf{J}^0(x,y) + \mathbf{J}^1(x,y)) dx dy \tag{6.142}$$

As a consequence, we can rework Equation 6.141 as an equation for the perturbed field, i.e.

$$\frac{j\beta\eta_0 \exp(-j\beta a)}{4\pi a} \hat{\mathbf{a}} \times \hat{\mathbf{a}} \times \int_{R^2} \exp(j\beta(\hat{a}_x x + \hat{a}_y y))(j\beta \hat{a}_z h(x,y)\mathbf{J}^0(x,y)$$
$$+ \mathbf{J}^1(x,y)) dx dy = C\mathbf{E}^1(\mathbf{a}) \tag{6.143}$$

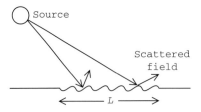

Figure 6.8 Scattering by a rough surface.

Define vector fields L and M by

$$\mathbf{L}(a_x, a_y) = -\frac{\beta^2 \eta_0}{4\pi} \int_{R^2} \exp(j\beta(\hat{a}_x x + \hat{a}_y y)) h(x,y) \mathbf{J}^0(x,y) dx dy \qquad (6.144)$$

and

$$\mathbf{M}(a_x, a_y) = \frac{j\beta \eta_0}{4\pi} \int_{R^2} \exp(j\beta(\hat{a}_x x + \hat{a}_y y)) \mathbf{J}^1(x,y) dx dy \qquad (6.145)$$

(Note that $\mathbf{L}$ only has horizontal components and that both $\mathbf{L}$ and $\mathbf{M}$ have no dependence on a_z.) We can rewrite Equation 6.141 as

$$\frac{\exp(-j\beta a)}{a} \hat{\mathbf{a}} \times \hat{\mathbf{a}} \times (\hat{a}_z \mathbf{L} + \mathbf{M}) = C\mathbf{E}^1(\mathbf{a}) \qquad (6.146)$$

Now consider Equation 6.141 at a point $\mathbf{a}$ above the plane and its conjugate point $\mathbf{a}_c = \mathbf{a} - 2a_z \hat{\mathbf{z}}$ below the plane. Let $\mathbf{J}$ be an arbitrary vector that has no dependence on $\hat{a}_z$, for points $\mathbf{a}$ and $\mathbf{a}_c$

$$\hat{\mathbf{a}} \times (\hat{\mathbf{a}} \times \mathbf{J}) = \mathbf{J} \cdot \hat{\mathbf{a}} \hat{\mathbf{a}} - \mathbf{J} \qquad (6.147)$$

and

$$\hat{\mathbf{a}}_c \times (\hat{\mathbf{a}}_c \times \mathbf{J}) = \mathbf{J} \cdot \hat{\mathbf{a}} \hat{\mathbf{a}} - 2\mathbf{J} \cdot \hat{\mathbf{a}} \hat{a}_z \hat{\mathbf{z}} - 2\hat{a}_z J_z \hat{\mathbf{a}} + 4\hat{a}_z^2 J_z \hat{\mathbf{z}} - \mathbf{J} \qquad (6.148)$$

respectively. If we consider Equation 6.146 at these conjugate points, then adding the vertical components

$$\mathbf{E}_v^1(\mathbf{a}) = \frac{\exp(-j\beta a)}{a} 2(\hat{a}_z^2 - 1) M_z \hat{\mathbf{z}} \qquad (6.149)$$

and subtracting the horizontal components

$$\mathbf{E}_h^1(\mathbf{a}) = \frac{\exp(-j\beta a)}{a} 2 M_z \hat{a}_z \hat{\mathbf{a}}_h \qquad (6.150)$$

The normal component of current will be zero on the surface and so, to the first order, $J_z^1 = \nabla h \cdot \mathbf{J}^0$ and, as a consequence, we find that

$$M_z(a_x, a_y) = \frac{j\beta \eta_0}{4\pi} \int_{R^2} \exp(j\beta(\hat{a}_x x + \hat{a}_y y)) \nabla h \cdot \mathbf{J}^0 dx dy \qquad (6.151)$$

We note that, provided the source is well separated from the plane, we can approximate the incident field as a plane wave with $\mathbf{H}^i = \mathbf{H}_0^i \exp(-j\beta \hat{\mathbf{p}} \cdot \mathbf{r})$. Since we assume the scattering plane to be PEC, $\mathbf{J}^0 = \hat{\mathbf{z}} \times \mathbf{H} \approx 2\hat{\mathbf{z}} \times \mathbf{H}^i$ and we have that $\nabla \mathbf{J}^0 \approx -j\beta \hat{\mathbf{p}} \cdot \mathbf{J}^0$. As a consequence of this,

$$M_z(a_x, a_y) = -\hat{\mathbf{a}} \cdot \mathbf{L}(a_x, a_y) + \hat{\mathbf{p}} \cdot \mathbf{L}(a_x, a_y) \qquad (6.152)$$

Since M_z is the only component of $\mathbf{M}$ that we require to calculate the scattered field $\mathbf{E}^1$, it is clear that we require a knowledge of $\mathbf{L}$ alone. From Equations 6.149 and 6.150, we will now obtain that

$$\mathbf{E}_v^1(\mathbf{a}) = 2\frac{\exp(-j\beta a)}{a} (\hat{a}_z^2 - 1)(\hat{\mathbf{p}} - \hat{\mathbf{a}}) \cdot \mathbf{L} \hat{\mathbf{z}} \qquad (6.153)$$

and
$$\mathbf{E}_h^1(\mathbf{a}) = 2\frac{\exp(-j\beta a)}{a}\hat{a}_z(\hat{\mathbf{p}} - \hat{\mathbf{a}}) \cdot \mathbf{L}\hat{\mathbf{a}}_h \tag{6.154}$$

If we know the form of $h(x, y)$, we can calculate the scattered field $\mathbf{E}^1$. Unfortunately, in many problems of interest, the surface roughness (i.e. $h(x, y)$) is only known in a statistical sense. In this case, one approach is to generate a *realization* of the surface. One of the more important applications of these kind of technique is in the study of scattering by a rough sea. The realization can be generated according to the following prescription (Ishimaru, 1997). First we pick a suitably large scale L and then assume the surface to be periodic in both the x and y directions with this length as a period (L can be chosen as a suitable multiple of the wavelength of greatest amplitude in the statistical description). The height is then generated according to

$$h(x, y) = \sum_{k=-\infty}^{\infty} \sum_{l=-\infty}^{\infty} P(i,j) \exp\left(\frac{2j\pi kx}{L} + \frac{2j\pi ly}{L}\right) \tag{6.155}$$

where the $P(i,j)$ are independent random variables with zero mean. In order for h to be real, we will need $P(i,j) = P^*(-i,-j)$. The autocorrelation will now take the form

$$\langle h(x,y)h^*(x',y')\rangle = \sum_{k=-\infty}^{\infty} \sum_{l=-\infty}^{\infty} \sum_{k'=-\infty}^{\infty} \sum_{l'=-\infty}^{\infty} \langle P(k,l)P^*(k',l')\rangle$$
$$\times \exp\left(\frac{2j\pi(kx - k'x')}{L} + \frac{2j\pi(ly - l'y')}{L}\right) \tag{6.156}$$

Noting the independence properties of $P(i,j)$, we then have

$$\langle h(x,y)h^*(x',y')\rangle = \sum_{k=-\infty}^{\infty} \sum_{l=-\infty}^{\infty} W_{k,l}$$
$$\times \exp\left(\frac{2j\pi k(x-x')}{L} + \frac{2j\pi l(y-y')}{L}\right)\left(\frac{2\pi}{L}\right)^2 \tag{6.157}$$

where $W_{k,l} = (L/2\pi)^2 \langle P(k,l)P^*(k,l)\rangle$. Let $W_{k,l} = W(p,q)$ with $p = 2\pi k/L$ and $q = 2\pi l/L$. Taking the limit $L \to \infty$,

$$\langle h(x,y)h^*(x',y')\rangle = \int_{-\infty}^{\infty}\int_{-\infty}^{\infty} W(p,q) \exp\left(jp(x-x') + jq(y-y')\right) dp\,dq \tag{6.158}$$

From the above expression, it will be noted that $W(p,q)$ is the spectral density of the irregularity and that the autocorrelation only depends on x, y, x' and y' through $\tilde{x} = x - x'$ and $\tilde{y} = y - y'$ (a surface satisfying this condition is said to be *stationary in the wide sense*). By means of Fourier's integral theorem, the spectral density can be expressed in terms of the autocorrelation by

$$W(p,q) = \frac{1}{4\pi^2} \int_{-\infty}^{\infty}\int_{-\infty}^{\infty} \langle hh^*\rangle \exp\left(-jp\tilde{x} - jq\tilde{y}\right) d\tilde{x}\,d\tilde{y} \tag{6.159}$$

We now need a method to generate a realization of the ground. We can do this by choosing coefficients $P(k,l)$ (for $k \geq 0$ and $l \geq 0$) according to $P(k,l) = \rho_{kl}(2\pi/L)\sqrt{W(p,q)}$

where $\rho_{kl} = (a_{kl} + jb_{kl})/\sqrt{2}$ with a_{kl} and b_{kl} independently distributed Gaussian random variables of zero mean and unit variance. Noting the properties of ρ_{kl}, the resulting $P(k, l)$ will have the correct independence and normalization properties. We consider the sums in Equation 6.155 to be suitably truncated (∞ is replaced by a suitably large number). Then, once we have generated $P(k, l)$, using the appropriate random number generator and spectrum, this will provide an effective realization of the ground with the correct spectral properties.

The ground will normally be defined through the autocorrelation function of the surface height. A simple example is the Gaussian autocorrelation $\langle h(x, y)h^*(x', y')\rangle = \langle h^2\rangle \exp\left(-|\mathbf{r}|^2/L^2\right)$ and for which $W(p, q) = \langle hh^*\rangle (L^2/4\pi) \exp\left(-(p^2 + q^2)L^2/4\right)$ where L is the correlation length. For the rough surface formed by the sea, several alternative representations have been developed, but that developed by Pierson and Moscowitz (1956) been found to be particularly effective.

If we only require an estimate of the order of the perturbations caused by the ground, an alternative is to calculate the variance of the perturbation fields, i.e. $\langle |\mathbf{E}_v^1|^2\rangle$ and $\langle |\mathbf{E}_h^1|^2\rangle$. From Equations 6.153 and 6.154,

$$\langle |\mathbf{E}_v^1|^2\rangle = \frac{4}{a^2}(\hat{a}_z^2 - 1)^2 \langle (\hat{\mathbf{p}} - \hat{\mathbf{a}}) \cdot \mathbf{LL}^* \cdot (\hat{\mathbf{p}} - \hat{\mathbf{a}})\rangle \qquad (6.160)$$

and

$$\langle |\mathbf{E}_h^1|^2\rangle = \frac{4\hat{a}_z^2}{a^2}\hat{\mathbf{a}}_h \cdot \hat{\mathbf{a}}_h \langle (\hat{\mathbf{p}} - \hat{\mathbf{a}}) \cdot \mathbf{LL}^* \cdot (\hat{\mathbf{p}} - \hat{\mathbf{a}})\rangle \qquad (6.161)$$

We need to study the matrix H with coefficients

$$H_{ij} = \langle L_i L_j^*\rangle = \frac{\beta^4 \eta_0^2}{16\pi^2} \int_A \int_A \exp(j\beta(\hat{a}_x(x - x') + \hat{a}_y(y - y'))) \qquad (6.162)$$

$$\times \langle h(x, y)h^*(x', y')\rangle J_i^0(x, y) J_j^{0*}(x', y') dx dy dx' dy' \qquad (6.163)$$

where A is the area over which the surface is rough. We assume the incident field can be approximated by a plane wave with $\mathbf{H}^i = \mathbf{H}_0^i \exp(-j\beta\hat{\mathbf{p}} \cdot \mathbf{r})$, $\mathbf{J}^0 \approx \tilde{\mathbf{J}} \exp(-j\beta\hat{\mathbf{p}} \cdot \mathbf{r})$ where $\tilde{\mathbf{J}} = 2\hat{\mathbf{z}} \times \mathbf{H}_0^i$. For this case

$$H_{ij} = \frac{\beta^4 \eta_0^2}{16\pi^2} \tilde{J}_i \tilde{J}_j^* \int_A \int_A \exp(j\beta\hat{\mathbf{a}} \cdot (\mathbf{r} - \mathbf{r}')) \qquad (6.164)$$

$$\times \exp(-j\beta\hat{\mathbf{p}} \cdot (\mathbf{r} - \mathbf{r}'))\langle h(x, y)h^*(x', y')\rangle dx dy dx' dy' \qquad (6.165)$$

We will assume that the simple representation 6.155 of the roughness applies and that the correlation length is much smaller than the dimensions of A. Then, in terms of the spectrum of irregularities,

$$H_{ij} = \frac{\beta^4 \eta_0^2}{4} \tilde{J}_i \tilde{J}_j^* W(\beta(\hat{\mathbf{p}} - \hat{\mathbf{a}}))A \qquad (6.166)$$

and, as a consequence

$$\langle |\mathbf{E}_v^1|^2\rangle = \frac{\beta^4 \eta_0^2}{a^2} W(\beta(\hat{\mathbf{p}} - \hat{\mathbf{a}}))A(\hat{a}_z^2 - 1)^2 (\hat{\mathbf{p}} - \hat{\mathbf{a}}) \cdot \tilde{\mathbf{J}}\tilde{\mathbf{J}}^* \cdot (\hat{\mathbf{p}} - \hat{\mathbf{a}}) \qquad (6.167)$$

and

$$\langle |E_h^1|^2 \rangle = \frac{\hat{a}_z^2 \beta^4 \eta_0^2}{a^2} W(\beta(\hat{\mathbf{p}} - \hat{\mathbf{a}})) A \hat{\mathbf{a}}_h \cdot \hat{\mathbf{a}}_h(\hat{\mathbf{p}} - \hat{\mathbf{a}}) \cdot \tilde{\mathbf{J}} \tilde{\mathbf{J}}^* \cdot (\hat{\mathbf{p}} - \hat{\mathbf{a}}) \qquad (6.168)$$

In terms of the amplitude of the incident electric field $\mathbf{E}_0^i$, we have $\tilde{\mathbf{J}} = 2(\hat{\mathbf{p}}\mathbf{E}_0^i \cdot \hat{\mathbf{z}} - \mathbf{E}_0^i \hat{\mathbf{p}} \cdot \hat{\mathbf{z}})/\eta_0$

A useful way of describing scattering is in terms of the bistatic scattering cross section per unit area of the surface; this is denoted by σ_0. This is a function of the incident direction and the scattered direction. If a unit amplitude electric field is incident upon a roughness patch of area A, then the cross section per unit area is defined to be $4\pi a^2 \langle |\mathbf{E}^s|^2 \rangle / A$ where $\mathbf{E}^s$ is the scattered electric field at distance a. In terms of the fields above, $\sigma_0 = 4\pi a^2 \langle |\mathbf{E}_v^1|^2 \rangle / A |\mathbf{E}_0^i|^2$. We will consider the important case of back scatter at grazing incidence, i.e. $\hat{a}_z \ll 1$ and $\hat{\mathbf{a}} = -\hat{\mathbf{p}}$. In this limit the vertical polarized scattered field dominates and $\sigma_0 = 64\pi \beta^4 W(2\beta\hat{\mathbf{p}})$. We can use this to calculate the back-scatter coefficient for a fully developed sea (Barrick, 1972). We use the spectrum that was developed by Phillips (1985). In terms of our definition of the spectrum

$$W(p,q) = \frac{0.005}{\pi(p^2 + q^2)^2} \quad \text{for } \sqrt{p^2 + q^2} > \frac{g}{U^2}$$
$$= 0 \qquad \text{otherwise} \qquad (6.169)$$

where g is the acceleration due to gravity in m/s² and U is the wind speed in m/s. This leads to a back-scatter coefficient of 0.02 (-17 dB).

In the above considerations, we have taken a perturbation approach to the problem of scattering from rough surfaces. There are, however, other approaches that will work when the surface perturbations become too large for this theory to be applied. In particular, Kirchhoff scattering theory is another approach that has been found particularly useful and can easily be extended to surfaces other than the PEC variety. If we consider Equation 6.141, the major problem is the estimation of the surface current **J**. In the Kirchhoff approach, we assume the surface can be treated as locally plane and the local current calculated according to $\mathbf{J} \approx 2\mathbf{n} \times \mathbf{H}^i$ where **n** is the unit normal to the local plane and $\mathbf{H}^i$ is the incident magnetic field (this approximation is sometimes known as *physical optics*). The theory works well providing that the radius of curvature of the surface at any point is greater than a few wavelengths (Ogilvy, 1991). Consequently, as the correlation length of the roughness decreases, the Kirchhoff theory will eventually break down. A further problem arises when a surface causes multiple scattering; i.e. several reflections are required before the radiation finally escapes from the surface (see Figure 6.9). In recent years, this problem has found effective solution in a method known as *shooting and bouncing rays* (Ling et al., 1989). In this approach, we use the Kirchhoff approach at each reflection to find the contribution to surface current and hence the total current distribution to be used in Equation 6.141. For a practical implementation, we would need to describe the incoming wave by a set of representative rays and follow each ray through its various reflections.

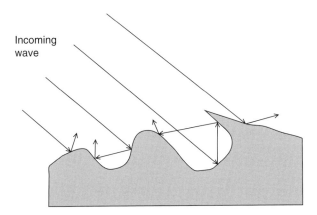

Figure 6.9 Multiple scatter at a rough surface.

6.9 References

D.E. Barrick, *First-Order Theory and Analysis of MF/HF/VHF Scatter from the Sea*, IEEE Transactions on Antennas and Propagation, AP-20, pp. 2–10, 1972.

H.G. Booker and W.E. Gordon, A theory of radio scattering in the troposphere, Proc. Inst. Radio Engineers, vol. 38, 1950.

K.G. Budden, *The Propagation of Radio Waves*, Cambridge University Press, Cambridge, 1988.

C.J. Coleman, Application of the extinction theorem to some antenna problems, IEE Proc. Microwaves, Antennas Propagation, vol. 143, pp. 471–474, 1996.

R.A. Dana, Propagation of RF signals through ionization, Rep DNA-TR-86-158, Defense Nuclear Agency, Washington, DC, May 1986.

K. Davies, *Ionospheric Radio*, IEE Electro-magnetic Waves Series, vol. 31, Peter Peregrinus, London, 1990.

L.B. Felsen and N. Marcuvitz, *Radiation and Scattering of Waves*, IEEE Press and Oxford University Press, Piscataway, NJ, 1994.

V.E. Gherm, N.N. Zernov and H.J. Strangeways, Propagation model for transionospheric fluctuating paths of propagation: Simulator of the transionospheric channel, Radio Sci., vol. 40, RS1003, doi:10.1029/2004RS003097, 2005.

R.F. Harrington, *Time Harmonic Electromagnetic Fields*, McGraw-Hill, New York, 1961.

D.S. Jones, *Methods in Electromagnetic Wave Propagation*, 2nd edition, IEEE/OUP series in Electromagnetic Wave Theory, IEEE Press and Oxford University Press, Oxford and New York, 1999.

A. Ishimaru, *Wave Propagation and Scattering in Random Media*, 2nd edition, IEEE/OUP series in Electromagnetic Wave Theory, IEEE Press and Oxford University Press, Oxford and New York, 1997.

D.L. Knepp, Analytic solution for the two-frequency mutual coherence function for spherical wave propagation, Radio Sci., vol. 18, pp. 535–549, 1983.

D.L. Knepp, Multiple phase-screen calculation of the temporal behaviour of stochastic waves, Proceedings IEEE, vol. 71, pp. 722–737, 1983.

D.L. Knepp and L.A. Wittwer, Simulation of wide bandwidth signals that have propagated through random media, Radio Sci., vol. 19, pp. 303–318, 1984.

H. Ling, R. Chou and S.W. Lee, Shooting and bouncing rays: Calculating the RCS of an arbitrarily shaped cavity, IEEE Tran. Antennas Propag., vol. 37, pp. 194–205, 1989.

R.M. Manning, The relationship between solutions of the parabolic equation method and the first Rytov approximation in stochastic wave propagation theory, Waves in Random Complex Media, vol. 18, pp. 615–621, doi: 10.1080/17455030802232737, 2008.

G.D. Monteath, *Applications of the Electro-magnetic Reciprocity Principle*, Pergamon Press, Oxford, 1973.

L.J. Nickisch, Non uniform motion and extended media effects on the mutual coherence function: An analytic solution for spaced frequency, position and time. Radio Sci., vol. 27, pp. 9–22, 1992.

L.J. Nickisch, G. St. John, S.V. Fridman. M.A. Hausman and C.J. Coleman, HiCIRF: A high-fidelity HF channel simulation, Radio Sci., vol. 47, RS0L11, doi:10.1029/2011RS004928, 2012.

M. Nieto-Vesperinas, Depolarisation of electromagnetic waves scattered from a slightly rough surface: A study by means of the extinction theorem, J. Opt. Soc. Am., vol. 72, pp. 539–547, 1982.

J.A. Ogilvy, *Theory of Wave Scattering from Random Rough Surfaces*, Institute of Physics Publishing, Bristol, 1991.

W.J. Pierson and L. Moskowitz, A proposed spectral form for fully developed wind sea based on the similarity theory of S. A. Kitaigorodski, J. Geophys. Res., vol. 69, pp. 5181–5190, 1956.

O.M. Phillips, *The Dynamics of the Upper Ocean*, Cambridge University Press, Cambridge, 1969.

V.I. Tatarskii, A. Ishimaru and V.U. Zavorotny (editors), Wave propagation in random media (scintillation), invited papers of a conference held 3–7 August 1992, Seattle, SPIE and Institute of Physics Publishing, 1993.

J.R. Wait, *Electromagnetic Waves in Stratified Media*, IEEE/OUP series in Electromagnetic Wave Theory, Oxford University Press, Oxford and New York, 1996.

K.C. Yeh and C.H. Liu, *Theory of Ionospheric Waves*, Academic Press, New York, 1972.

7 The Approximate Solution of Maxwell's Equations

In the study of propagation, we are often confronted by electromagnetic problems for which the boundary geometry, and/or the propagation medium, are extremely complex. We can rarely find an analytic solution to such problems and often need to resort to approximate methods. Asymptotic methods, such as the geometric optics and perturbation series, are often used and are so important that we have already studied them in some detail in the two previous chapters. In the current chapter, however, we will study some other approximate procedures that have proven useful in propagation studies. Of particular importance, we will study the paraxial approximation. We use this approximation to derive the parabolic equations for propagation and then consider their solution through finite difference (FD) methods. In addition, we use the paraxial approximation to derive the Kirchhoff integral equations of propagation and consider their solution through fast Fourier transform (FFT) techniques. The problem of boundary conditions is also considered and, in particular, those arising from irregular terrain. The chapter ends with a description of the finite difference time domain (FDTD) technique as a means of solving transient propagation problems.

7.1 The Two-Dimensional Approximation

In general, the electric field $\mathbf{E}$ will satisfy the vector equation

$$\nabla \times \nabla \times \mathbf{E} - \omega^2 \mu \epsilon \mathbf{E} = 0 \tag{7.1}$$

but we will assume the medium to have the permeability of free space ($\mu = \mu_0$) as this is normally the case for the propagation problems we will study. We note that $\nabla \times \nabla \times \mathbf{E} = -\nabla^2 \mathbf{E} + \nabla(\nabla \cdot \mathbf{E})$ and $\nabla \cdot (\epsilon \mathbf{E}) = 0$, so that

$$\nabla^2 \mathbf{E} + \omega^2 \mu_0 \epsilon_0 \epsilon_r \mathbf{E} - \nabla\left(\frac{\nabla \epsilon}{\epsilon}\right) = 0 \tag{7.2}$$

Further, we assume that the length scales of the medium are much larger than a wavelength and so we can neglect the last term on the left-hand side and obtain

$$\nabla^2 \mathbf{E} + \omega^2 \mu_0 \epsilon \mathbf{E} = 0 \tag{7.3}$$

which is a vector Helmholtz equation. We note that ϵ can be tensor in nature, an important example of this being the propagation medium formed by Earth's ionosphere.

For the moment, however, we will consider the medium to be isotropic (ϵ a scalar) and so will study the generic scalar Helmholtz equation

$$\nabla^2 E + \omega^2 \mu_0 \epsilon E = 0 \tag{7.4}$$

As we have seen in earlier chapters, Equation 7.4 has well-developed solution techniques when ϵ is a constant. For nonhomogeneous media, however, we normally have to resort to approximate methods (perturbation series and geometric optics, for example) or numerical techniques. Some limited progress can be made in the case of two-dimensional (2D) fields. Although these require the unrealistic proposition of infinite line sources, they can form a good approximation far away from real sources where the curvature of the wavefront, in at least one dimension, has become very large. This is the assumption for the canonical solutions that are used in the GTD approach. We take the 2D coordinates to be x and y then the Helmholtz equation will have the form

$$\frac{\partial^2 E}{\partial x^2} + \frac{\partial^2 E}{\partial y^2} + \omega^2 \mu_0 \epsilon E = 0 \tag{7.5}$$

If we introduce the complex variable $z = x + jy$, we can rewrite the above equation as

$$\frac{\partial^2 E}{\partial z \partial z^*} + \omega^2 \mu_0 \epsilon E = 0 \tag{7.6}$$

where epsilon is now a function of z and z^*. We now consider a permeability of the form

$$\epsilon = \epsilon_0 R^2(\Im\{g(z)\}) |g'(z)|^2 \tag{7.7}$$

(see Mikaelian (1980) for the study of such media) and apply the conformal transformation $Z = g(z)$ to Equation 7.6, where $Z = X + jY$. As a result

$$\frac{\partial^2 E}{\partial X^2} + \frac{\partial^2 E}{\partial Y^2} + \omega^2 \mu_0 \epsilon_0 R^2(Y) E = 0 \tag{7.8}$$

The new propagation medium is now horizontally stratified and this greatly simplifies the solution procedure. Indeed, through Airy functions, we can make some progress analytically when the variation of R^2 is linear in the variable Y.

Consider the generic 2D Helmholtz equation

$$\frac{\partial^2 E}{\partial x^2} + \frac{\partial^2 E}{\partial y^2} + \beta_0^2 N^2 E = 0 \tag{7.9}$$

where N is a function of y alone and assume solutions of the form

$$E(x, y) = E_0(y) \exp(-j\beta_0 \gamma x) \tag{7.10}$$

In this case, Equation 7.9 will reduce to

$$\frac{d^2 E_0}{dy^2} + \beta_0^2 \left(N^2 - \gamma^2\right) E_0 = 0 \tag{7.11}$$

7.1 The Two-Dimensional Approximation

We now assume that N has the simple form $N = N_0^2 + \alpha y$ and then Equation 7.11 will reduce to

$$\frac{d^2 E_0}{dy^2} + \beta_0^2 \left(y - \frac{\gamma^2 - N_0^2}{\alpha} \right) \alpha E_0 = 0 \qquad (7.12)$$

Equation 7.12 is essentially a Stokes equation and so its solution can be expressed in terms of Airy functions (Levy, 2000)

$$E_0(y) = E_a Ai\left((-\alpha\beta_0^2)^{\frac{1}{3}}\left(y - \frac{\gamma^2 - N_0^2}{\alpha}\right)\right) + E_b Bi\left((-\alpha\beta_0^2)^{\frac{1}{3}}\left(y - \frac{\gamma^2 - N_0^2}{\alpha}\right)\right) \qquad (7.13)$$

where we have implicitly assumed that α is negative. In the limit $|z| \to \infty$, $Ai(z) \approx \exp\left(-2z^{3/2}/3\right)/2\sqrt{\pi}z^{1/4}$ and $Bi(z) \approx \exp\left(2z^{3/2}/3\right)/\sqrt{\pi}z^{1/4}$. Consequently, we set $E_b = 0$ as this part of the solution is nonphysical. If E represents a vertically polarized electric field above a PMC boundary that consists of the plane $y = 0$, we require that $E_0(0) = 0$. Consequently, we will need to determine permissible γ from the equation $Ai\left(-(-\alpha\beta_0^2)^{1/3}(\gamma^2 - N_0^2)/\alpha\right) = 0$. The zeros $a_1, a_2, a_3, \ldots$ of Ai form a countable infinity that are all real and negative (Abramowitz and Stegun, 1970), so the permissible γ are given by $\gamma_i = \pm\sqrt{N_0^2 - a_i\alpha(-\alpha\beta_0^2)^{-1/3}}$. For each γ_i, Equation 7.10 will yield a *mode* and a general solution is obtained as a sum of these modes

$$E_0(y) = \sum_i E_a^i Ai\left((-\alpha\beta_0^2)^{\frac{1}{3}}\left(y - \frac{\gamma_i^2 - N_0^2}{\alpha}\right)\right) \exp\left(-j\beta_0\gamma_i x\right) \qquad (7.14)$$

for waves traveling in the positive x direction. It should be noted, however, that each γ_i will become imaginary for a sufficiently low frequency and so each mode will have a *cut-off frequency* below which the solution will become exponentially decaying (*evanescent*).

We have seen that conformal transformations can be extremely useful in simplifying the nature of the propagation medium. They can, however, also significantly change the boundary and boundary conditions. Indeed, we may have simplified the medium only to find that the boundary conditions are now difficult to implement. Of course, the converse can be true and it might be possible to simplify the boundaries and throw the complexity into the properties of the medium. An example of this is a wedge with exterior angle α. The transformation $z = Z^{\alpha/\pi}$ opens out the angle π in the Z-plane into angle α in the z-plane (Figure 7.1). In the Z coordinate system we can solve a problem where the boundary is simply the $\Im\{z\}$ axis, but the equation for E now has a more complex permittivity

$$\frac{\partial^2 E}{\partial Z \partial Z^*} + \left(\frac{\alpha\omega|Z|^{\left(\frac{\alpha}{\pi}-1\right)}}{\pi}\right)^2 \mu_0\epsilon E = 0 \qquad (7.15)$$

In reality, the above transformation approach can only provide analytic results for simple boundaries; more complex boundaries will need to be handled by a numerical approach. In numerical approaches, however, the boundary can often be the cause of

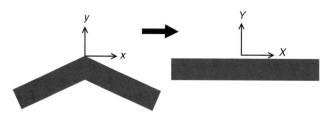

Figure 7.1 Conformal mapping that flattens angle.

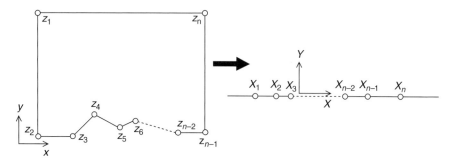

Figure 7.2 Schwarz–Christoffel conformal mapping.

great difficulty and it is still useful to simplify the boundary. In general, numerical techniques have far less difficulties with a complex medium than with a tortuous boundary. A general transform that can greatly simplify boundaries (Figure 7.2) is known as the *Schwarz–Christoffel transform* (Carrier et al., 1966). A statement of the Schwarz–Christoffel theorem is as follows:

Let P be the interior of a polygon in the z-plane with vertices $z_1, z_2, \ldots, z_n$ (numbered in the anticlockwise direction) with right turns through angles $\alpha_1, \alpha_2, \ldots, \alpha_n$, respectively. Consider a conformal mapping $z = z(Z)$ from the upper half Z plane to P, then

$$\frac{dz}{dZ} = K \prod_{i=1}^{n} (Z - X_i)^{\frac{\alpha_i}{\pi}} \tag{7.16}$$

for some K and $X_1, X_2, \ldots, X_n$ on the $\Im\{Z\} = 0$ axis satisfying $z_1 = z(X_1), z_2 = z(X_2), \ldots, z_n = z(X_n)$.

We can integrate Equation 7.15 to obtain

$$z(Z) = L + M \int_0^Z \prod_{i=1}^{n} (W - X_i)^{\frac{\alpha_i}{\pi}} dW \tag{7.17}$$

where L and M are arbitrary constants. In order to complete the transformation, however, we need to ascertain the values of $X_1, X_2, \ldots, X_n$, L and M. It might seem that we have n conditions through $z_1 = z(X_1), z_2 = z(X_2), \ldots, z_n = z(X_n)$, but it should be noted that, as we have the turning angles at the vertices, the last condition is redundant. Consequently, we only have $n - 1$ conditions for $n + 2$ unknowns. We could choose convenient values for X_1, X_2 and X_n or any other triplet of Z corresponding to vertices. It is customary to choose X_n to be the point at infinity ($z_n = z(\infty)$), and then we only need two additional conditions. With such a choice for z_n, the effect of this vertex can be absorbed into L and we can write

$$z(Z) = L + M \int_0^Z \prod_{i=1}^{n-1} (W - X_i)^{\frac{\alpha_i}{\pi}} dW \tag{7.18}$$

Although the determination is fairly straightforward for simple polygons, it can become quite involved for a large number of vertices (e.g. in the representation of ground topography as a polygon). The integral in Equation 7.17 will need to be performed numerically and the equations for the constraints solved numerically. Nevertheless, there are good numerical techniques for implementing these processes and some good computer software for the transform is available. Once we have converted the domain of solution into the upper half plane (the complex boundary now becoming the $\Im\{Z\} = 0$ axis), Equation 7.6 reduces to

$$\frac{\partial^2 E}{\partial Z \partial Z^*} + \left|\frac{dz}{dZ}\right|^2 \omega^2 \mu_0 \epsilon E = 0 \tag{7.19}$$

Noting that $\mu_0 \epsilon = \mu_0 \epsilon_0 N^2$ where N is the refractive index of the medium, we see the effect of the conformal transformation is to change the refractive index to a *modified* refractive index $\hat{N} = N|dz/dZ|$. Consequently, we have traded a complex boundary geometry for a possibly more complex medium. As far as boundary conditions are concerned, it will be noted that the two most important conditions, $E_t = 0$ for a PEC boundary and $E_n = 0$ for a PMC boundary, remain unaltered under a conformal mapping (note that the subscripts t and n refer to tangential and normal components, respectively).

A much simpler conformal transformation is given by $Z = a \ln(z/a)$, which, in terms of suitably defined polar coordinates (see Figure 7.3), yields $Z = a \ln(r/a) + ja\theta$. We will assume that we are studying propagation in a plane that goes through the center of

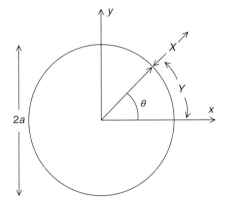

Figure 7.3 Earth flattening conformal mapping.

the Earth and that a is Earth's radius. For $|r-a| \ll a$ we find that $Z \approx (r-a)+ja\theta$ and so we can interpret Y as the range along the surface of the Earth and X as the altitude above the surface of the Earth. Since $z = a\exp(Z/a)$, we find that $dz/dZ = \exp(Z/a)$ and so the modified refractive index is given by $\hat{N} = N\exp(X/a)$. In the case that the refractive index only deviates slightly from 1 (this is the case for most tropospheric calculations), and the altitude is low, $\hat{N} \approx N + X/a$. The use of a modified refractive index is fairly common in calculations of terrestrial propagation and we will use it extensively in the rest of this chapter.

The 2D Cartesian Helmholtz equation is more than an academic curiosity as we can rework some quite realistic models into this form. Consider a problem with cylindrical symmetry. If we choose cylindrical polar coordinates (r, θ, z), then the electric field will be independent of the azimuthal coordinate θ and will satisfy

$$\frac{\partial^2 E}{\partial z^2} + \frac{1}{r}\frac{\partial}{\partial r}\left(r\frac{\partial E}{\partial r}\right) + \omega^2 \mu_0 \epsilon E = 0 \tag{7.20}$$

Cylindrical symmetry is the simplest idealization that includes the lateral spread in a wave field as it moves out from an isolated source. We can often use this as a first realistic approximation in modeling problems for which information is required at a particular azimuth. Since the permittivity needs to be independent of azimuth angle θ, we therefore need to take the behavior of boundary geometry and permittivity to be that in the direction of the required azimuth. Assuming a source at $r = 0$, we introduce the normalized electric field $\hat{E} = \sqrt{r}E$ and then Equation 7.20 will reduce to

$$\frac{\partial^2 \hat{E}}{\partial z^2} + \frac{\partial^2 \hat{E}}{\partial r^2} + \left(\omega^2 \mu_0 \epsilon + \frac{1}{4r^2}\right)\hat{E} = 0 \tag{7.21}$$

that is, we have produced a 2D Cartesian Helmholtz equation, but with slightly more complex permittivity. At large distances from the source (in comparison with a wavelength) the modified permittivity will reduce to that of our original 2D problem. The above 2D equations are elliptical in nature and, as such, are amenable to the finite element (FE) and FD methods. However, even after reduction to a 2D problem, the solution can still prove very demanding on computational resources due to the large distances over which propagation can take place and the need to capture detail at the wavelength level. Further, in any of the above numerical procedures, we will need to truncate the boundaries, and this raises the possibility of wave reflections at the artificial boundaries that we introduce. Of course, we could move the boundaries far enough way from the sources in order to sufficiently reduce their influence on the region of interest. This, however, would greatly increase the demand on computer resources. An alternative, as mentioned in Chapter 2, is to introduce absorbing boundaries (Senior and Volakis, 1995) that prevent any artificial reflections from returning to the region of interest. In this case, we could introduce some artificial conductivity into the medium near the boundary in order to damp out these reflections and then discard results in this region. FD methods have proven useful for propagation calculations in the form of the FDTD method, and this will be discussed at the end of the chapter.

7.2 The Paraxial Approximation

The elliptic nature of the time harmonic electromagnetic equations means that, even with a reduction in dimension, the computational requirement can still be excessive. Consequently, we will look at another approach that can greatly reduce this requirement. Consider our generic Helmholtz equation

$$\nabla^2 E + \omega^2 \mu_0 \epsilon_0 N^2 E = 0 \tag{7.22}$$

where N is the refractive index. Assume the propagation to be predominantly in one direction (restricted to narrow angles about this direction) and, without loss of generality, choose this direction to be that of the x axis. We define a new field E_a through

$$E(\mathbf{r}) = \exp(-j\beta_0 x) E_a(\mathbf{r}) \tag{7.23}$$

where E_a is slowly varying on the scale of a wavelength, and substitute this into Equation 7.22. The resulting equation is

$$-\beta_0^2 E_a - 2j\beta_0 \frac{\partial E_a}{\partial x} + \frac{\partial^2 E_a}{\partial x^2} + \nabla_T^2 E_a + \omega^2 \mu_0 \epsilon_0 N^2 E_a = 0 \tag{7.24}$$

where $\nabla_T^2 = \partial^2/\partial y^2 + \partial^2/\partial z^2$ is the Laplace operator in the transverse direction. In the paraxial approximation, we assume that the $\exp(-j\beta_0 x)$ factor embodies the dominant variations of E in the x direction and that E_a varies in this direction on a scale much greater than it varies in the transverse directions. Consequently, we ignore the $\partial^2 E_a/\partial x^2$ in comparison to all other terms and Equation 6.81 reduces to

$$-2j\beta_0 \frac{\partial E_a}{\partial x} + \nabla_T^2 E_a + \omega^2 \mu_0 \epsilon_0 \left(N^2 - 1\right) E_a = 0 \tag{7.25}$$

The character of this equation is parabolic and it is far less computationally expensive to solve than an elliptic equation. Another way of viewing the paraxial approximation is through the formal factorization of the operator in Equation 7.22, i.e.

$$\left(\frac{\partial}{\partial x} + j\beta_0 \sqrt{N^2 + \frac{1}{\beta_0^2} \nabla_T^2}\right) \left(\frac{\partial}{\partial x} - j\beta_0 \sqrt{N^2 + \frac{1}{\beta_0^2} \nabla_T^2}\right) E = 0 \tag{7.26}$$

The first operator of the factorization represents forward propagation and the second operator backward propagation. Assuming only forward propagation we have

$$\left(\frac{\partial}{\partial x} + j\beta_0 \sqrt{N^2 + \frac{1}{\beta_0^2} \nabla_T^2}\right) E = 0 \tag{7.27}$$

Under the assumptions of the paraxial approximation, operator $Q = N^2 - 1 + \beta_0^{-2} \nabla_T^2$ will be small in some sense and we can expand the square root operator as $\sqrt{N^2 + \beta_0^{-2} \nabla_T^2} \approx 1 + Q/2$ so that Equation 7.27 reduces to

$$\frac{\partial E}{\partial x} + j\beta_0 E + \frac{j\beta_0}{2} QE = 0 \tag{7.28}$$

Then, substituting Equation 7.23 into Equation 7.28, we obtain the standard parabolic equation 7.25. We have used the simplest approximation to the operator in Equation 7.27, but there are other approximations that can extend the range of applicability of the parabolic equation (Levy, 2000). In particular, $\sqrt{1+Q} \approx (4+3Q)/(4+Q)$ is a rational approximation (Collis, 2011; Lin et al., 2012) that is valid for propagation over much wider angles. This approximation 7.27 reduces to

$$\left(1+\frac{Q}{4}\right)\frac{\partial E}{\partial x} + j\beta_0\left(1+\frac{3Q}{4}\right)E = 0 \qquad (7.29)$$

Then, substituting Equation 7.23 into 7.29, we obtain the *generalized parabolic equation*

$$\left(1+\frac{Q}{4}\right)\frac{\partial E_a}{\partial x} + \frac{j\beta_0}{2}QE_a = 0 \qquad (7.30)$$

Although more difficult to discretize, this provides effective modeling of propagation over much wider angles than the basic parabolic equation.

In two dimensions, the standard parabolic equation takes the form

$$-2j\beta_0\frac{\partial E_a}{\partial x} + \frac{\partial^2 E_a}{\partial y^2} + \omega^2\mu_0\epsilon_0(N^2-1)E_a = 0 \qquad (7.31)$$

We will first look at discretizing Equation 7.31 through a FD approach and for this purpose consider a rectangular grid of sample points (see Figure 7.4). The $\partial^2 E_a/\partial y^2$ derivative can be discretized using the approximation

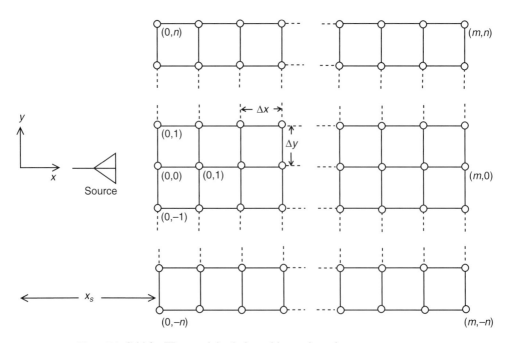

Figure 7.4 Grid for FD paraxial solution without a boundary.

7.2 The Paraxial Approximation

$$\frac{\partial^2 E_a}{\partial y^2} \approx \frac{E_a^{i(k-1)} + E_a^{i(k+1)} - 2E_a^{ik}}{\Delta y^2} \tag{7.32}$$

where $E_a^{ik} = E_a(x_s + i\Delta x, k\Delta y)$. Further, a simple discretization of the x derivative is given by $\partial E_a/\partial x \approx (E_a^{(i+1)k} - E_a^{ik})/\Delta x$. Equation 7.31 can now be written in the discretized form

$$E_a^{(i+1)k} = E_a^{ik} - \frac{j\Delta x}{2\beta_0} \frac{E_a^{i(k-1)} + E_a^{i(k+1)} - 2E_a^{ik}}{\Delta y^2} - \frac{j\Delta x}{2\beta_0} \omega^2 \mu_0 \epsilon_0 (N_{ik}^2 - 1) E_a^{ik} \tag{7.33}$$

where $N_{ik} = N(x_s + i\Delta x, k\Delta y)$. Given that we know E_a on the initial surface (distance x_s from the source), we can then use Equation 7.33 to progress the solution through the grid. The only difficulty is the question of what value we assign at the top and bottom of the grid. If the radiation is traveling outward unimpeded, we would need to apply a radiation condition. This is problematic at a finite boundary and our alternative is to introduce some artificial conductivity near the boundary in order to absorb any reflected waves (the conductivity should blend smoothly into the real propagation medium in order to minimize artificial reflections). The boundary condition would then be immaterial (the field could simply be set to zero here). An alternative, however, is to consider the source to be an antenna that produces a narrow beam (a Gaussian beam pattern is often used) and fix the boundaries such that the field does not reach them before the end of the propagation region of interest. The field on the initial boundary at $x = x_s$ can be obtained from the field of the antenna that acts as the field source.

The FD approach allows us to move in steps of Δx with very little requirement on computer memory. It is, however, by no means the most efficient method, since, in order for the method to be stable, it requires that the step Δx satisfies $\Delta x/\beta_0 \Delta y^2 \leq 1$. This constraint can make the step size Δx quite small and require a large amount of computer resource to reach the end of the propagation region. Our problem with stability arises from that fact that the above scheme is explicit. If we evaluate the term $\partial^2 E_a/\partial y^2$ at $x_s + (i+1)\Delta x$ instead of $x_s + i\Delta x$, we obtain an implicit scheme (i.e. a linear set of equations needs to be solved at each step)

$$E_a^{(i+1)k} + \frac{j\Delta x}{2\beta_0} \frac{E_a^{(i+1)(k-1)} + E_a^{(i+1)(k+1)} - 2E_a^{(i+1)k}}{\Delta y^2} = E_a^{ik} - \frac{j\Delta x}{2\beta_0} \omega^2 \mu_0 \epsilon_0 (N_{ik}^2 - 1) E_a^{ik} \tag{7.34}$$

which is unconditionally stable. We can now make the steps in Δx as large as we like but, due to the low order of approximation in the $\partial E_a/\partial x$ derivative, the solution will be very inaccurate if the steps are too large. We can, however, increase the order of accuracy by taking the approximation to $\partial^2 E_a/\partial y^2$ and E_a at the midpoint between $x_s + (i+1)\Delta x$ and $x_s + i\Delta x$. This can be done by taking the average of the approximations at points $x_s + (i+1)\Delta x$ and $x_s + i\Delta x$. We now have an implicit scheme

$$E_a^{(i+1)k} + \frac{j\Delta x}{4\beta_0} \frac{E_a^{(i+1)(k-1)} + E_a^{(i+1)(k+1)} - 2E_a^{(i+1)k}}{\Delta y^2} + \frac{j\Delta x}{4\beta_0} \omega^2 \mu_0 \epsilon_0 \left(N_{(i+1)k}^2 - 1\right) E_a^{(i+1)k}$$

$$= E_a^{ik} - \frac{j\Delta x}{4\beta_0} \frac{E_a^{i(k-1)} + E_a^{i(k+1)} - 2E_a^{ik}}{\Delta y^2} - \frac{j\Delta x}{4\beta_0} \omega^2 \mu_0 \epsilon_0 \left(N_{ik}^2 - 1\right) E_a^{ik} \tag{7.35}$$

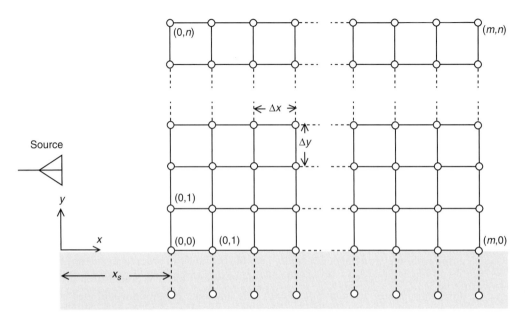

Figure 7.5 Grid for FD paraxial solution with a boundary.

which is unconditionally stable and has a high degree of accuracy, even for quite large steps Δx. This numerical procedure is known as the Crank–Nicolson scheme (Press et al., 1992) and was applied to the propagation problem by Lee and McDaniel (1987). (It should be noted that a scheme of the Crank–Nicolson variety would be required in the solution of the extended parabolic equation 7.29.)

Up to now we have considered the solution of the paraxial equations for propagation through an infinite region, only truncated by artificial boundaries in order to facilitate a numerical solution. More often than not, however, it is necessary to consider a real boundary, especially for propagation close to the ground. Figure 7.5 shows the discretization for such a problem. In the implementation of Crank–Nicolson scheme with only artificial boundaries, Eqaution 7.35 is used to develop the solution through the mesh, except for the top and bottom rows where the value is provided by the boundary value (probably taken to be zero). However, in the case of propagation over ground, the bottom row is now the surface of the ground. At this surface either E_a or $\partial E_a/\partial y$ will be zero (we assume the surface to be PEC or PMC). The implementation of condition $E_a = 0$ is straightforward, but, if $\partial E_a/\partial y = 0$, we have a major problem with the discretization of the boundary condition. With the points available, the discretization will need to be of the form $E_a^{i1} - E_a^{i0} = 0$. This is a low accuracy implementation of the boundary condition and will hence compromise the high accuracy of the Crank–Nicolson scheme. The solution is to introduce some phantom points below the bottom of the mesh and introduce the more accurate central difference approximation to the boundary condition, i.e. $E_a^{i1} - E_a^{i(-1)} = 0$. Consequently, we now apply Equation 7.35 on the bottom row of the mesh with the value at the phantom point replaced by E_a^{i1}. That is, on the bottom row we use

$$E_a^{(i+1)0} + \frac{j\Delta x}{2\beta_0} \frac{E_a^{(i+1)1} - E_a^{(i+1)0}}{\Delta y^2} + \frac{j\Delta x}{4\beta_0}\omega^2\mu_0\epsilon_0(N_{(i+1)0}^2 - 1)E_a^{(i+1)0}$$
$$= E_a^{i0} - \frac{j\Delta x}{2\beta_0}\frac{E_a^{i1} - E_a^{i0}}{\Delta y^2} - \frac{j\Delta x}{4\beta_0}\omega^2\mu_0\epsilon_0(N_{i0}^2 - 1)E_a^{i0} \qquad (7.36)$$

From Equations 7.35 and 7.36, it will be noted that the Crank–Nicolson scheme does not preclude us from changing the Δx step size at each stage. This option can be very useful over regions where the boundary topography, boundary conditions and refractive index change only very slowly with range. In this case, we can greatly increase the step length in order to reduce computational requirement.

7.3 Kirchhoff Integral Approach

Another approach to progressing a solution through the solution region is to use some of the integral results developed in Chapter 3. The following development is based on Coleman (2005, 2010). Let $(\mathbf{H}, \mathbf{E})$ be a field that is generated by sources outside a closed surface S, then, from Equation 3.15,

$$\hat{\mathbf{s}} \cdot \mathbf{E} = \int_S (\mathbf{E}_0 \times \mathbf{H} - \mathbf{E} \times \mathbf{H}_0) \cdot \mathbf{n}\, dS \qquad (7.37)$$

where $(\mathbf{H}_0, \mathbf{E}_0)$ is the field generated by a unit electric dipole at point A (current density of the form $\mathbf{J}_0 = \delta(\mathbf{r} - \mathbf{r}_0)\hat{\mathbf{s}}$ where $\hat{\mathbf{s}}$ is an arbitrary unit vector). This expression allows us to calculate $\mathbf{E}$ at points away from the surface when $(\mathbf{H}, \mathbf{E})$ is known on S. If we assume the paraxial limit, that is, we take the propagation to be predominantly in one direction, then $\mathbf{H}_0 \approx \hat{\mathbf{z}} \times \mathbf{E}_0/\eta$ and $\mathbf{H} \approx -\hat{\mathbf{z}} \times \mathbf{E}/\eta$ when the propagation direction is the z direction. If the surface S is the plane orthogonal to the z direction (note that contributions from any closing surface at infinity will be zero for bounded sources), Equation 7.37 will reduce to

$$\hat{\mathbf{s}} \cdot \mathbf{E} = -\int_S \frac{2}{\eta} \mathbf{E}_0 \cdot \mathbf{E}\, dS \qquad (7.38)$$

(Coleman, 2005, 2010). If we have the field $\mathbf{E}$ on a surface S, the above integral equation will allow us to develop the field forward of the surface in the manner of a *Kirchhoff integral*. In general, however, the medium will be inhomogeneous and the requisite dipole field $\mathbf{E}_0$ will be difficult to derive. If we assume that any inhomogeneity is a perturbation $\delta\epsilon = (N^2 - 1)\epsilon_0$ of a homogeneous background permittivity ϵ_0, we could use the Rytov approximation to find an approximate form of $\mathbf{E}_0$. For a homogeneous medium, $\mathbf{E}_0$ is given by

$$\mathbf{E}_0 = -\frac{j\omega\mu_0}{4\pi}\hat{\mathbf{s}}\frac{\exp(-j\beta_0|\mathbf{r}_0 - \mathbf{r}|)}{|\mathbf{r}_0 - \mathbf{r}|} \qquad (7.39)$$

where $\mathbf{r}_0$ is the position of the point where the field is evaluated and $\mathbf{r}$ is the position of the dipole. We take the dominant propagation direction to be the z direction and then, in the paraxial limit, we can further approximate this to obtain

$$\mathbf{E}_0 \approx -\frac{j\omega\mu_0}{4\pi}\hat{\mathbf{s}}\frac{\exp\left(-j\beta_0\left(|z_0-z|+\frac{(x_0-x)^2+(y_0-y)^2}{2|z_0-z|}\right)\right)}{|z_0-z|} \qquad (7.40)$$

which we use as the zeroth-order solution in the Rytov approximation. Assuming that the scale of lateral variation in $\delta\epsilon$ is much larger than $\sqrt{\lambda|z_0-z|}$, the next Rytov approximation Equation 6.39 yields

$$\mathbf{E}_0 \approx -\frac{j\omega\mu_0}{4\pi}\hat{\mathbf{s}}\frac{\exp\left(-j\beta_0\left(|z_0-z|+\frac{(x_0-x)^2+(y_0-y)^2}{2|z_0-z|}\right)\right)}{|z_0-z|}$$
$$\times \exp\left(-\frac{j\beta_0}{2\epsilon_0}\int_z^{z_0}\delta\epsilon(\gamma(x_0-x)+x,\gamma(y_0-y)+y,\zeta)d\zeta\right) \qquad (7.41)$$

where $\gamma = (\zeta-z)/(z_0-z)$. We can now write our integral equation as

$$E(\mathbf{r}_0) = \int_S K(\mathbf{r}_0,\mathbf{r})E(\mathbf{r})dS \qquad (7.42)$$

where

$$K(\mathbf{r}_0,\mathbf{r}) = \frac{j\beta_0}{2\pi}\frac{\exp\left(-j\beta_0\left(z_0-z+\frac{(x_0-x)^2+(y_0-y)^2}{2|z_0-z|}\right)\right)}{|z_0-z|}\exp\phi \qquad (7.43)$$

with

$$\phi = -\frac{j\beta_0}{2\epsilon_0}\int_z^{z_0}\delta\epsilon(\gamma(x_0-x)+x,\gamma(y_0-y)+y,\zeta)d\zeta \qquad (7.44)$$

This integral equation applies to all components of $\mathbf{E}$ and allows us to develop the electric field through a series of intermediate parallel surfaces in the propagation direction (see Figure 7.6). For intermediate surfaces that are close, we can approximate ϕ by

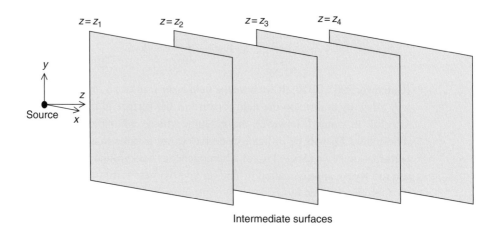

Figure 7.6 Intermediate surfaces for developing a field away from the source.

7.3 Kirchhoff Integral Approach

$\phi \approx -(j\beta_0/2\epsilon_0)\delta\epsilon(x,y,z)\Delta z$ or, more accurately, by $\phi \approx \phi_-(x,y,z) + \phi_+(x_0,y_0,z_0) = -(j\beta_0/4\epsilon_0)\delta\epsilon(x,y,z)\Delta z - (j\beta_0/4\epsilon_0)\delta\epsilon(x_0,y_0,z_0)\Delta z$ where $\Delta = z_0 - z$ is the distance between surfaces. With these approximations, it will be noted that the solution process can be split into several steps (see Levy (2000), Dockerty and Kuttler (1996) and Kuttler and Dockerty (1991) for some other approaches to split step algorithms). For $\phi \approx -(j\beta_0/2\epsilon_0)\delta\epsilon(x,y,z)\Delta z$, we calculate $E(\mathbf{r}_0)$ according to

$$E(\mathbf{r}_0) = \int_S K_0(\mathbf{r}_0, \mathbf{r}) \exp \phi E(\mathbf{r}), dS \qquad (7.45)$$

where

$$K_0(\mathbf{r}_0, \mathbf{r}) = \frac{j\beta_0}{2\pi} \frac{\exp\left(-j\beta_0\left(\Delta z + \frac{(x_0-x)^2 + (y_0-y)^2}{2\Delta z}\right)\right)}{\Delta z} \qquad (7.46)$$

i.e. we first apply a phase shift ϕ to E and then progress the result using the free space kernel K_0. The split step approach can be made more accurate using the improved approximation to ϕ. In this case

$$E(\mathbf{r}_0) = \exp\phi_+ \int_S K_0(\mathbf{r}_0, \mathbf{r}) \exp\phi_- E(\mathbf{r}) dS \qquad (7.47)$$

from which it will be noted that there are phase corrections both before and after the integral over the free space kernel. Further, the free space kernel is of the displacement variety and so the integral is amenable to integral transform techniques.

If ground is present (see Figure 7.7), we can remove the ground by treating the field below as the reflection of the field above the ground, multiplied by a suitable reflection coefficient. In mathematical terms, $E(x, -y, z) = RE(x, y, z)$ where R is the reflection coefficient (the relationship to surface impedance can be found in Chapter 2). For a PEC material $R = 1$ and for a PMC material $R = -1$ (note that the PMC material is a good approximation to the ground at frequencies above a few hundred MHz). In general, we will need the angle of incidence θ^i (the *grazing angle*) to calculate the reflection coefficient (see Chapter 2). In the paraxial approximation, this can be estimated by $\theta^i \approx (\partial E/\partial y)/j\beta_0 E$ where the calculation is performed at a point just above the ground.

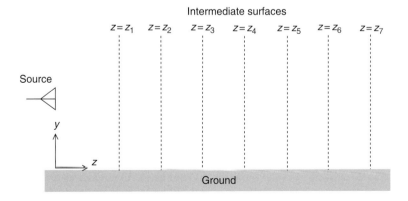

Figure 7.7 Intermediate surfaces for developing a field in the case of a ground.

Consequently, to progress the field from one intermediate surface to the next, we take the field above the ground for the initial surface and then calculate the reflected field below the ground by the above process. The field above the ground on the next surface is then calculated using Equation 7.45 or 7.47. After this, the process is repeated until the solution has reached the desired range. In a practical implementation, the intermediate surfaces will need to be truncated, and, to prevent reflections from this artificial boundary, the solutions will require some artificial attenuation close to the boundary (this can be achieved by suitably windowing the solution at each stage). In addition, we will need to discretize the field on each surface by a grid of representative values and then evaluate the integrals in the above integral equations by means of a suitable numerical quadrature. The displacement nature of the free space kernel introduces the possibility of using transform techniques and, in particular, FFT techniques (Press et al., 1992).

In many situations the lateral variations in topography, boundary conditions and the propagation medium are negligible in comparison with their variation in altitude and range (this is especially so when distant from the source). Under such circumstances, we can assume that the field is approximately constant laterally and perform the x integral in the integral equations (7.45 or 7.47). From Equation 7.47, we now find that the equation for progressing the solution from the intermediate surface at $z = z_i$ to that at $z = z_{i+1}$ is

$$E^{i+1}(y_0) = \exp \phi_+ \int_{-\infty}^{\infty} \hat{K}_0(y_0, y, z_0, z) \exp \phi_- E^i(y) \, dy \tag{7.48}$$

where

$$\hat{K}_0(y_0, y, z_0, z) = \sqrt{\frac{j\beta_0}{2\pi}} \frac{\exp\left(-j\beta_0 \left(z_{i+1} - z_i + \frac{(y_0-y)^2}{2|z_{i+1}-z_i|}\right)\right)}{\sqrt{|z_{i+1} - z_i|}} \tag{7.49}$$

and $E^i(y) = E(y, z_i)$. In order to discretize this equation, we first truncate the simulation region at height H_{max}. After the addition of the reflected field, the y domain for simulation will then consist of the interval from $-H_{max}$ to H_{max}. We divide this domain into N_y shorter intervals of length $\Delta y = 2H_{max}/N_y$ and take sample values of the field, and kernel, at the midpoints of these intervals. The integral equation can now be discretized using the midpoint rule, the integral being simply a sum over the integrand at the interval midpoints, multiplied by Δy. Unfortunately, this is a very computationally expensive procedure. However, since the kernel $\hat{K}_0$ is of the displacement invariant variety, we can apply FFT techniques and greatly accelerate the algorithm. (This requires us to make N_y a power of 2, a small sacrifice considering the massive increase in computational efficiency.) In terms of FFTs, Equation 7.48 becomes

$$E^{i+1} = \exp(-j\beta_0\phi_+) \, \text{FFT}^{-1} \left\{ \text{FFT} \left\{ \hat{K}_0 \right\} \text{FFT} \left\{ \exp(-j\beta_0\phi_-) E^i \right\} \right\} A \tag{7.50}$$

where $A = \Delta y$. It should be noted that the FFT is circular in nature and this means that the field values near to H_{max} and $-H_{max}$ have the potential to contaminate each other. As a consequence, it is necessary to window the samples at these extremes, something we need to do anyway in order to prevent contamination by reflections from these artificial boundaries. A suitable window is given by $W(y) = (1/2 + \cos(\pi y/H_{max})/2)^\alpha$ where α has a small positive value ($\alpha = 1$ yields the Hanning window).

7.3 Kirchhoff Integral Approach

The FFT of the kernel $\hat{K}_0$ can be calculated from the Fourier transform (FT) of this kernel, i.e. FT$\{\hat{K}_0\} = \exp(-j\beta_0 \Delta z) \exp(j\alpha^2 \Delta z / 2\beta_0)$ where α is the transform variable and $\Delta z = z_{i+1} - z_i$. We can estimate the required FFT from the FT by sampling at points $\alpha_i = (2i - N_y - 1)\pi/2H_{max}$ where i runs from 1 to N_y. However, the truncation of the transform domain requires that there be an appropriate windowing of the kernel in the spatial domain. An appropriate window is that of Hanning, and this can be implemented in the transform domain by applying the moving average $\{f_{i-1}/4 + f_i/2 + f_{i+1}/4\}$ to the transform samples $\{f_i\}$. This new sequence constitutes the desired FFT$\{\hat{K}_0\}$ to be used in Equation 7.50.

Thus far we have assumed a parabolic approximation to distance in the kernel $\hat{K}_0$, and this is equivalent to the paraxial approximation to the Helmholtz equation (i.e. it is only effective for propagation that deviates by narrow angles). For wider angles, we need to use a more accurate approximation, at least for the phase part of the kernel. An improved kernel for wider angles is given by

$$\text{FT}\left\{\hat{K}_0(y,z)\right\} \approx \frac{k}{\beta_0} \exp(-jk\Delta z) \tag{7.51}$$

where $k = \sqrt{\beta_0^2 - \alpha^2}$. (It will be noted that the narrow angle kernel is obtained in the small α limit of the above kernel.) The FFT for the above wide angle kernel can now be obtained in the same fashion as for the narrow angle kernel.

Important choices in the implementation of the above algorithms are the range interval Δz, the maximum height and the number of height quadrature points. Essentially, between the intermediate surfaces, we use a GO approximation and this needs to remain valid between the surfaces. For this to be the case, we will need $\Delta z \ll D^2/\lambda$ where D is a typical scale on which the environment varies (the height of variations in topography or the scale of variations in permittivity) and this limit will usually be satisfied if $\Delta z = D^2/5\lambda$. It is usually sufficient to choose H_{max} to be the maximum height of the region of interest, plus a few Fresnel scales D_{Fres} $(=\sqrt{\lambda z_{max}}$ where z_{max} is the maximum propagation range) to allow room for windowing. To ensure the discretization can capture expected variations in the electric field, N_y is chosen to be a power of 2 such that $H_{max}/N_y < \lambda \Delta z / 10D$. A further consideration in any implementation of the above algorithms is the field that is used on the initial intermediate surface. Providing this surface is close enough to the source, this can be the geometric optics field of the source antenna. In many situations, the field of this antenna for a homogeneous medium is adequate.

As mentioned earlier, the effect of the curvature of Earth's curvature can be modeled by employing the modified refractive index $\hat{N} = N + y/a$ ($a = 6371$ for y in kilometers) where y is the altitude above Earth's surface. Figure 7.8 shows the result of some simulations using the integral approach and the modified refractive index $\hat{N} = 1 + 0.000125z$ (this represents a *standard atmosphere*). The figure shows propagation losses (power received at a point divided by the power transmitted) for a source that is 30 meters above the ground and operating on a frequency of 400 MHz. It will be noted that the propagation peels away from Earth's surface (altitude 0) as a result of Earth's curvature. Figure 7.9 shows some propagation with modified refractive index

144 **The Approximate Solution of Maxwell's Equations**

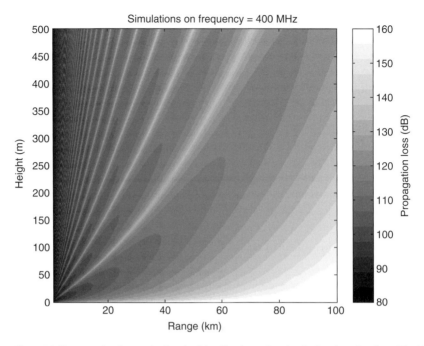

Figure 7.8 Propagation loss calculated with effective refractive index that simulates Earth's curvature.

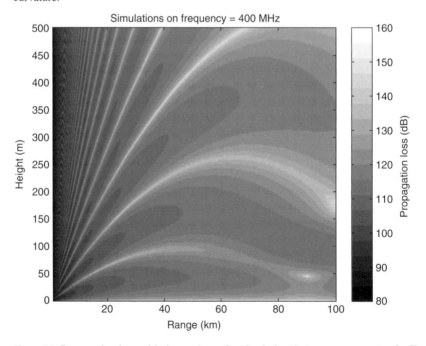

Figure 7.9 Propagation loss with decreasing refractive index that overcompensates for Earth's curvature.

$\hat{N} = 1 - 0.00013z$. The real refractive index now overcompensates for Earth's curvature and it will be noted that the energy is refracted back to the ground.

7.4 Irregular Terrain

Consider the problem of calculating propagation over irregular terrain. We could, of course, use a Schwarz–Christoffel transform to convert the boundary, but the paraxial approximation

$$-2j\beta_0 \frac{\partial E_a}{\partial z} + \frac{\partial^2 E_a}{\partial y^2} + \beta_0^2 \left(N^2 - 1\right) E_a = 0 \tag{7.52}$$

introduces the possibility of a much simpler transform (Barrios, 1992). Suppose we have calculated the solution on some intermediate surface S_A and we want to calculate the solution on a nearby intermediate surface S_B (distance Δz away) where the height of the ground has changed by Δh. We assume the height changes linearly between surfaces with slope α $(= \Delta h/\Delta z)$. Between the surfaces we replace the y coordinate by the new coordinate $Y = y - \alpha z$, which has the effect of flattening the ground. In addition, we introduce the new field $\hat{E}_a$ through $E_a(y, z) = \exp(-j\beta_0 \alpha Y)\hat{E}_a(Y, z)$ and then

$$\frac{\partial E_a}{\partial y} = -j\beta_0\alpha \exp(-j\beta_0\alpha Y)\hat{E}_a + \exp(-j\beta_0\alpha Y)\frac{\partial \hat{E}_a}{\partial Y} \tag{7.53}$$

from which

$$\frac{\partial^2 E_a}{\partial y^2} = -\beta_0^2\alpha^2 \exp(-j\beta_0\alpha Y)\hat{E}_a - 2j\beta_0\alpha \exp(-j\beta_0\alpha Y)\frac{\partial \hat{E}_a}{\partial Y} + \exp(-j\beta_0\alpha Y)\frac{\partial^2 \hat{E}_a}{\partial Y^2} \tag{7.54}$$

For the z partial derivative, we have

$$\frac{\partial E_a}{\partial z} = \exp(-j\beta_0\alpha Y)\frac{\partial \hat{E}_a}{\partial z} - \alpha \exp(-j\beta_0\alpha Y)\frac{\partial \hat{E}_a}{\partial Y} + j\beta_0\alpha^2 \exp(-j\beta_0\alpha Y)\hat{E}_a \tag{7.55}$$

If we substitute these expressions into the parabolic equation for E_a, we obtain

$$-2j\beta_0 \frac{\partial \hat{E}_a}{\partial z} + \frac{\partial^2 \hat{E}_a}{\partial Y^2} + \beta_0^2 \left(N^2 + \alpha^2 - 1\right) \hat{E}_a = 0 \tag{7.56}$$

that is, we have the same parabolic equation as for E_a, except for a modified refractive index. Before we proceed, we also need to consider what happens to the boundary conditions under the above transformation. We consider the general impedance boundary condition

$$n_z \frac{\partial E}{\partial z} + n_y \frac{\partial E}{\partial y} - j\beta_0 ZE = 0 \tag{7.57}$$

which includes the cases of boundaries at PEC and PMC materials (Senior and Volakis, 1995). Assuming the paraxial approximation (i.e. $\partial E_a/\partial z \approx -j\beta_0 E_a$), and ground with slope α, this condition yields

$$j\beta\alpha E_a + \frac{\partial E_a}{\partial y} - j\beta_0 ZE_a(1 + \alpha^2) = 0 \tag{7.58}$$

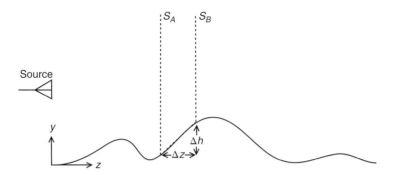

Figure 7.10 Intermediate surfaces for developing field in the case of irregular terrain ground.

and, on applying the above transformations, this yields

$$\frac{\partial \hat{E}_a}{\partial Y} - j\beta_0 Z \hat{E}_a (1 + \alpha^2) = 0 \qquad (7.59)$$

which is the boundary condition for flat terrain, but with the surface impedance Z modified by factor $1 + \alpha^2$. (It should be noted that α must be small for the paraxial approximation to remain valid and so we can usually ignore the α^2 terms.)

It is clear that we can use the above ideas to progress a solution between intermediate surfaces S_A and S_B with differing ground heights. Given the field E_a on intermediate surface S_A, we first multiply by factor $\exp(j\beta_0 \alpha y)$ to form the field $\hat{E}_a$ on S_A. We then progress field $\hat{E}_a$ across the now flat ground to surface S_B using Equation 7.56 and the boundary condition 7.59. On surface S_B we obtain E_a from the product of $\hat{E}_a$ and the phase factor $\exp(-j\beta_0 \alpha Y)$. We then transform back to the original y coordinates using $y = Y + \alpha \Delta z$ (Δz is the distance between intermediate surfaces). In this way, we can progress the field from intermediate surface to intermediate surface across irregular terrain (Figure 7.10).

The above procedure can also be used with the Kirchhoff integral approach. In this case, we form a field $\hat{E}$ on S_A through the product of E and the phase factor $\exp(j\beta_0 \alpha y)$. We then use the integral equation approach to progress this field to surface S_B. At surface S_B we obtain E by multiplying the new $\hat{E}$ by phase factor $\exp(-j\beta_0 \alpha Y)$ and then transform back to the original y coordinates.

In the case of the split step integral approach, however, there is an alternative procedure available (Coleman, 2010). In propagating the solution from one intermediate surface to the next using the integral approach, we effectively use the kernel

$$K_0(\mathbf{r}_0, \mathbf{r}) = \frac{j\beta_0}{2\pi} \left(\frac{\exp(-j\beta_0 |\mathbf{r}_0 - \mathbf{r}|)}{|\mathbf{r}_0 - \mathbf{r}|} + R \frac{\exp(-j\beta |\mathbf{r}_0 + \mathbf{r}|)}{|\mathbf{r}_0 + \mathbf{r}|} \right) \qquad (7.60)$$

with the Kirchhoff integral performed over that part of the intermediate surface above the ground. The reflection coefficient needs to be chosen such that the kernel $\hat{K}$ satisfies the impedance boundary condition $\partial K_0 / \partial n = j\beta_0 Z K_0$, where n is a coordinate normal to the ground and Z is the relative surface impedance (Senior and Volakis, 1995)

7.4 Irregular Terrain

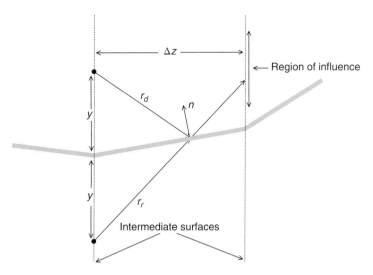

Figure 7.11 Geometry for calculating effective reflection coefficient.

(Figure 7.11). ($Z = \eta_r$ for vertical polarization and $Z = 1/\eta_r$ for horizontal polarization with η_r the relative impedance of the ground.) To the leading order in β (the GO limit), this is satisfied when

$$R = -\frac{\hat{\mathbf{r}}_d \cdot \mathbf{n} + Z\,|\mathbf{r}_r|}{\hat{\mathbf{r}}_r \cdot \mathbf{n} + Z\,|\mathbf{r}_d|} \exp\left(j\beta_0(|\mathbf{r}_r| - |\mathbf{r}_d|)\right) \tag{7.61}$$

where $\mathbf{n}$ is unit normal to the surface, $\mathbf{r}_d = \mathbf{r}_0 - \mathbf{r}$ and $\mathbf{r}_d = \mathbf{r}_0 + \mathbf{r}$ (a hat denotes a unit vector in the same direction as the vector under the hat). For ground with a small constant slope, $(|\mathbf{r}_r|/|\mathbf{r}_d|)\exp(j\beta_0(|\mathbf{r}_r|-|\mathbf{r}_d|)) \approx \exp(2j\beta_0\alpha y)$ and so, for a PMC surface ($Z \to \infty$), $R \approx -\exp(2j\beta_0\alpha y)$ (for a PEC material $R \approx \exp(2j\beta_0\alpha y)$). It is now possible to apply the methods of the previous section with the above reflection coefficient. At an intermediate surface the artificial field below the ground can be generated through $E(x, -y, z) = RE(x, y, z)$ with the above reflection coefficient (using the slope of the ground between surfaces). The split step integral approach is used to calculate the field above the ground on the next intermediate surface. On this surface, we then repeat the procedure and so on. Figure 7.12 shows results of simulations using this technique for propagation over a Gaussian hill. The figure shows propagation losses (power received at a point divided by the power transmitted) for a source that is 30 m above the ground and operating on a frequency of 400 MHz. It will be noted that there is significant diffraction into the rear of the hill.

The reflection coefficient approach can be applied when the ground is nonperfectly conducting, but in this case the impedance boundary condition on $\hat{K}$ can only be satisfied at a representative point. In general, this point will need to represent the points at which the reflected field has most influence on the next intermediate surface. Typically, for point at height y, the points on the next surface with greatest influence will be plus or minus a few Fresnel scales ΔF ($\Delta F = \sqrt{\lambda \Delta z}$ where Δz is the distance between surfaces) either side of the height y. (It will be noted that this region will be truncated

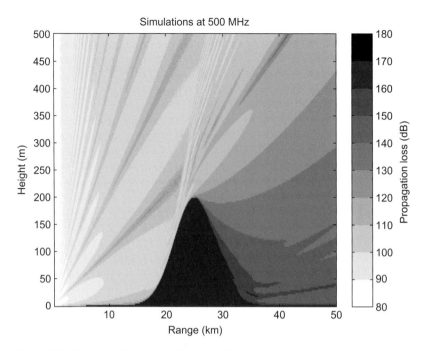

Figure 7.12 2D propagation over a Gaussian hill.

on the lower side for y that are close to the ground.) Consequently, R is best chosen to be its average over the reflection points for a few Fresnel scales about height y.

It is possible to apply the terrain flattening approach of Barrios to the more general terrain $y = h(z)$ by introducing the new coordinate $Y = y - h(z)$ and the new field $E_a(y,z) = \exp\left(-j\beta_0 h'(z) Y\right) \hat{E}_a(Y,z)$ (Barrios, 1992). The parabolic equation now reduces to

$$-2j\beta_0 \frac{\partial \hat{E}_a}{\partial z} + \frac{\partial^2 \hat{E}_a}{\partial Y^2} + \beta_0^2(N^2 + h'^2 - 2h''Y - 1)\hat{E}_a = 0 \qquad (7.62)$$

which has a far more complex refractive index. The above transformation can be used to derive the effective refractive index that models the curvature of the Earth. In this case, $h(z) = -z^2/2a$, where a is the radius of the Earth. Assuming $z \ll a$, we can ignore the h'^2 term in the effective refractive index and then find, as expected, that $\hat{N} \approx N + Y/a$ (we have implicitly assumed that N does not differ greatly from 1). The coordinate Y is now the altitude above the ground and z is the distance along the ground. (For some other approaches to propagation over irregular terrain, the reader should consult Donohue and Kuttler (2000) and Eibert (2002).)

7.5 3D Kirchhoff Integral Approach

Although we can use plane surfaces to develop the 3D electromagnetic field away from an isolated source, this can only be done over a limited range of azimuth and poses

7.5 3D Kirchhoff Integral Approach

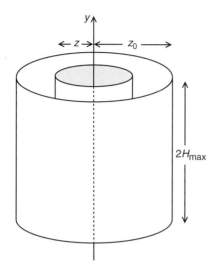

Figure 7.13 Cylindrical surfaces for developing the 3D solution.

problems for the field at the edge of this range. A more natural proposition is to develop the field through a set of intermediate concentric cylindrical surfaces that surround the source (Coleman, 2010). We will use cylindrical polar coordinates in which z is the radial coordinate, ϕ the azimuthal coordinate and y as the altitude coordinate (see Figure 7.13). Equation 7.47 is still applicable, i.e.

$$E(z_0, \theta_0, y_0) = \exp\phi_+ \int_{-\infty}^{\infty} \int_{-\pi}^{\pi} K_0(z_0, z, \theta_0, \theta, y_0, y) \exp\phi_- E(z, \theta, y) z d\theta dy \quad (7.63)$$

but with S an intermediate cylindrical surface. The problem is now one of finding an appropriate kernel for this integral equation. In the paraxial limit, this can be derived by translating Equation 7.46 into cylindrical coordinates

$$K_0(z_0, z, \theta_0, \theta, y_0, y) \approx \frac{j\beta_0}{2\pi} \frac{\exp\left(-j\beta_0\left(\Delta z + \frac{zz_0(\theta_0-\theta)^2 + (y_0-y)^2}{2\Delta z}\right)\right)}{\Delta z} \quad (7.64)$$

where $\Delta z = z_0 - z$. In this approximation, the kernel is of the displacement variety in y and θ coordinates and so Equation 7.63 is amenable to transform techniques in both of these dimensions. The 2D FT of the kernel is given by

$$\text{FT2D}\{K_0\} = \frac{1}{\sqrt{z_0 z}} \exp(-j\beta_0 \Delta z) \exp\left(-j\frac{\alpha^2 \Delta z}{2\beta_0}\right) \exp\left(-j\frac{\kappa^2 \Delta z}{2z_0 z \beta_0}\right) \quad (7.65)$$

where α and κ are the transform variables with respect to the y and θ coordinates, respectively. It will be noted that the 2D FT is the product of separate 1D transforms in the coordinates y and θ and this greatly reduces the storage and computational requirements when calculating the FFT of the kernel. We now need to transform Equation 7.65 into a 2D FFT. In the α variable, we can achieve this in exactly the same fashion as previously. In the case of the θ coordinate, we divide the range of integration $[-\pi, \pi]$

into N_θ (a power of 2) subintervals of length $2\pi/N_\theta$ and sample at the midpoints. Since the field will be periodic in this coordinate, we have no need to filter these samples. We produce the FFT in κ from the FT by sampling at points $\kappa_i = i - N_\theta/2 - 1/2$ where i runs from 1 to N_θ. The fields can now be developed between the cylindrical intermediate surfaces using the algorithm

$$E(z_0, \theta_0, y_0) = \exp(\phi_+)$$
$$\times \text{FFT2D}^{-1}\left\{\text{FFT2D}\left\{\hat{K}_0\right\}\text{FFT2D}\{\exp(\phi_-)E(z,\theta,y)\}\right\}A \quad (7.66)$$

where $A = z\Delta\theta\Delta y$ is the area of a discretization cell. Apart from the use of 2D FFTs, the solution proceeds as for the 2D case.

We can also adapt the techniques of the previous section to allow consideration of propagation over irregular terrain. At each azimuth sample, we flatten the terrain according to the radial slope at that azimuth. After calculating $\hat{E}$ for each azimuth (including the image field), we progress this field to the next intermediate surface using Equation 7.66. At the new intermediate surface, we convert $\hat{E}$ back to E using the appropriate slope and ground height at each azimuth. Figure 7.14 shows the simulation of propagation over a Gaussian hill. When compared with the 2D calculation of Figure 7.12, it will be noted that there is additional diffraction to the rear of the hill. This arises due to the energy that can diffract around the sides of the hill. Both simulations, however, give fairly close results and this tends to justify the use of 2D calculations as a first approximation.

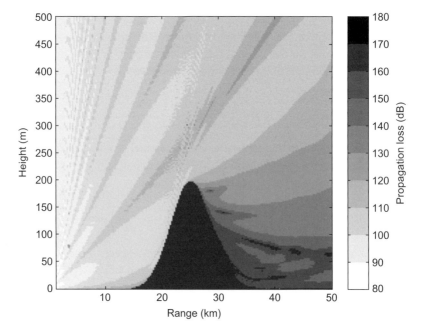

Figure 7.14 3D propagation over a Gaussian hill.

7.6 Time Domain Methods

Thus far, we have only considered techniques that are applicable to time harmonic fields. In studying propagation, however, there are circumstances where we need to consider transient behavior in a signal (the propagation of a pulse, for example) or a transient medium (atmospheric turbulence, for example). Although, in theory, we could construct a transient solution out of harmonic solutions over a suitable range of frequencies, this is difficult in practice. As a consequence, we look at some time domain methods in the current section. We consider a 2D field (y and z dependence alone) that is transverse magnetic (TM) (i.e. $\mathcal{E} = (0, \mathcal{E}_y, \mathcal{E}_z)$ and $\mathcal{H} = (\mathcal{H}_x, 0, 0)$). Then, Maxwell's equations imply that

$$\mu_0 \frac{\partial \mathcal{H}_x}{\partial t} = -\frac{\partial \mathcal{E}_z}{\partial y} + \frac{\partial \mathcal{E}_y}{\partial z} - \mathcal{M}_x \qquad (7.67)$$

$$\epsilon \frac{\partial \mathcal{E}_y}{\partial t} = \frac{\partial \mathcal{H}_x}{\partial z} - \mathcal{J}_y \qquad (7.68)$$

$$\epsilon \frac{\partial \mathcal{E}_z}{\partial t} = -\frac{\partial \mathcal{H}_x}{\partial y} - \mathcal{J}_z \qquad (7.69)$$

The equation $\nabla \cdot \mathcal{H} = 0$ is identically satisfied and $\nabla \cdot \mathcal{D} = 0$ implies $\partial \mathcal{E}_y / \partial y + \partial \mathcal{E}_z / \partial z = 0$ (we assume that the scale of permittivity variations is much larger than a wavelength). From Equations 7.68 and 7.69 we obtain that $\partial / \partial t \nabla \cdot \mathcal{E} = 0$ and so, providing $\mathcal{E}$ is initially divergence free, it will remain so. Consequently, Equations 7.67 to 7.69 will describe the development of the field with time.

The numerical solution of the above equations has been investigated by Yee (1966) in a technique that has become known as the FDTD method. Consider a step length of Δt in the time domain, Δy in the y direction and Δz in the z direction. Spatially, we consider a grid of the form shown in Figure 7.15 (points (x_l, y_m)) where $x_l = x_s + l \Delta x$ and $y_m = m \Delta y$. In the FDTD technique, we consider a pulse that is emitted by the source and travels through this grid. Consequently, at any one time, the pulse is confined to a limited region of the grid. The above equations can be discretized (Yee, 1966) to yield

$$\mathcal{H}_x^{n+1}(i+1, j+1) = \mathcal{H}_x^{n-1}(i+1, j+1) - \frac{\Delta t}{\Delta y \mu_0} \left(\mathcal{E}_z^n(i+2, j+1) - \mathcal{E}_z^n(i, j+1) \right)$$
$$+ \frac{\Delta t}{\Delta z \mu_0} \left(\mathcal{E}_y^n(i+1, j+2) - \mathcal{E}_y^n(i+1, j) \right) - \frac{2\Delta t}{\mu_0} \mathcal{M}_x^{n-1}(i+1, j+1) \qquad (7.70)$$

$$\mathcal{E}_y^{n+2}(i+1, j) = \mathcal{E}_y^n(i+1, j) + \frac{\Delta t}{\epsilon(i+1, j) \Delta z} \left(\mathcal{H}_x^{n+1}(i+1, j+1) - \mathcal{H}_x^{n+1}(i+1, j-1) \right)$$
$$- \frac{2\Delta t}{\epsilon(i+1, j)} \mathcal{J}_y^n(i+1, j) \qquad (7.71)$$

and

$$\mathcal{E}_z^{n+2}(i, j+1) = \mathcal{E}_z^n(i, j+1) - \frac{\Delta t}{\epsilon(i, j+1) \Delta y} \left(\mathcal{H}_x^{n+1}(i+1, j+1) - \mathcal{H}_x^{n+1}(i-1, j+1) \right)$$
$$- \frac{2\Delta t}{\epsilon(i, j+1)} \mathcal{J}_z^n(i, j+1) \qquad (7.72)$$

(note that, for function $A(y, z, t)$, $A^n(i, j)$ is used to denote $A(i\Delta y, j\Delta z, n\Delta t)$). Given $\mathcal{H}$ at $t = 0$ and $\mathcal{E}$ at time $t = \Delta t$, Equations 7.70 to 7.72 allow us to calculate $\mathcal{H}$ at $t = \Delta t$ and then $\mathcal{E}$ at $t = 2\Delta t$. We can next calculate $\mathcal{H}$ at $t = 3\Delta t$ and then $\mathcal{E}$ at $t = 4\Delta$ and so on until we have calculated the propagation of the pulse to the desired distance. Not all quantities are calculated at all mesh points; the values are staggered in a way that each quantity is evaluated at alternate mesh points (in both time and space). This is illustrated in Figure 7.15. We assume that the magnetic field is known at $n = -1$ and the electric field at $n = 0$ (obviously subject to the divergence-free condition). The major problem with the above procedure is that it says nothing about what we do at the boundaries, both real (the ground, for example) and artificial (the upper truncation of the mesh, for example). In the case of propagation over ground (the situation shown in Figure 7.15), we could assume the ground to be a PMC material (a good approximation in many situations). We would then start the $\mathcal{E}_z$ samples at the first grid point above the ground ($m = 1$) and the $\mathcal{E}_y$ samples at the second grid point above the ground ($m = 2$). The first $\mathcal{H}_x$ sample would start at the ground $m = 0$ where its value would be zero according to the PMC condition. The other boundaries of the mesh are artificial and we need to ensure that they do not reflect power back into the region of interest. This is normally achieved by some sort of absorbing layer at these boundaries and such layers can be composed of material with both electrical

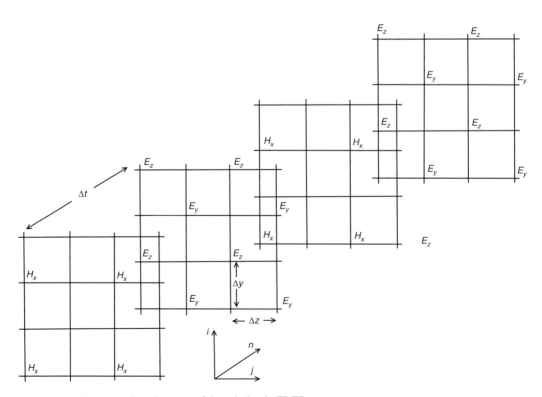

Figure 7.15 Development of the solution in FDTD.

σ_e and magnetic σ_m conductivity. A particularly effective approach to this problem is the *perfectly matched layer* (PML) that was introduced by Berenger (1994). In this approach, the conductivities are non-isotropic with electric current flowing in a direction normal to the boundary and magnetic current parallel to the boundary. To implement this scheme, we consider $\mathcal{H}_x$ to consist of two components ($\mathcal{H}_x = \mathcal{H}_x^y + \mathcal{H}_x^z$) and likewise $\mathcal{M}_x$. Equations 7.67 to 7.69 now take the form

$$\mu_0 \frac{\partial \mathcal{H}_x^y}{\partial t} = -\frac{\partial \mathcal{E}_z}{\partial y} - \mathcal{M}_x^y \tag{7.73}$$

$$\mu_0 \frac{\partial \mathcal{H}_x^z}{\partial t} = \frac{\partial \mathcal{E}_y}{\partial z} - \mathcal{M}_x^z \tag{7.74}$$

$$\epsilon \frac{\partial \mathcal{E}_y}{\partial t} = \frac{\partial \mathcal{H}_x}{\partial z} - \mathcal{J}_y \tag{7.75}$$

$$\epsilon \frac{\partial \mathcal{E}_z}{\partial t} = -\frac{\partial \mathcal{H}_x}{\partial y} - \mathcal{J}_z \tag{7.76}$$

We consider the case where we introduce a PML layer at an upper horizontal boundary (see Figure 7.16) and so $\mathcal{J}_y = \sigma_e \mathcal{E}_y$, $\mathcal{J}_z = 0$, $\mathcal{M}_x^z = \sigma_m \mathcal{H}_x^z$ and $\mathcal{M}_x^y = 0$. If we require that $\sigma_e/\epsilon = \sigma_m/\mu$, this ensures that impedance in the absorbing layer will be that of the propagation medium and no energy will be reflected back from this layer (see Appendix H). Equations 7.73 to 7.76 are now discretized in the same fashion as Equations 7.67 to 7.69 and, consistent with this, the currents are discretized according to $\mathcal{J}_y^n(i+1,j) = \sigma_e(i+1,j)\mathcal{E}_y^n(i+1,j)$, $\mathcal{J}_z^n(i+1,j) = 0$, $\mathcal{M}_x^{z(n-1)}(i+1,j+1) = \sigma_m(i+1,j+1)\mathcal{H}_x^{z(n-1)}(i+1,j+1)$ and $\mathcal{M}_x^{y(n-1)}(i+1,j+1) = 0$.

Although the initial field is arbitrary, it is common to assume that this takes the form of a pulse. A plane wave with pulse-like behavior is given by $\mathcal{E} = \mathcal{E}_0 \exp\left(-(z-ct)^2/L^2\right)$. This, however, is usually modulated by a lateral window in order to mimic the effect

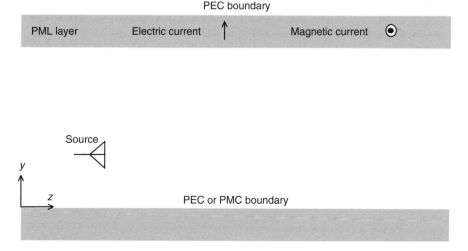

Figure 7.16 Propagation with upper PML layer.

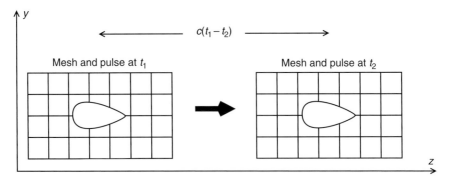

Figure 7.17 Moving mesh for FDTD wave propagation.

of a pulse transmitted by a beam antenna. The reason for a pulse is that, in the limit $L \to 0$, this will have a uniform distribution over all frequencies. Consequently, by taking FTs, we can obtain the effect of propagation on harmonic waves at all frequencies. Obviously, because of practical constraints, we cannot make L too small. We can, however, make it sufficiently small to obtain a useful range of frequency coverage. In order to reduce the computational requirement, Akleman and Sevgi (2000) have introduced the idea of a moving mesh. Since we are dealing with pulses of limited length, the region of activity is likely to remain relatively compact throughout the propagation. Consequently, after the pulse has passed, a mesh point need only be kept active for a limited time. Therefore, as new mesh points ahead of the pulse are activated, others to the rear may be deactivated. Effectively, we have a mesh that moves with the pulse and this greatly reduces computational and storage requirement. Obviously, it is once again important to ensure that there are no reflections at the now moving artificial boundaries (Figure 7.17).

Akleman and Sevgi (2000, 2003) have considered the application of FTDT to propagation problems, including those with irregular boundaries. However, the approach poses serious challenges when dispersive media are involved. In the case of dispersive impedance boundary conditions, Beggs et al. (1992) have developed an equivalent time domain condition that can be used with the FDTD approach. In the case of a dispersive medium consisting of a plasma, Nickisch and Franke (1992, 1996) have developed time domain equations for the polarization vector that facilitate an FDTD approach to the problem of radio wave propagation in a plasma. In particular, Nickisch and Franke (1996) have used this approach to consider propagation in an irregular ionospheric plasma.

As discussed so far, the FDTD algorithm is explicit and we must restrict the time step if the algorithm is to remain stable. Stability is guaranteed by the Courant condition $(1/\Delta y)^2 + (1/\Delta z)^2 < (1/\Delta t c_{max})^2$ where c_{max} is the maximum propagation speed. This restriction can severely reduce the efficiency of the algorithm, and we would like an alternative algorithm in which we can vary the time step to any level consistent with the time scale of the problem at hand. Such an algorithm is given by alternating direction implicit (ADI) schemes (Press et al., 1992) and Zheng and Chen (1999) have considered an FDTD scheme based on such an approach. In such schemes, derivatives in one spatial dimension are approximated at the new time step (the rest at the old time step), but this dimension alternates between steps. As a consequence, the solution procedure is far more efficient than for a fully implicit scheme, but is obviously a bit more complex than an explicit scheme. Importantly, such algorithms are unconditionally stable. We now must solve for all quantities at each stage, but alternate stages (A and B) are implicit in different directions. Stage A is implicit in the z direction, with

$$\mathcal{H}_x^{n+1}(i+1,j+1) = \mathcal{H}_x^n(i+1,j+1) - \frac{\Delta t}{2\Delta y \mu_0}\left(\mathcal{E}_z^n(i+2,j+1) - \mathcal{E}_z^n(i,j+1)\right)$$
$$+ \frac{\Delta t}{2\Delta z \mu_0}\left(\mathcal{E}_y^{n+1}(i+1,j+2) - \mathcal{E}_y^{n+1}(i+1,j)\right)$$
$$+ \frac{\Delta t}{2\mu_0}\left(\mathcal{M}_x^{n+1}(i+1,j+1) + \mathcal{M}_x^n(i+1,j+1)\right) \quad (7.77)$$

$$\mathcal{E}_y^{n+1}(i+1,j) = \mathcal{E}_y^n(i+1,j) + \frac{\Delta t}{2\epsilon(i+1,j)\Delta z}\left(\mathcal{H}_x^{n+1}(i+1,j+1) - \mathcal{H}_x^{n+1}(i+1,j-1)\right)$$
$$- \frac{\Delta t}{2\epsilon(i+1,j)}\left(\mathcal{J}_y^{n+1}(i+1,j) + \mathcal{J}_y^n(i+1,j)\right) \quad (7.78)$$

and

$$\mathcal{E}_z^{n+1}(i,j+1) = \mathcal{E}_z^n(i,j+1) - \frac{\Delta t}{2\epsilon(i,j+1)\Delta y}\left(\mathcal{H}_x^n(i+1,j+1) - \mathcal{H}_x^n(i-1,j+1)\right)$$
$$- \frac{\Delta t}{2\epsilon(i,j+1)}\left(\mathcal{J}_z^{n+1}(i,j+1) + \mathcal{J}_z^n(i,j+1)\right) \quad (7.79)$$

Stage B is implicit in the y direction, with

$$\mathcal{H}_x^{n+2}(i+1,j+1) = \mathcal{H}_x^{n+1}(i+1,j+1) - \frac{\Delta t}{2\Delta y \mu_0}\left(\mathcal{E}_z^{n+2}(i+2,j+1) - \mathcal{E}_z^{n+2}(i,j+1)\right)$$
$$+ \frac{\Delta t}{2\Delta z \mu_0}\left(\mathcal{E}_y^{n+1}(i+1,j+2) - \mathcal{E}_y^{n+1}(i+1,j)\right)$$
$$+ \frac{\Delta t}{2\mu_0}\left(\mathcal{M}_x^{n+2}(i+1,j+1) + \mathcal{M}_x^{n+1}(i+1,j+1)\right) \quad (7.80)$$

$$\mathcal{E}_y^{n+2}(i+1,j) = \mathcal{E}_y^{n+1}(i+1,j) + \frac{\Delta t}{2\epsilon(i+1,j)\Delta z}\left(\mathcal{H}_x^{n+1}(i+1,j+1) - \mathcal{H}_x^{n+1}(i+1,j-1)\right)$$
$$- \frac{\Delta t}{2\epsilon(i+1,j)}\left(\mathcal{J}_y^{n+2}(i+1,j) + \mathcal{J}_y^{n+1}(i+1,j)\right) \quad (7.81)$$

and

$$\mathcal{E}_z^{n+2}(i,j+1) = \mathcal{E}_z^{n+1}(i,j+1) - \frac{\Delta t}{2\epsilon(i,j+1)\Delta y}\left(\mathcal{H}_x^{n+2}(i+1,j+1) - \mathcal{H}_x^{n+2}(i-1,j+1)\right)$$
$$- \frac{\Delta t}{2\epsilon(i,j+1)}\left(\mathcal{J}_z^{n+2}(i,j+1) + \mathcal{J}_z^{n+1}(i,j+1)\right) \quad (7.82)$$

The above scheme allows us to step in time, but at each stage we must solve a linear system of equations. With suitable rearrangement, however, this can be reduced to a simple tridiagonal system that is easily solved.

In the case of an isolated source, it is more useful to consider cylindrical coordinates (ρ, ϕ, z) where ϕ is the azimuthal angle, and ρ is a radial coordinate. We consider a 2D field (ρ and z dependence alone) that is transverse magnetic (TM) (i.e. $\mathcal{E} = (\mathcal{E}_\rho, 0, \mathcal{E}_z)$ and $\mathcal{H} = (0, \mathcal{H}_\phi, 0)$), then Maxwell's equations imply that

$$\mu_0 \frac{\partial \mathcal{H}_\phi}{\partial t} = -\frac{\partial \mathcal{E}_\rho}{\partial z} + \frac{\partial \mathcal{E}_z}{\partial \rho} - \mathcal{M}_\phi \qquad (7.83)$$

$$\epsilon \frac{\partial \mathcal{E}_\rho}{\partial t} = -\frac{\partial \mathcal{H}_\phi}{\partial z} - \mathcal{J}_\rho \qquad (7.84)$$

and

$$\epsilon \frac{\partial \mathcal{E}_z}{\partial t} = \frac{1}{\rho}\frac{\partial (\rho \mathcal{H}_\phi)}{\partial \rho} - \mathcal{J}_z \qquad (7.85)$$

The discretized version of these equations for the explicit scheme of Lee (1966) is then

$$\mathcal{H}_\phi^{n+1}(i+1,j+1) = \mathcal{H}_\phi^{n-1}(i+1,j+1) - \frac{\Delta t}{\Delta z \mu_0}\left(\mathcal{E}_\rho^n(i+2,j+1) - \mathcal{E}_\rho^n(i,j+1)\right)$$

$$+ \frac{\Delta t}{\Delta \rho \mu_0}\left(\mathcal{E}_z^n(i+1,j+2) - \mathcal{E}_z^n(i+1,j)\right) - \frac{2\Delta t}{\mu_0}\mathcal{M}_\phi^{n-1}(i+1,j+1) \qquad (7.86)$$

$$\mathcal{E}_\rho^{n+2}(i+1,j) = \mathcal{E}_\rho^n(i+1,j) - \frac{\Delta t}{\epsilon(i+1,j)\Delta z}\left(\mathcal{H}_\phi^{n+1}(i+1,j+1) - \mathcal{H}_\phi^{n+1}(i+1,j-1)\right)$$

$$- \frac{2\Delta t}{\epsilon(i+1,j)}\mathcal{J}_\rho^n(i+1,j) \qquad (7.87)$$

and

$$\mathcal{E}_z^{n+2}(i,j+1) = \mathcal{E}_z^n(i,j+1) + \Delta t\left(\frac{\rho_{i+1}\mathcal{H}_\phi^{n+1}(i+1,j+1) - \rho_{i-1}\mathcal{H}_\phi^{n+1}(i-1,j+1)}{\epsilon(i,j+1)\rho_i \Delta \rho}\right)$$

$$- \frac{2\Delta t}{\epsilon(i,j+1)}\mathcal{J}_z^n(i,j+1) \qquad (7.88)$$

where $\rho_j = j\Delta z$.

7.7   References

F. Akleman and L. Sevgi, A novel time-domain wave propagator, IEEE Trans. Antennas Propagat., vol. 48, pp. 839–841, 2000.

F. Akleman and L. Sevgi, Realistic surface modelling for a finite-difference time-domain wave propagator, IEEE Trans. Antennas Propagat., vol. 51, pp. 1675–1679, 2003.

A.E. Barrios, Parabolic equation modelling in horizontally inhomogeneous environments, IEEE Trans. Antennas Propagat., vol. 40, pp. 791–797, 1992.

A.E. Barrios, A terrain parabolic equation model for propagation in the troposphere, IEEE Trans. Antennas Propagat., vol. 42, pp. 90–98, 1994.

J.H. Beggs, R.J. Luebbers, K.S. Yee and K.S. Kunz, Finite-difference time-domain implementation of surface boundary conditions, IEEE Trans. Antennas Propagat., vol. 40, pp. 49–56, 1992.

7.7 References

J.B. Berenger, A perfectly matched layer for the absorption of electromagnetic waves, J. Comput. Phys., vol. 114, pp. 185–200, 1994.

G.F. Carrier, M. Krook and C.E. Pearson, *Functions of a Complex Variable*, McGraw-Hill, New York, 1966.

C.J. Coleman, A Kirchhoff integral approach to estimating propagation in an environment with nonhomogeneous atmosphere and complex boundaries, IEEE Trans. Antennas Propag., vol. 53, pp. 3174–3179, 2005.

C.J. Coleman, An FFT based Kirchhoff integral technique for the simulation of radio waves in complex environments, Radio Sci., vol. 45, RS2002, doi:10.1029/2009RS004 197, 2010.

J.M. Collis, Three-dimensional underwater sound propagation using a split-step Pade parabolic equation solution, J. Acoust. Soc. Am., vol. 130, p. 2528, 2011.

G.D. Dockery and J.R. Kuttler, An improved impedance boundary algorithm for Fourier split-step solutions of the parabolic wave equation, IEEE Trans. Antennas Propag., vol. 44, pp. 1592–1599, 1996.

D.J. Donohue and J.R. Kuttler, Propagation modelling over terrain using the parabolic wave equation, IEEE Trans. Antennas Propag., vol. 48, pp. 260–277, 2000.

T.F. Eibert, Irregular terrain wave propagation by a Fourier split-step wide-angle parabolic wave equation technique for linearly bridged knife edges, Radio Sci., vol. 37, pp. 5-1–5-11, 2002.

R.F. Harrington, *Time Harmonic Electromagnetic Fields*, McGraw-Hill, New York, 1961.

J.R. Kuttler and G.D. Dockery, Theoretical description of the parabolic approximation/Fourier split-step method of representing electromagnetic propagation in the troposphere, Radio Sci., vol. 26, pp. 381–393, 1991.

D. Lee and S.T. McDaniel, Ocean acoustics propagation by finite difference methods, Comput. Math. Applic., vol. 14, pp. 305–423, 1987.

M. Levy, Parabolic equation methods for electromagnetic wave propagation, IEE Electromagnetic Wave Series 45, London, 2000.

J.H. Mathews, *Numerical Methods for Computer Science, Engineering and Mathematics*, Prentice Hall, Englewood Cliffs, NJ, 1987.

A.L. Mikaelian, Self-focusing media with variable index of refraction, in *Progress in Optics*, vol. XVII, edited by E. Wolf, North-Holland, New York.

F.A. Milinazzo, C.A. Zala and G.H. Brook, Rational square-root approximations for parabolic equation algorithms, J. Acoust. Soc. Am., vol. 101, pp. 760–766, 1997.

L.J. Nickisch and P.M. Franke, Finite-difference time-domain solution of Maxwell's equations for the dispersive ionosphere, IEEE Antennas Propag. Mag., vol. 34, pp. 33–39, 1992.

L.J. Nickisch and P.M. Franke, Finite difference tests of random media propagation theory, Radio Sci., vol. 31, pp. 955–963, 1996.

M.O. Ozyalcin, F. Akleman and L. Sevgi, A novel TLM-based time-domain wave propagator, IEEE Trans. Antennas Propagat., vol. 51, pp. 1680–1682, 2003.

W.H. Press, S. Teukolsky, W. Vetterling and B. Flannery, *Numerical Recipes in FORTRAN: The Art of Scientific Computing*, 2nd edition, Cambridge University Press, Cambridge, 1992.

T.B.A. Senior and J.L. Volakis, *Approximate Boundary Conditions*, IEE Electromagnetic Wave Series 41, The Institution of Electrical Engineers, London, 1995.

J.R. Smith, *Introduction to the Theory of Partial Differential Equations*, Van Nostrand, London, 1967.

K.S. Yee, Numerical solution of initial boundary value problems involving Maxwell's equations, IEEE Trans. Antennas Propagat., vol. 14, pp. 302–307, 1996.

F. Zheng and Z. Chen, A finite-difference time-domain method without the Courant stability conditions, IEEE Microw. Guided Wave Lett., vol. 9, pp. 441–443, 1999.

8 Propagation in the Ionospheric Duct

As mentioned previously, the ionosphere can cause radio waves to be refracted back to the ground. At the ground, these waves can then be reflected back to the ionosphere where they are again refracted back to the ground. In this manner, the ionosphere can form a duct that is capable of transmitting significant amounts of radio energy around the globe. Such propagation (known as *sky wave* propagation) normally takes place at frequencies below about 30 MHz, although, under extreme ionospheric conditions, there can be propagation at frequencies of 50 MHz and more. In this chapter, we apply some of the techniques described in previous chapters to the study of such propagation. In addition, we look at some of the issues (such as loss and noise) that are required when studying such propagation for practical purposes. Due to the advent of artificial satellites and fiber optic cables, communications via the ionospheric duct have become less common. The ionospheric duct, however, is still important in providing back-up communications for both civil and military purposes. In particular, it is most important in locations where satellite coverage is lacking. Additionally, some important surveillance technologies, such as skywave over the horizon radar, are reliant on the ionospheric duct. In order that the reader can appreciate the nature of the ionospheric duct, we start this chapter with a description of the basic physics of the ionosphere. In addition, we discuss some important ionospheric phenomena that can severely affect the propagation of radio waves in the ionospheric duct.

8.1 The Benign Ionosphere

Radiation from the sun causes the molecules in the atmosphere to ionize. This ionization is dependent on the molecular structure and tends to occur in layers at heights of 80 km for the D layer, 110 km for the E layer, 170 km for the F1 layer and 320 km for the F2 layer. The F2 layer is by far most complex and its height can vary considerably across the globe. The simplest model of the atmosphere gives a number density $n = n_0 \exp(-(h - h_0)/H)$ of neutral molecules where H is known as the scale height, h is the altitude and h_0 is some reference altitude. Due to the change in molecular composition and temperature with height, H can vary with a value of about 5 km in the D layer, 10 km in the E layer, 40 km in the F1 layer and 60 km in the F2 layer (Figure 8.1). Consequently, it is more accurate to write $n = n_0 \exp\left(-\int_{h_0}^{h} H^{-1} dh\right)$.

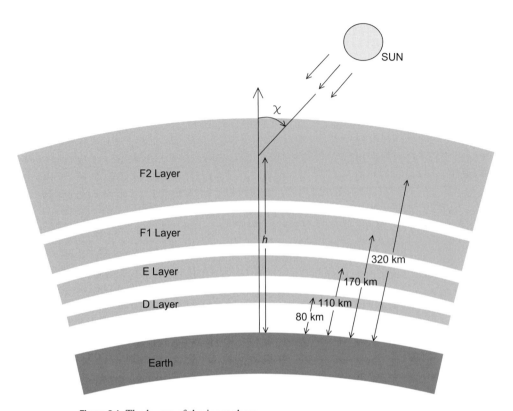

Figure 8.1 The layers of the ionosphere.

The intensity I of the radiation striking Earth's atmosphere will initially have a strength I_∞ and will then decay as it is steadily absorbed by the atmosphere as it travels to the ground. Consequently,

$$\frac{dI}{dh} = -\sigma n \sec \chi I \tag{8.1}$$

where n is the density of molecules, σ is the absorption cross section and χ is the angle from the vertical at which the solar radiation strikes the ionosphere. From this,

$$I(h) = I_\infty \exp\left(-\sigma \sec \chi \int_h^\infty n(h) dh\right) \tag{8.2}$$

Not all the radiation generates ionization, and, as a consequence, the rate of ion production is given by

$$q(h) = \eta \sigma I_\infty n(h) \exp\left(-\sigma \sec \chi \int_h^\infty n(h) dh\right) \tag{8.3}$$

where η is known as the ionization efficiency. The peak production rate will occur at the altitude h_{peak} for which $dq/dh = 0$ and from which $n_{\text{peak}} \sigma \sec \chi H_{\text{peak}} = 1$. The scale height is slowly varying and so we can take $n = n_{\text{peak}} \exp(-(h - h_{\text{peak}})/H_{\text{peak}})$ around the peak. As a consequence, we have

$$q_{\text{peak}} = \frac{\eta I_\infty}{H_{\text{peak}}} \cos \chi \tag{8.4}$$

and
$$q(h) = q_{\text{peak}} \exp(1 - U - \exp(-U)) \tag{8.5}$$

where $U = (h - h_{\text{peak}})/H_{\text{peak}}$. The height of peak ionization h_{peak} will be given by

$$h_{\text{peak}} = h_0 + H_{\text{peak}} \ln(\sec \chi) \tag{8.6}$$

where reference height h_0 is the peak height when $\chi = 0$. The above distribution of ionization is known as a Chapman layer (Davies, 1990).

Due to the complex structure of the atmosphere, there will be several peaks in production corresponding to the different ionospheric layers mentioned above. Consequently, the ionization distribution in a real ionosphere will consist of a combination of Chapman layers. After production, ions can be lost by a process of recombination. For ionization in the D, E and F1 layers, the loss L will be given by

$$L(h) = \alpha N_e^2 \tag{8.7}$$

and for the F2 layer

$$L(h) = \beta N_e \tag{8.8}$$

where N_e is the number density of electrons produced by the ionization. The simplest model of the ionosphere is to assume that production and recombination balance each other (a good approximation in the lower layers for much of the time). From this, we find that

$$N_e = \frac{q_{\text{peak}}}{\beta} \exp(1 - U - \sec \chi \exp(-U)) \tag{8.9}$$

for the F2 layer and

$$N_e = \sqrt{\frac{q_{\text{peak}}}{\alpha}} \exp\left(\frac{1}{2}(1 - U - \sec \chi \exp(-U))\right) \tag{8.10}$$

for the D, E and F1 layers. The dependence on χ will lead to a seasonal and geographic dependence of N_e due to the fact that

$$\cos \chi = \sin \phi \sin \delta + \cos \phi \cos \delta \cos \hat{\theta} \tag{8.11}$$

where ϕ is the latitude, δ is the solar declination, $\hat{\theta} = \theta - 2\pi t/24$ (rad), t is UT time (h) and θ is longitude (rad). (To a reasonable degree of accuracy (Davies, 1990), $\delta = 23.4 \sin(0.9856(Y - 80.7))$ where Y is the number of the day in the year.) The intensity of radiation from the sun I_∞ varies with an 11-year cycle that is known as the *sunspot cycle* and the cycle is well correlated with the *sunspot number R* ($I_\infty \approx I_0 + I_1 R$ to a high degree of accuracy).

A plasma is often described in terms of its plasma frequency f_p where $f_p^2 = N_e^2/4\pi^2 \epsilon_0 m = 80.61 N_e$ (in terms of Hz) and N_e is the electrons per cubic meter. Figure 8.2 shows a typical distribution of plasma frequency with height (distributions at local noon and midnight are shown). The above models work well for the lower layers, but the F2 layer is far more complex. This can be seen from Figure 8.3, which shows a typical global distribution of *peak plasma frequency* for the F2 layer (designated

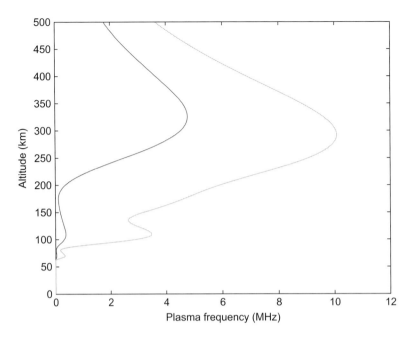

Figure 8.2 Variation of mid-latitude plasma frequency between day and night (at equinox with $R = 100$).

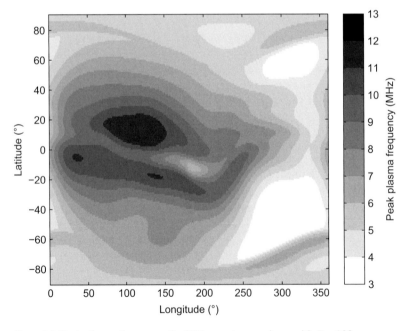

Figure 8.3 Peak plasma frequency for F2 layer at an equinox with $R = 100$.

foF2) at 0 UT. Our simple model would give a distribution of plasma that is minimum around the poles and peaks around the equator (latitude 0) and midday (longitude 180° in the figure). It will be noted, however, that the maximum snakes around the magnetic equator and that there are areas of significant depletion and enhancement. All of these effects owe their origin to the Earth's magnetic field, a field that is mainly the result of the dynamics of Earth's core. To a first approximation, it can be regarded as a dipole whose axis is slightly offset from Earth's axis (magnetic north is approximately 85°N and 227°W at present). In terms of a coordinate system based on the Earth's center, the magnetic field is given by

$$\mathbf{B}_0 = -\nabla\left(\frac{\mathbf{M}\cdot\mathbf{r}}{r^3}\right) \tag{8.12}$$

where $\mathbf{M}$ is a vector in the direction of magnetic north with magnitude $M = 0.31 \times 10^{-4} a^3$ tesla where a is the radius of Earth. Complicating matters, however, the sun emits a continuous stream of electrons and protons that travel to Earth as the *solar wind*. Earth, however, is shielded from this wind by its magnetic field. The interaction forms a shock wave and causes the wind to travel around the Earth. The interaction causes Earth's magnetic field lines to be compressed and so the total field is, to some extent, a function of solar activity. The net effect is shown in Figure 8.4.

Around the equator, there is a depletion of electron density that is known as the *equatorial anomaly*, which is explained as follows. In the upper atmosphere, around the E layer level, the neutral winds drag the electrons and ions across the magnetic field lines and this causes an east to west electric field by a dynamo process. In the ionosphere, the electrons are largely constrained to move along the magnetic field lines that, as a consequence, act like conducting wires. The electric field generated in the E

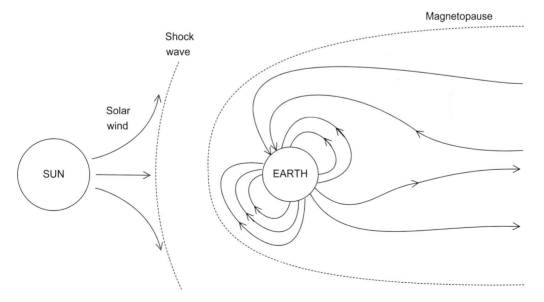

Figure 8.4 The effect of the solar wind on Earth's magnetic field.

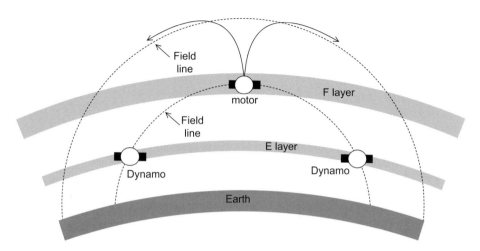

Figure 8.5 Mechanism for the generation of the equatorial anomaly.

layer will be transmitted along these conductors to the equatorial F layer and here it will then cause an upward drift of electrons (an electron velocity $\mathbf{v} = \mathbf{E} \times \mathbf{B}_0/B_0^2$ that results from the balance between electric and magnetic forces) since the magnetic field will be horizontal there. The electrons thus driven to higher altitudes will eventually fall back along the field lines and hence deposit themselves on either side of the magnetic equator. Consequently, the F layer at the equator will become depleted while, either side, it will become enhanced (see Figure 8.5). This results in the equatorial structure that is seen in Figure 8.3.

The Earth's polar regions provide additional zones of anomalous behavior. In particular, around the magnetic poles, there is a ring of increased plasma density that is known as the *auroral oval*. At the poles, the magnetic field lines are perpendicular to the Earth's surface and stretch out far into the magnetosphere (the region contained by the magnetopause shown in Figure 8.4). As previously mentioned, the field lines act like conductors, and charged particles from the solar wind can enter the auroral oval via this route (under extreme circumstances, this causes a visible aurora). These conductors also transfer, into the polar region, the electric fields that are generated by the interaction of the solar wind with the interplanetary magnetic field (IMF). In the polar regions this results in an electric field pointing from dawn to dusk and causes a drift of ionization from the day side to the night side. As a consequence, there is strong nighttime auroral ionization, as illustrated in Figure 8.3. An additional phenomenon associated with the oval is known as the *mid-latitude trough* and occurs between dusk and midnight. This is a region of depleted electron density just on the equator side of the auroral oval and is associated with a stagnation point in the flow of plasma across the oval. Figure 8.6 depicts the oval with the center of the figure located at the geomagnetic pole. The outer extent of the oval is approximately a circle of radius $18 + 1.7K_p$ degrees with center displaced from the magnetic pole by $5°$ in a direction opposite to the solar direction. The index K_p is a measure of geomagnetic activity and is affected by solar activity through the influence of the sun on Earth's magnetic field. Under normal circumstances, K_p has a value around 3 but, with solar storm activity, its value can rise to as much as 9.

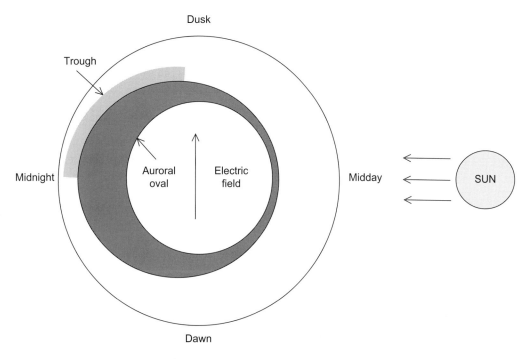

Figure 8.6 The auroral oval.

For the modeling of radio wave propagation, we normally use an ionospheric description that is derived from measured plasma frequencies or from a global model. The most comprehensive global model is known as the International Reference Ionosphere (IRI). This model (Bilitza, 1990; Bilitza et al., 2014) is under continual development and contains the best of our current knowledge concerning the ionosphere. In the present work we use IRI layer parameters (foF2, etc.), together with Chapman layers, in order to provide an ionospheric representation with sufficient differentiability for ray tracing purposes.

8.2 The Disturbed Ionosphere

As mentioned above, the behavior of the ionosphere can become disturbed by solar storms and this is usually indicated by large values of the magnetic indices. The main consequence is an increase in the size of the auroral oval. Importantly, however, there can also be a large increase of the ionization in the D layer, and, as we shall see later, this can lead to much increased absorption of the radio waves that travel through this layer. In addition to disturbances of solar origin, there are also disturbances of terrestrial origin. In particular, there can be traveling ionospheric disturbances (TIDs) that are the result of gravity waves driving plasma along the magnetic field lines. The gravity waves have their origin in disturbances in the lower atmosphere (earthquakes, for example) and can be quite complex in character. In addition to TIDs, there are other disturbances that arise as a result of instabilities in the plasma itself. In particular, after sunset, the equatorial F2 layer considerably rises in height and this results in a large-scale irregular structure that can persist through the night. Additionally, an irregular structure will also be present in the auroral oval. All of the above disturbances can have marked impact on radio wave propagation and need to be taken into consideration in the analysis, and modeling, of ionospheric radio wave propagation.

TIDs are frequently present and have been observed with periods from tens of minutes to hours and wavelengths from tens to thousands of kilometers (Friedman, 1966; Baulch and Butcher, 1985). TIDs are formed when gravity waves drive the ionospheric plasma along Earth's magnetic field lines (Hines, 1960) and the connection was established theoretically by Hooke (1968). In this theory, a simple gravity wave with velocity field

$$\mathbf{U} = \mathbf{U}_0 \exp\left(j\left(\omega t - k_x x - k_y y\right)\right) \qquad (8.13)$$

will result in a variation

$$\delta N = \Re\left\{\frac{\hat{\mathbf{I}}_H \cdot \mathbf{U}}{\omega}\left(\mathbf{k} \cdot \hat{\mathbf{I}}_H N + j\hat{\mathbf{I}}_H \cdot \nabla N\right)\right\} \qquad (8.14)$$

of the plasma density N where $\hat{\mathbf{I}}_H$ is a unit vector in the direction of Earth's magnetic field.

It is clear that an understanding of gravity waves is crucial to the understanding of TIDs. The horizontal wavelength λ_x, vertical wavelength λ_y and the frequency ω are related by the dispersion relation for gravity waves

$$\frac{\omega^4}{C^2} - \omega^2\left(\left(\frac{2\pi}{\lambda_x}\right)^2 + \left(\frac{2\pi}{\lambda_y}\right)^2 + \frac{\omega_A^2}{C^2}\right) + \omega_B^2\left(\frac{2\pi}{\lambda_x}\right)^2 = 0 \qquad (8.15)$$

where ω_B is the Brunt–Vaisala frequency, C is the speed of sound and ω_A is the acoustic cut-off frequency. The above parameters are related to the physical quantities γ (the ratio of specific heats), H (the scale height) and C (the speed of sound) through $\omega_A = C/2H$, $\omega_B^2 = (g/H)(\gamma - 1/\gamma)$ and $C = \sqrt{\gamma g H}$ where g is the acceleration due to gravity. The wave numbers are given by $k_x = 2\pi/\lambda_x$ and $k_y = -2\pi/\lambda_y + j/2H$ (the wave direction is normally observed to point downward) and from which it will be noted that a small velocity perturbation in the lower atmosphere can grow to be quite large in the upper atmosphere. This growth will be moderated by diffusion (both thermal and viscous), but, for the longer wavelengths, the perturbations can reach significant strength at F-layer heights.

The above dispersion relation is correct for a plane gravity wave in the thermosphere, but gravity waves can have very complex structures stretching from the ground to the upper reaches of the thermosphere. In particular, for a given frequency, gravity waves are found to occur in modes with distinct phase speeds. As noted previously, the scale height varies with altitude and, as a consequence, so will the sound speed C (Figure 8.7 shows the variation of C with height). Consequently, in a similar fashion to radio waves

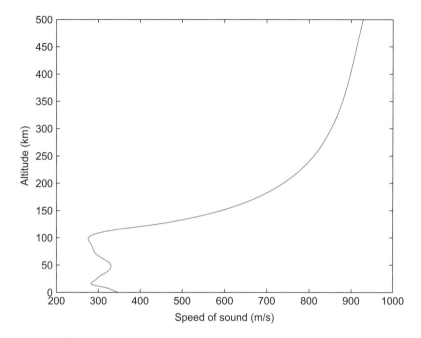

Figure 8.7 The variation of the speed of sound with altitude.

in the ionosphere, the propagation speed increases with height and this can cause the gravity wave paths to bend and hence cause them to be trapped in the lower layers. Friedman (1966) and Francis (1973) have carried out detailed numerical studies of gravity wave structure and have confirmed that these do indeed occur at distinct phase speeds for a given frequency. While many of these waves are fully trapped, some will leak energy into the upper atmosphere and will consequently decay after traveling a distance of a few wavelengths. Figure 8.8 shows an example of a gravity wave driven TID with a horizontal wavelength of 251 km.

8.3 Vertical and Quasi-Vertical Propagation

Reflection occurs at a discontinuity in refractive index, but can also occur when the gradient of the refractive index is discontinuous. Such a discontinuity will occur when $N = 0$ (N the refractive index). If the magnetic field of Earth is ignored, a wave of frequency f that is propagated upward will therefore be reflected downward at a height h_v where the plasma frequency is equal to f. For a pulse, the delay in traveling from the ground to height h_v will be the group delay τ_g, which is related to the plasma frequency through $\tau_g = c_0^{-1} \int_0^{h_v} N^{-1} dz = c_0^{-1} \int_0^{h_v} (1 - f_p^2/f^2)^{-1/2} dz$ where z is the height above the ground. The delay τ_g is a function of frequency, and a plot of this function is known as a *vertical ionogram*. Figure 8.9 shows a typical variation of plasma frequency with height and Figure 8.10 shows the corresponding vertical ionogram (note that we

Propagation in the Ionospheric Duct

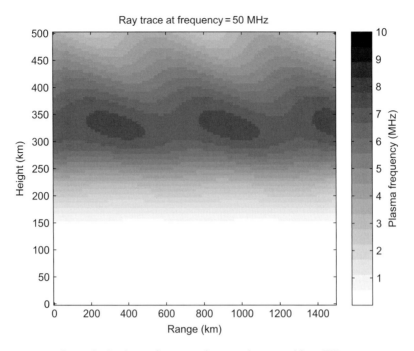

Figure 8.8 Ionospheric plasma frequency for a gravity wave driven TID.

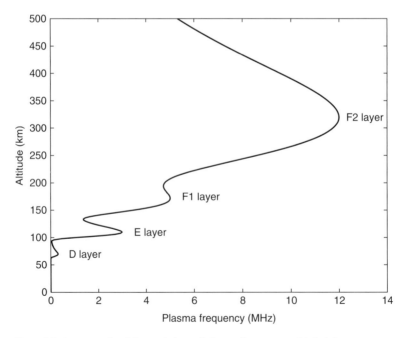

Figure 8.9 An example of the variation of plasma frequency with height.

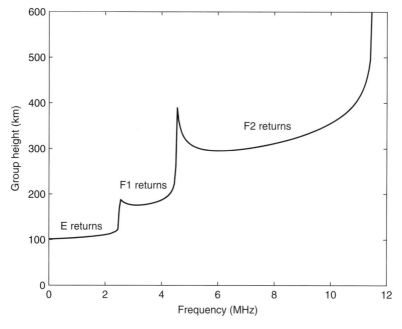

Figure 8.10 Vertical ionogram for the ionosphere of Figure 8.9.

have used group height $c_0\tau_g$ instead of group delay to label the vertical axis). The function $\tau_g(f)$ can be measured using an instrument known as an *ionosonde* and then the equation $\tau_g(f) = c_0^{-1} \int_0^{h_v(f)} (1 - f_p^2/f^2)^{-1/2} dz$ inverted to obtain an ionospheric profile (the variation of plasma frequency with height). This will only be an approximation since, in reality, the propagation will be affected by Earth's magnetic field. An O mode, however, is reflected at the same height as for the field-free case and so inversions based on an O trace will usually provide a reasonable estimate of the ionospheric plasma distribution.

For oblique propagation over short ranges (a few hundred kilometers), we can glean considerable information from a knowledge of vertical propagation (i.e. an ionogram) through results that are known, respectively, as the Breit and Tuve theorem, Martyn's theorem and the secant law. For propagation at frequency f_o, Snell's law gives us that $\sin\phi_0 = \sin\phi\sqrt{1 - f_p^2/f_o^2}$ where ϕ is the angle between the propagation direction and the vertical at a general point on the ray with ϕ_0 this angle at the ground. Let the ray be turned back toward the ground at a height h and so $\phi = 90°$ at this height. Consequently, $\sin\phi_0 = \sqrt{1 - f_v^2/f_o^2}$ where f_v is the plasma frequency at height h. We can rearrange this to obtain that $f_o = f_v \sec\phi_0$, a result that is known as the *secant law*. Since f_v is the frequency at which a vertical ray would be reflected at the same height as the oblique ray on frequency f_o, it is known as the *equivalent vertical frequency*.

We now consider the delay τ_g of a pulse in traveling between A and B along the oblique path ATB (see Figure 8.11)

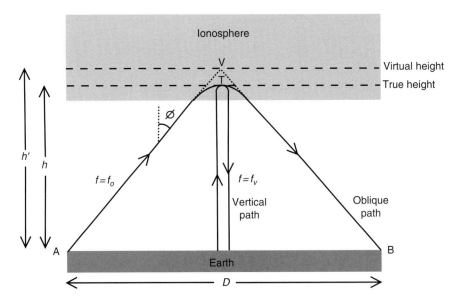

Figure 8.11 Vertical and oblique path for Breit and Tuve's theorem.

$$\tau_g = \frac{1}{c_0} \int_A^B \frac{ds}{N} = \frac{1}{c_0} \int_A^B \frac{ds}{\sqrt{1 - \frac{f_p^2}{f_o^2}}} \quad (8.16)$$

where s is the distance along the path and x is a the distance along the ground. From Snell's law, $\sin \phi_0 = (dx/ds)\sqrt{1 - f_p^2/f_o^2}$ and so

$$\tau_g = \frac{1}{c_0 \sin \phi_0} \int_A^B dx = \frac{D}{c_0 \sin \phi_0} \quad (8.17)$$

where D is the distance along the ground. Consequently, the time of transit is that which would occur in free space with a reflection at virtual height $D \tan \phi_0/2$, a result that is known as the *Breit and Tuve theorem*. On noting that $ds^2 = dx^2 + dz^2$ where z is the vertical coordinate, Snell's law implies that $(dz/dx)^2 \tan^2 \phi_0 = 1 - f_p^2/f_o^2 \cos^2 \phi_0$ and, using this to eliminate dx in Equation 8.17, we obtain

$$\tau_g = \frac{2}{c_0 \cos \phi_0} \int_0^h \frac{dz}{\sqrt{1 - \frac{f_p^2}{f_o^2 \cos^2 \phi_0}}} \quad (8.18)$$

where h is the physical height of reflection for the oblique propagation. It will now be noted that the above integral is the group height h' for the equivalent vertical frequency $f_v = f_o \cos \phi_0$. Consequently, we have *Martyn's theorem*, i.e.

$$\tau_g = \frac{2h'}{c_0 \cos \phi_0} \quad (8.19)$$

8.3 Vertical and Quasi-Vertical Propagation

The Breit and Tuve theorem, together with Martyn's theorem, now yield that $\tan \phi_0 = D/2h'$ from which we deduce that the virtual height of reflection for the oblique propagation is that given by the group height for vertical propagation on the equivalent vertical frequency. This is an important result as it means that we can obtain information about oblique propagation directly from a vertical ionogram. From the secant law we have

$$f_o = f_v \sec \phi_0 = f_v \sqrt{1 + \left(\frac{D}{2h'}\right)^2} \tag{8.20}$$

and so, given a range D and a frequency f_o for an oblique propagation, Equation 8.20 will provide a relation between the equivalent vertical frequency f_v and the group height h' for vertical propagation on this frequency. Obviously, a vertical ionogram (measured or possibly simulated) will provide an additional relationship between these quantities. Consequently, solving between these two relationships, we can find the f_v and h' pairs that correspond to the possible oblique propagation modes (both low and high rays). This process is illustrated graphically in Figure 8.12, which, for a range $D = 500$ km and sample values of 6.5 and 12.5 MHz for f_o, shows the curves defined by Equation 8.20 (the *transmission curves*) plotted over an ionogram. The points of intersection then identify the oblique propagation modes. It will be noted that, as frequency rises, we will eventually reach a transmission curve that only touches the ionogram, and the corresponding frequency is known as the MUF (*maximum usable frequency*). Above the MUF, the ionosphere cannot support oblique propagation over the given range.

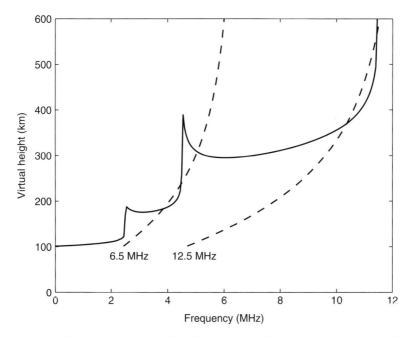

Figure 8.12 Transmission curves plotted over a vertical ionogram for a range of 500 km.

In some circumstances, it is not possible to measure vertical propagation, and one must resort to oblique propagation measurements in which the receiver and transmitter are well separated. If we plot transmitter frequency against measured group delay (or distance) we will have what is known as an *oblique ionogram*. Denoting the group distance for oblique propagation by s', we will have the relation $s' = \sqrt{4h'^2 + D^2}$ and, from the secant law, the relationship $f_o = (f_v/2h')\sqrt{4h'^2 + D^2}$ will follow. It is clear that we can use these two relations to transform vertical ionograms into oblique ionograms and vice versa. In theory, this allows us to use techniques developed for inverting vertical ionograms on oblique ionograms. Figure 8.13 shows three oblique traces derived from the vertical ionogram of Figure 8.10. The ionograms are for the ranges 100, 500 and 1000 km, respectively. It will be noted that the trace for a range of 100 km is very close to that of the vertical ionogram. As the range increases, however, there is a distinct pattern of two returns for frequencies above the peak plasma frequency. These are the high and low rays that we have seen in previous ray tracing. It should be noted that, as the range increases, the above approach becomes less and less effective due to the implicit assumption that the Earth is flat. For ranges of a few hundred kilometers, however, the approach can be quite effective. In this case, it is more correct to make D the chordal distance between the path end points and to make a suitable correction to heights.

When the magnetic field of the Earth is included, vertical propagation becomes much more complex. The propagation is still reflected at a point where $N = 0$ and, from the Appleton–Hartree formula, we find that this occurs when $X = 1$ for an O mode and when $X = 1 \pm Y$ for an X mode. For an X mode traveling upward in a benign

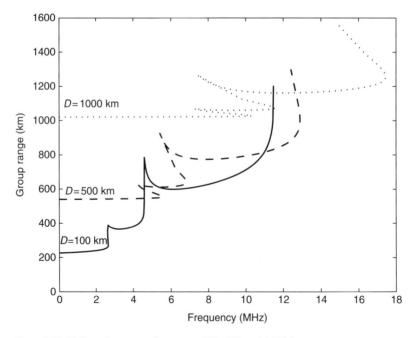

Figure 8.13 Oblique inograms for ranges 100, 500 and 1000 km.

ionosphere, and with $Y < 1$, the wave will usually reach a height where $X = 1 - Y$ before it reaches one where $X = 1 + Y$. Under some circumstances, however, an O mode can be converted into an X mode above the height where $X = 1 - Y$ and so reach an altitude where $X = 1 + Y$. For frequencies on which $Y > 1$ (medium and low frequencies), propagation with $X = 1 - Y$ is not possible and so the X mode can reach a height where $X = 1 + Y$. Due to Earth's magnetic field, waves propagated upward will not remain traveling in a vertical direction, but can deviate considerably. Figure 8.14 shows an example of vertical propagation, for both O and X rays, at a hight latitude location in the Northern Hemisphere. It will be noted that the O and X modes deviate in different directions, toward magnetic north in the case of the O ray (positive direction in the figure) and toward magnetic south (negative direction in the figure) in the case of the X ray. In the case of the O mode, the longitudinal component of the **Y** approaches zero as the reflection point is approached. In order to study vertical and quasi-vertical propagation in more detail, we will need some further results.

We first define the *refractive index surface*. This is a surface in **p** (the wave vector) space and is defined by $|\mathbf{p}| = N$ for a constant X. This is illustrated in Figure 8.15a, which shows a slice through some typical refractive index surfaces. The surfaces are concentric, the largest being that for which $X = 0$ and the other surfaces reducing in size to a point as $X \to 1$. The surfaces do not always reduce to a point, and Figure 8.15b shows the special case of a slice through a magnetic meridian plane for the refractive index surface of an O mode. In this case, the surfaces reduce to a finite length line in the direction of Earth's magnetic field and not a point.

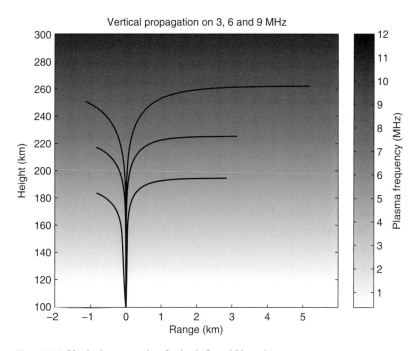

Figure 8.14 Vertical propagation for both O and X modes.

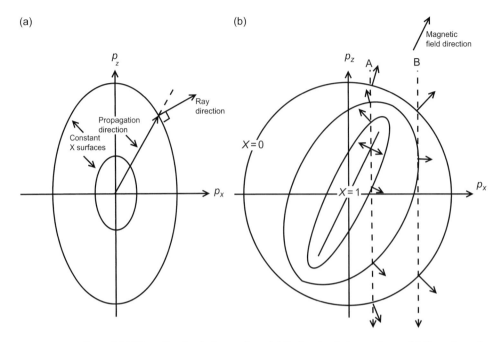

Figure 8.15 The refractive index surface. (a) Refractive index surfaces. (b) O mode surfaces in the meridian plane.

An extremely important result is that ray direction will always be orthogonal to the refractive index surface. We first note that, for a given X, a refractive index surface satisfies $\mathbf{p} \cdot \mathbf{p} - N^2 = 0$ where

$$N^2 = 1 - \frac{2X(1-X)}{2(1-X) - \left(Y^2 - \left(\mathbf{Y} \cdot \frac{\mathbf{p}}{p}\right)^2\right) \pm \sqrt{\left(Y^2 - \left(\mathbf{Y} \cdot \frac{\mathbf{p}}{p}\right)^2\right)^2 + 4(1-X)^2 \left(\mathbf{Y} \cdot \frac{\mathbf{p}}{p}\right)^2}} \quad (8.21)$$

(the $+$ sign corresponds to an ordinary mode and the $-$ sign to an extraordinary mode). A vector $\mathbf{n}$, normal to this surface, will have components

$$n_i = 2p_i - 2N\frac{\partial N}{\partial p_i} \quad (8.22)$$

From Fermat's principle

$$\delta \int_A^B \frac{N}{p} \mathbf{p} \cdot \dot{\mathbf{r}} \, dt = 0 \quad (8.23)$$

(dot indicates a derivative with respect to variable t) we obtain the Euler-Lagrange equation

$$\frac{N\dot{r}_i}{p} - \frac{\mathbf{p} \cdot \dot{\mathbf{r}}}{p}\left(\frac{Np_i}{p^2} - \frac{\partial N}{\partial p_i}\right) = 0 \quad (8.24)$$

8.3 Vertical and Quasi-Vertical Propagation

Noting that $p = N$, 8.22 and 8.24 imply that $\mathbf{n} \propto \dot{\mathbf{r}}$. In the case of a horizontally stratified ionosphere, this result is the basis of an alterative technique for finding ray paths (Poeverlein, 1948, 1949, 1950).

Consider a plane wave that is incident upon a plane stratified ionosphere (the plasma only depends on the vertical coordinate z), then

$$\mathbf{E}(x, y, z) = \mathbf{E}_0 \exp\left(-j\beta_0(xp_x + yp_y + zp_z)\right) \qquad (8.25)$$

as the wave enters the ionosphere. If Earth's magnetic field is sufficiently constant over the propagation region, we can assume a solution of the form

$$\mathbf{E}(x, y, z) = \mathbf{E}_0 \exp\left(-j\beta_0(xp_x + yp_y + \int_0^z p_z dz)\right) \qquad (8.26)$$

where $p_z = \pm\sqrt{N^2 - p_x^2 - p_y^2}$.

The Poeverlein approach consists of the following procedure. At the bottom of the ionosphere the propagation vector components p_x and p_y will be known and will remain constant throughout propagation. We now consider a sequence of relatively close refractive index surfaces with $X = X_0 = 0, X = X_1, X = X_2$, etc. (see Figure 8.16). In $\mathbf{p}$ space, we consider an axis that passes through point $(p_x, p_y, 0)$ and is parallel to the p_z axis. The intersection of this axis with surface $X = X_0$ yields two values of p_y, one corresponding to an upward propagating wave and one to a downward propagating wave. When the wave initially reaches the ionosphere, we obviously choose upward propagation (i.e. we follow the axis downward) and find the normal direction at the point of intersection. This is the ray direction, and, from our point at the bottom of the ionosphere we move in this direction through $\mathbf{r}$ space (the space in which the ray is traced out) until we reach the level of the ionosphere where $X = X_1$. We now repeat the procedure and move to

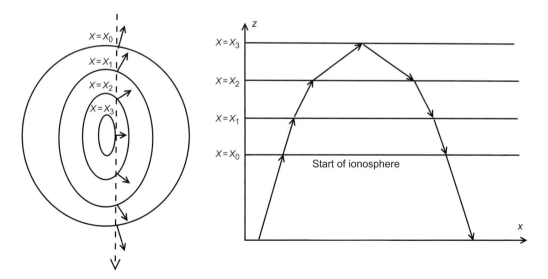

Figure 8.16 Poeverlein's ray tracing method.

the level where $X = X_2$ and continue in this manner until the height where the ray levels out ($X = X_3$ in Figure 8.16). At this level, we continue in **p** space (along the dotted line) until we reach the intersection with $X = X_2$ again and then use the corresponding ray direction to advance in **r** space from the $X = X_3$ level to the $X = X_2$ level. We now continue in **p** space to the $X = X_1$ surface and use the corresponding ray direction to move in **r** space from the $X = X_2$ to the $X = X_1$ level. We then continue in this manner until we reach the ground. The process is normally straightforward, but a special case must be differentiated. Figure 8.15b shows a refractive index surface slice that passes through the magnetic meridian in the case of an O mode. The Poeverlein procedure proceeds normally for the incident wave corresponding to axis B, but there is a problem for axis A when it intersects the $X = 1$ line. The axis no longer touches a surface at any point (the surface at which the ray turns back toward the ground), but only intersects surfaces. When the ray reaches the $X = 1$ level, there is a singularity in the refractive index and the ray will be reflected (i.e. it will rapidly reverse its direction). Importantly, the ray just before and after reflection will be orthogonal to Earth's magnetic field. The wave vector, however, will be in the direction of the magnetic field and will be continuous through the reflection process. This abrupt change in the behavior of the ray forms a cusp that was termed a *Spitze* by Poeverlein. The rays in Figure 8.17 exhibit propagation where a *Spitze* occurs. At lower initial elevations the ray is refracted back to the ground in the normal fashion, but as this elevation increases a *Spitze* forms until we have vertical propagation. It will be noted that there is always a *Spitze* in vertical propagation since this is the case where the axis A goes through the origin of **p** space.

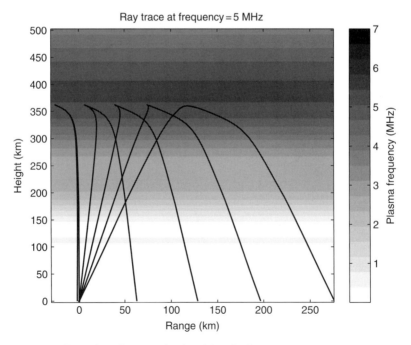

Figure 8.17 Examples of propagation involving the *Spitze*.

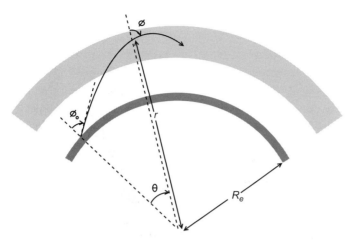

Figure 8.18 2D ray tracing in polar coordinates.

8.4 Oblique Propagation over Long Ranges

The methods described in the previous section are extremely useful over short ranges (a few hundred kilometers). At greater distances, however, the effect of Earth's curvature, and of gradients in the ionosphere, become increasingly difficult to ignore. If we can ignore Earth's magnetic field, and the ionosphere only varies slowly with respect to longitude and latitude, we can employ 2D ray tracing (much of the ionosphere can be dealt with in this fashion) (see Figure 8.18). In this case, we can use the ray tracing Equations 5.39, 5.40 and 5.41, that is

$$\frac{dQ}{dg} = \frac{1}{2}\frac{\partial N^2}{\partial r} + \frac{N^2 - Q^2}{r} \tag{8.27}$$

$$\frac{dr}{dg} = Q \tag{8.28}$$

and

$$\frac{d\theta}{dg} = \frac{\sqrt{N^2 - Q^2}}{r} \tag{8.29}$$

The parameter g that is the group distance along the ray ($dg = ds/N$) and, if required, the phase distance P can be found from the additional equation

$$\frac{dP}{dg} = N^2 \tag{8.30}$$

In 2D ray tracing, we assume that the propagation between two points is in the plane contained by the center of the Earth and these two points (the projection of the path onto the ground is thus a great circle). Consequently, θR_e measures the distance along the ground and r is the height above the center of the Earth (R_e is the radius of the Earth). The above system can be solved by the Runge–Kutta–Fehlberg method (described in Appendix B). This solution approach adjusts the integration steps to allow for significant

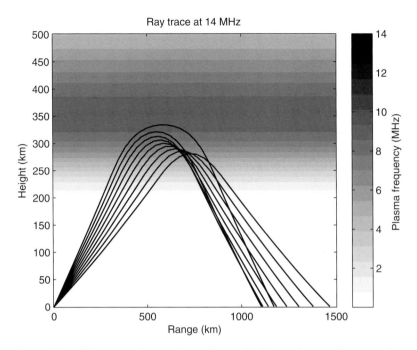

Figure 8.19 A 2D ray trace for a horizontally stratified ionosphere on frequency above peak plasma frequency.

changes in the ionosphere and/or the ray path. Starting at the ground, the initial values of r, θ and Q will be R_e, 0 and $\cos \phi_0$, respectively (ϕ is the angle between the vertical and the ray direction with ϕ_0 its initial value). For long ranges, a 2D ray trace can give a reasonable representation of the propagation. However, as we shall see shortly, there are circumstances where the 2D assumption becomes untenable.

Figure 8.19 shows some 2D ray tracing for a horizontally stratified ionosphere. The figure shows rays traced out for a variety of initial elevations and at a constant frequency that is above the peak plasma frequency. This ray trace exhibits some important features that we have already noted in the chapter on geometric optics. First, for an operating frequency above the peak plasma frequency, there is a region around the source that cannot be reached by the rays (the *skip zone*). Second, beyond the skip zone, all points can be reached by two rays (the *high ray* and the *low ray*). As previously noted, the low rays have *focal points* (i.e. they cross over each other) and the geometric optics solution breaks down around these points It has been shown (Jones, 1994), however, that the geometric optics solution can still be used providing the phase is advanced by $\pi/2$ after passage through the focal point. Figure 8.20 shows some rays for a frequency below the peak plasma frequency. It will now be noted that the skip zone has disappeared and that all rays have a focal point. Figure 8.21 shows a ray trace in an ionosphere with strong horizontal structure in the propagation direction. The propagation is northward from a point south of the equator, the propagation ending north of the equator. The propagation is for an equinox evening during a high sunspot year and clearly shows the effect of the

8.4 Oblique Propagation over Long Ranges

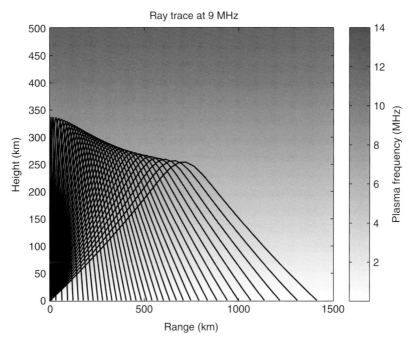

Figure 8.20 A 2D ray trace for a horizontally stratified ionosphere on frequency below peak plasma frequency.

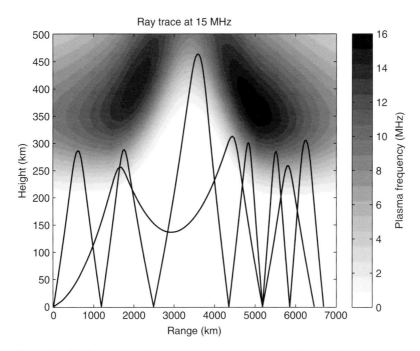

Figure 8.21 A 2D ray trace of propagation across the equatorial anomaly.

equatorial anomaly. While most of the propagation across the anomaly requires several hops (i.e. there are reflections at the ground), there is also propagation that can reach these long ranges via one hop alone. This transequatorial propagation is only possible because of the structure of the anomaly and includes a long section between the anomaly peaks that is outside the ionosphere. (Note that the linear paths between the two anomaly peaks are curved due to our use of the non-Cartesian coordinates of ground range and height.)

In many situations, we require information about a ray between given end points, i.e. we require *point to point ray tracing*. Up to now, however, we have considered ray tracing for which the initial start point and direction are given. We can use such ray tracing to solve the point to point problem through a process known as shooting (or sometimes known as *ray homing*). Essentially, we guess an initial direction and shoot out a ray and see where it lands. We then attempt to improve this initial direction through a minimization process. The function to be minimized is the distance of the landing point from the desired landing point with minimization performed with respect to the initial direction. There are many minimization techniques that can be used, but all will require a good estimate of the initial direction. In the case of 2D ray tracing, the initial bearing is known and we can therefore make a rough search in elevation in order to find elevation pairs with landing points that bracket the desired landing point. Once a suitable pair of bracketing elevations has been found, a suitable minimization technique can be used to refine the brackets to a point where the landing error is small enough. Figures 8.22 and 8.23 show some point to point ray tracing that has been calculated by

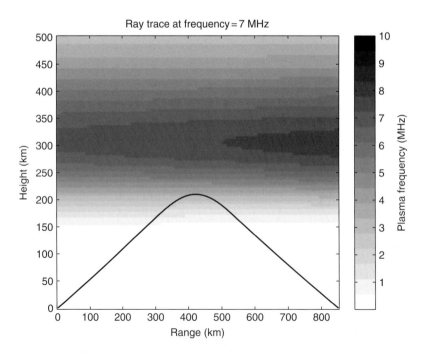

Figure 8.22 A 2D point to point ray trace on a frequency below the peak plasma frequency.

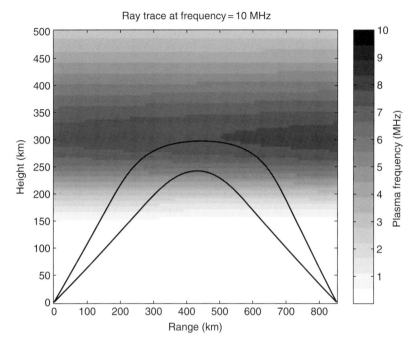

Figure 8.23 A 2D point to point ray trace on a frequency above the peak plasma frequency.

means of the above approach. Figure 8.22 shows some ray tracing for a frequency below the peak plasma frequency and Figure 8.23 for a frequency above. As expected, it will be noted that both high and low rays are present for the wave frequency above the peak plasma frequency.

If horizontal gradients in plasma become large, it becomes necessary to use full 3D ray tracing equations such as (see Chapter 5)

$$\frac{d^2\mathbf{r}}{dg^2} = \frac{1}{2}\nabla N^2 \tag{8.31}$$

where $\mathbf{r}$ is now a 3D position vector. We will, however, to go one step further and include the effect of Earth's magnetic field and hence the split the propagation into the O and X modes. For ordinary rays, as we have already seen in the previous section, the vertical ray is reflected at the same height as the field free ray and so the above ray tracing can be expected to give a reasonable approximation to the ordinary ray. In the case of the extraordinary ray, however, the vertical ray is reflected at a height where $f_p^2/f^2 = 1 - f_H/f$ where f_H is the gyro frequency, f_p is the plasma frequency and f is the wave frequency. As a consequence, this suggests that an approximation to the X ray in which we employ the above field-free ray tracing at an effective frequency is given by $f_{\text{eff}} = \sqrt{f^2 - f f_H}$. Such an approach is found to give a reasonable first approximation to the X ray in many cases. A more correct approach, however, is to employ the Haselgrove equations (Haselgrove, 1954) that were derived in Chapter 5. The relevant equations are 5.77, 5.81 and 5.80, i.e.

$$\frac{dx_i}{dt} = 4p_i(q+1)(1-X-Y^2) + 2p_iY^2 + 2p_i(\mathbf{p}\cdot\mathbf{Y})^2 + 2Y_iXq\mathbf{p}\cdot\mathbf{Y} \quad (8.32)$$

and

$$\frac{dp_i}{dt} = \left((1-Y^2)(q+1)^2 - 2(1-X-Y^2)(q+1) - Y^2\right)\frac{\partial X}{\partial x_i}$$
$$+ 2q(q+1)X\mathbf{Y}\cdot\frac{\partial \mathbf{Y}}{\partial x_i} - 2qX(\mathbf{p}\cdot\mathbf{Y})\mathbf{p}\cdot\frac{\partial \mathbf{Y}}{\partial x_i} \quad (8.33)$$

where independent variable t parameterizes the path and q is a solution of the equation

$$q^2(1-X-Y^2) + q((\mathbf{p}\cdot\mathbf{Y})^2 + 2 - 2X - Y^2) + 1 - X = 0 \quad (8.34)$$

The above equation is of the form $\alpha q^2 + 2\beta q + \gamma = 0$, which has solutions of the form $q = (-\beta \pm \sqrt{\beta^2 - \alpha\gamma})/\alpha$ where we choose $+$ for the O mode and $-$ for the X mode. An alternative form of this solution is $q = \gamma/(-\beta \mp \sqrt{\beta^2 - \alpha\gamma})$ and this can be useful if α approaches zero. We now have a system of equations that can be solved by numerical techniques such as the Runge–Kutta–Fehlberg method. For a ray that starts outside the ionosphere, the initial value of $\mathbf{p}$ is a unit vector in the direction of propagation. Figures 8.24 and 8.25 show some examples of ray tracing northward across the equatorial anomaly. Figure 8.24 shows the vertical behavior of the ray traces and Figure 8.25 shows the horizontal behavior. The ordinary rays are shown as dashed curves and the extraordinary rays as solid curves. It will be noted that the ray whose first hop lands before the equator has a considerably wider deviation from the great circle path.

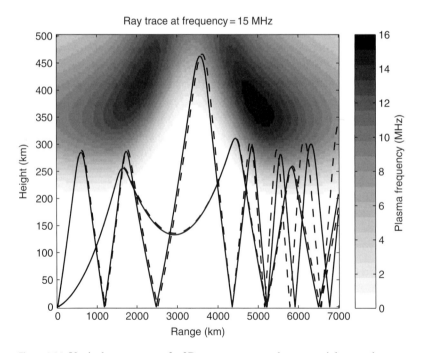

Figure 8.24 Vertical component of a 3D ray trace across the equatorial anomaly.

8.4 Oblique Propagation over Long Ranges

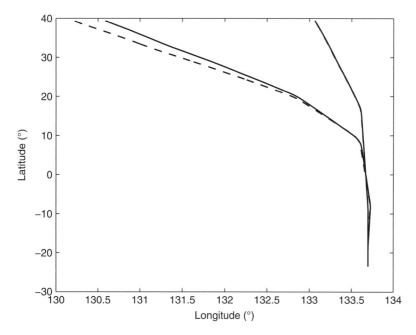

Figure 8.25 Horizontal component of a 3D ray trace across the equatorial anomaly. Deviations from great circle path

The Haselgrove equations are best suited to problems where we know the initial position and direction of the propagation. Consequently, if we require the propagation between two given points (point to point ray tracing), we will need to implement a shooting (homing) method. Starting from one of the end points, we make an initial guess of the direction, possibly derived from the easier 2D point to point problem (discussed earlier in this section). We then use the Haselgrove equations to trace the ray to find its end point and hence find the landing error (distance of the calculated end point from the desired end point). We now iteratively refine our initial guess to minimize the error (the process homes in on the desired end point). One approach (Strangeways, 2000) is to use the Nelder–Mead simplex algorithm to home in on the desired end point by minimizing the landing error with respect to the initial direction. We use our initial guess to form a triangle that brackets the initial direction (a point in elevation and azimuth space) and then calculate the landing point of the vertices and hence the error corresponding to these vertices. The Nelder–Mead algorithm then steadily refines the vertices so that the triangle shrinks in a way that steadily reduces the landing error at the vertices until an acceptable level of error is achieved. As an example, consider the particularly demanding example of propagation in an ionosphere that is perturbed by a TID of wavelength 300 km traveling orthogonal to the propagation direction. Figure 8.26 shows the vertical aspect of propagation at a frequency of 5 MHz for a short path (70 km); the figure shows rays at various times during a period of the TID. Figure 8.27 shows the horizontal aspect of propagation for two snapshots (separated by half a TID period). It will be noted that the propagation is well removed from the great

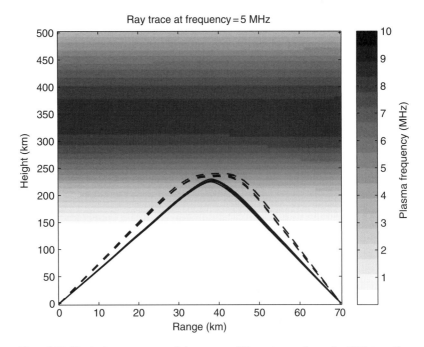

Figure 8.26 Vertical components of short-range 3D ray traces through a TID traveling orthogonal to the propagation direction.

circle path of the 2D approximation (the dotted line) and that there is considerable variation over the period of the gravity wave. This serves to emphasize the fact that purely 2D ray tracing is inadequate when there are significant horizontal gradients. Figures 8.28 and 8.29 show propagation through a TID that travels in the direction of the radio wave; propagation is now over a much longer range and at a frequency close to the MUF. The figures show propagation separated in time by half a TID period. First, it will be noted that there are now both high and low rays since the propagation is well above the background peak plasma frequency and, second, that the paths have far more complex vertical behavior. Propagation of this sort can pose serious difficulties for homing methods and, in particular, for high rays. Due to this, the propagation displayed in Figures 8.28 and 8.29 was calculated using the direct variational method of Chapter 5. This method is better suited to long range point to point ray tracing but, as with the homing approach, requires a good initial estimate of the propagation (this can once again be provided by a 2D ray trace). Figures 8.28 and 8.29 show the vertical component of propagation at times separated by half a TID period, while Figures 8.30 and 8.31 show the corresponding horizontal deviations of the various rays (the great circle path is shown as a dotted line). In these figures, the low rays deviate very little from the great circle path, while the high rays deviate substantially more. The deviation, however, is far less than for the previous example of short ranges. The effect of TIDs on HF propagation has been studied by several authors; for more detail the reader should consult the papers by Stocker et al. (2000) and Earl (1975).

8.4 Oblique Propagation over Long Ranges

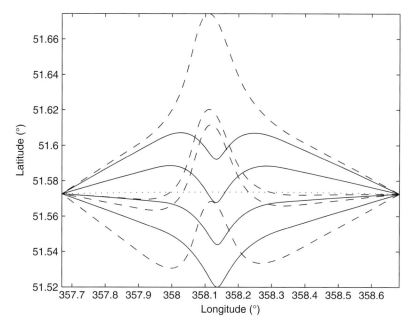

Figure 8.27 Horizontal component of short-range 3D ray traces through a TID. Deviations from great circle path.

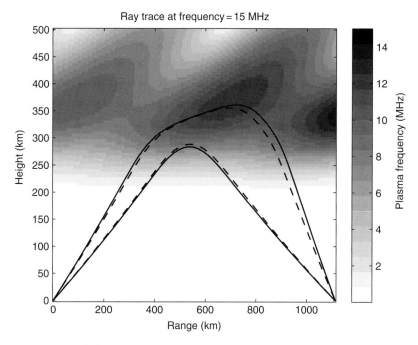

Figure 8.28 Vertical components of a 3D ray trace through a TID traveling in the same direction as the radio wave.

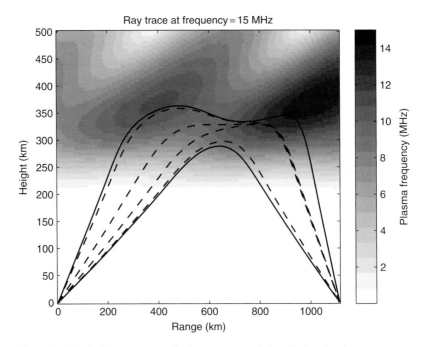

Figure 8.29 Vertical components of a 3D ray trace at half a TID period later.

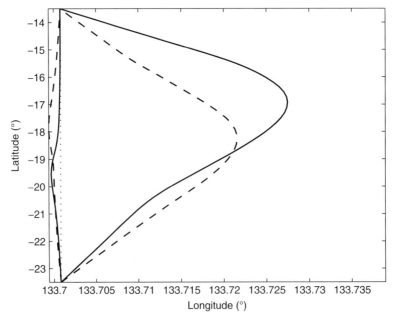

Figure 8.30 Horizontal components of the 3D ray trace. Deviations from great circle path.

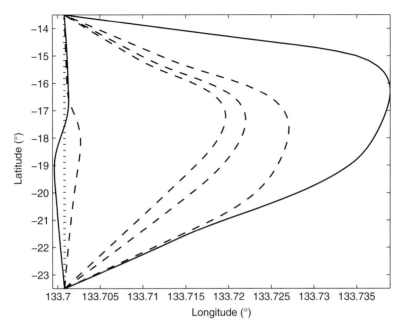

Figure 8.31 Horizontal components of the 3D ray trace at half a TID period later. Deviations from great circle path.

8.5 Propagation Losses

Ray tracing provides us with the path of propagation, but a full description of the field will require knowledge of the way in which the polarization and amplitude change along a ray. The major change in amplitude occurs as the rays spread out from their source, and, in the case of an isotropic homogenous medium, the amplitude will fall away in proportion to the inverse of distance traveled. In more complex media, this will only provide a rough estimate of changes in amplitude. Refractive effects can cause focusing and de-focusing and so we must look at this issue in further detail. Importantly, assuming we can ignore ionospheric collisions, the ray direction is also the direction in which energy propagates. Consequently, energy within a small bundle of rays will remain confined within that bundle. The power will fall off within the ray bundle as the inverse of cross sectional area (the amplitude falls off as the inverse of the square root of this area). We can consider the change in amplitude on a ray by considering how the rays in the immediate vicinity deviate from that ray. This could be achieved by looking at a small bundle of rays instead of a single ray, but an alternative is to use the equations for ray deviations about the ray of interest. We will first consider the simpler problem of the case where Earth's magnetic field can be ignored and 2D ray tracing can be used. For 2D ray tracing in polar form, we can employ the differential equations (5.37 and 5.37) for the deviations δr and δQ in the quantities r and Q. These are

$$\frac{d(\delta Q)}{dg} = \frac{\delta r}{2}\left(\frac{1}{r}\frac{\partial N^2}{\partial r} + \frac{\partial^2 N^2}{\partial r^2}\right) - \left(\frac{1}{2(N^2 - Q^2)}\frac{\partial N^2}{\partial r} - \frac{1}{r}\right)\left(\frac{\delta r}{2}\frac{\partial N^2}{\partial r} - Q\delta Q\right)$$

(8.35)

and

$$\frac{d(\delta r)}{dg} = \frac{\delta r Q}{r} + \delta Q - \frac{Q}{N^2 - Q^2}\left(\frac{\delta r}{2}\frac{\partial N^2}{\partial r} - Q\delta Q\right) \tag{8.36}$$

and these can be solved alongside the equations for r and Q. We take the initial deviation in ray elevation to be 1, and then the initial value of δQ will then be $N_0 \sin\phi_0$ (the initial deviation of δr will be 0). The above equations provide us with information about the way in which rays deviate in altitude, but we also need to know how the rays deviate laterally. In the 2D approach, we assume that the rays remain on a great circle path and so nearby rays will initially diverge laterally and then come together at the antipodal point (the observed phenomenon of antipodal focusing). Putting all of this together, the power will fall off as the inverse square of *effective distance* s_eff where

$$s_\text{eff}^2 = R_e \sin\theta \delta r \frac{\sin\phi}{\sin\phi_0} \tag{8.37}$$

and R_e is the radius of Earth. In terms of dB, the *spreading loss* (as it is known) is given by

$$L_\text{sprd} = -20\log_{10}\left(\frac{\lambda}{4\pi s_\text{eff}}\right) \tag{8.38}$$

When horizontal variations become significant, we must resort to full 3D ray tracing. In this case, assuming Earth's magnetic field can be ignored, the effective distance can be calculated from $s_\text{eff}^2 = s_+ s_-$ where s_+ and s_- satisfy (Coleman, 2002)

$$\frac{d^2 s_\pm}{dg^2} = -\left(\gamma^{11} + \gamma^{22} \pm \sqrt{(\gamma^{11} - \gamma^{22})^2 + 4\gamma^{12^2}}\right) s_\pm \tag{8.39}$$

with boundary conditions $ds_\pm/dg = 1$ and $s_\pm = 0$ at the source. (Note that in solving Equation 8.39 it is advisable to use the factorization

$$\sqrt{(\gamma^{11} - \gamma^{22})^2 + 4\gamma^{12^2}} = \sqrt{\gamma^{11} - \gamma^{22} + 2j\gamma^{12}}\sqrt{\gamma^{11} - \gamma^{22} - 2j\gamma^{12}} \tag{8.40}$$

to avoid problems with change of branch that can occur on some rays.) The coefficients γ^{ij} are given by

$$\gamma^{ij} = \sum_{m=1}^{3}\sum_{l=1}^{3} P_l^i P_m^j \left(-\frac{1}{4}\frac{\partial^2 N^2}{\partial x_m \partial x_l} + \frac{3}{8N^2}\frac{\partial N^2}{\partial x_m}\frac{\partial N^2}{\partial x_l}\right) \tag{8.41}$$

where $\mathbf{P}^1$ and $\mathbf{P}^2$ are vectors orthogonal to each other and to the ray path (the polarization vectors). $\mathbf{P}^1$ and $\mathbf{P}^2$ will be unit vectors that satisfy

$$\frac{d\mathbf{P}^i}{dg} = -\frac{1}{2N^2}\frac{d\mathbf{r}}{dg}\mathbf{P}^i \cdot \nabla N^2 \tag{8.42}$$

With suitably chosen initial orientation, these vectors will trace out the directions of the electric and magnetic fields. It should be noted that s_+ and s_- are the wavefront principal radii of curvature and, as such, provide invaluable information concerning the distortions caused by the propagation medium.

Thus far, all considerations of loss have ignored magneto-ionic effects and we need to now consider their inclusion. As in the magnetic field free case, we can trace out a tube of rays and calculate the change in cross section. Alternatively, we can calculate ray deviations using the appropriate accessory equation (see Appendix C). (The approach based on ray deviations has been considered by Nickisch (1988) for the situation where magneto-ionic effects can be ignored.) In the direct variational approach of Chapter 5, the form of the equations for calculating the refinement of a solution at each iteration of the Newton–Raphson method turn out to be identical to the equations for calculating the deviations. Consequently, the calculation of spreading loss can be a byproduct of the direct variational method. In the homing approach, the Nelder–Mead optimization algorithm develops a triangular tube of rays that converges on the true ray. Consequently, we can use this tube to find out how the cross section changes. Once again, the spreading loss is a byproduct of the ray tracing.

As already mentioned, a primary loss mechanism in ionospheric propagation occurs through electron collisions. We have seen that such collisions can be incorporated into the field equations ($U \neq 1$) and hence into the ray tracing equations. When the loss is relatively weak, the collision-free ray tracing equations can be used and the effect of collisions introduced through a reduction in field amplitude. To simplify matters we will ignore magneto-ionic effects and consider the case of a horizontally stratified ionosphere (vertical axis in the z direction). For a plane wave that is incident upon the ionosphere, we can take the components p_x and p_y of the wave vector $\mathbf{p}$ to be constant (see Section 8.8). We note that $\mathbf{p} \cdot \mathbf{p} = N^2$ and so, ignoring Earth's magnetic field, we obtain that

$$\mathbf{p} \cdot \mathbf{p} = 1 - \frac{X}{U} = 1 - X \frac{\omega^2 + j\nu\omega}{\omega^2 + \nu^2} \tag{8.43}$$

It is obvious that p_z will become complex in the ionosphere, i.e. $p_z = p_z^r + jp_z^i$ and so

$$p_x^2 + p_y^2 + p_z^{r2} - p_z^{i2} = 1 - X\frac{\omega^2}{\omega^2 + \nu^2} \tag{8.44}$$

and

$$2p_z^r p_z^i = -X\frac{\nu\omega}{\omega^2 + \nu^2} \tag{8.45}$$

If $\nu \ll \omega$ then $p_z^i \approx -X\nu/2p_z^r\omega$ and, from Equation 8.26, we find that the wave amplitude reduction due to collision is given by $\exp(\beta_0 \int_0^z p_z^i dz)$. Noting that $(1 - X)dz \approx p_z^r ds$, we find that the reduction in amplitude will be $\exp(-\int_0^s \kappa ds)$ where the integral is along the ray path and $\kappa = X\nu/2c_0(1 - X)$ is the absorption constant. A more accurate expression for the absorption coefficient κ, incorporating magneto-ionic effects, is (Davies, 1990)

$$\kappa = \frac{X\nu}{2c_0(1 - X)} \frac{\omega^2}{(\omega \pm \omega_L)^2 + \nu^2} \tag{8.46}$$

where $\omega_L = \omega Y_L$ with Y_L the component of $\mathbf{Y}$ in the propagation direction (+ refers to the ordinary ray and − refers to the extraordinary ray). In terms of dB, the collision loss is given by

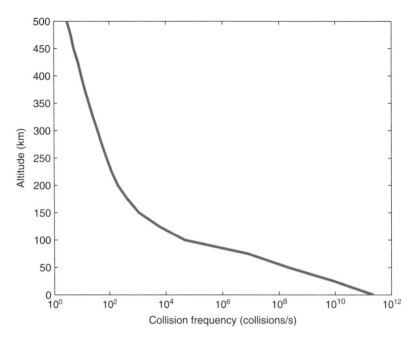

Figure 8.32 Electron-neutral collision frequency as a function of height.

$$L_{coll} = 8.69 \int_{\text{ray path}} \kappa \, ds \tag{8.47}$$

where the integral is taken along the ray path. The collision frequency ν (collisions/s) is related to temperature T (K) and neutral density n (particles/m³) through $\nu = 5.4 \times 10^{-16} n \sqrt{T}$. Figure 8.32 shows a typical distribution of collision frequency with the largest values below the level of the E layer. There are two major regions on a ray that contribute to the above loss. First, loss will be large in the D layer where both ν and N_e are relatively large (N being close to 1 here). Second, close to the peak of a ray, N can be quite small and hence give rise to appreciable losses. These two losses are sometimes termed *nondeviative* and *deviative*, respectively.

For a horizontally stratified ionosphere, Equation 8.47 can be rewritten as

$$L_{coll} = 8.69 \int_{\text{ray path}} \frac{X\nu}{2c_0 p_z^r} \, dz \tag{8.48}$$

If we consider a ray that is confined to the xz plane, then $p_z^r \approx \sqrt{C^2 - X}$ where $C = \cos\phi_0$ and ϕ_0 is the initial angle of the ray to the vertical. Consequently,

$$L_{coll} = 8.69 \int_0^{z_0} \frac{X\nu}{c_0\sqrt{C^2 - X}} \, dz = 8.69 C \int_0^{z_0} \frac{\frac{f_p^2}{f_v^2}\nu}{c_0\sqrt{1 - \frac{f_p^2}{f_v^2}}} \, dz \tag{8.49}$$

where z_0 is the height of reflection and $f_v = Cf$ is the equivalent vertical frequency. Let $L_{coll}^v(f)$ denote the collision loss for vertical propagation on frequency f, then 8.49

will imply that $L_{\text{coll}}(f) = CL^v_{\text{coll}}(Cf)$. This result is known as *Martyn's theorem for attenuation* (Martyn, 1935).

For multi-hop propagation, we will also need to take account of the losses as the wave is reflected from the ground. The loss depends on polarization, which is difficult to predict in a realistic ionosphere due to the effects of Faraday rotation. Consequently, we often use a reflection loss that is an average of the horizontal and vertical loss, i.e.

$$L_{\text{rfl}} = -10 \log_{10} \left(\frac{|R_V|^2 + |R_H|^2}{2} \right) \quad (8.50)$$

in terms of *dB* where R_V and R_H are the vertical and horizontal reflection coefficients, respectively (see Chapter 2). In terms of dB, the total loss over any propagation path will be the sum of the spreading, collision and reflection losses, i.e.

$$L_{\text{total}} = \underbrace{\sum L_{\text{sprd}}}_{\text{hops}} + \underbrace{\sum L_{\text{coll}}}_{\text{hops}} + \underbrace{\sum L_{\text{rfl}}}_{\text{reflections}} \quad (8.51)$$

8.6 Fading

In the ionospheric duct, a signal can often propagate between a transmitter and a receiver by several modes. These can consist of the high- and low-ray modes and also multi-hop combinations of these (we will assume the frequency is high enough for magneto-ionic effects to be dealt with through Faraday rotation). If a signal $s_{Tx}(t) = m(t) \exp(-2\pi j f_c t)$ is transmitted, then the signal

$$s_{Rx}(t) = \sum_i \frac{\cos \phi_i}{\sqrt{L_i}} \exp(2\pi j f_c (t - \tau_p^i)) m(t - \tau_g^i) \quad (8.52)$$

will be received. The sum will be over all modes with τ_p^i the phase delay of the *i*th mode, τ_g^i the group delay and L_i the loss (absorption, ground reflection, etc.). If a signal propagates via mode *i*, there will also be a Faraday rotation through angle ϕ_i and the factor $\cos \phi_i$ will express the reduction in amplitude due to polarization mismatch between the wave and receiver antenna as a result of this rotation.

It will be noted that a received signal consists of a linear combination of copies of the original signal that have been time shifted with respect to each other. As a consequence, signal processing will be required to extract the original signal. This, however, is not the only problem. The ionosphere can be an extremely dynamic medium, especially with the passage of a TID. This can cause the received signal to vary quite considerably in amplitude over time, a phenomena that is known as *fading*. In particular, the amplitude can vary significantly due to variations in the Faraday rotation ϕ_i, this being known as *polarization fading*. In addition, the spatial variations caused by TIDs can cause the spreading loss to vary in time (focusing and defocusing), resulting in *amplitude fading*. Both of the above types of fading will occur whatever the number of modes. When there is more than one mode, however, there is a third kind of fading that is known as *multipath fading*. If there is more than one propagation path, there is the possibility of interference between the modes. The paths can pass through considerably different regions of the ionosphere, and so a disturbance, such as a TID, can lead to differential phase delays between modes that vary with time and hence cause variations in the interference. Fading is a big problem in HF systems and its effects need to be included in any realistic modeling.

8.7 Noise

In a radio system, a desired signal must compete with undesired signals, or *noise*. There will always be noise generated within the electronics of the receiver, but this is normally under the control of the radio designer. This *internal noise* is most affected by the first stages of a radio receiver, and modern components have brought its intensity down to very low levels. The designer, however, has no control over the noise that enters the receiver, along with the desired signal, through the antenna. This *external noise* can arise from man-made sources (ignition interference, for example) and from natural sources (lightning strikes and galactic noise). Man-made noise will usually propagate from source to receiver via surface wave propagation, lightning noise by both surface and ionospheric duct propagation and galactic noise by transionospheric propagation. Figure 8.33 shows the contributions to noise for HF (3–30 MHz) around dusk for a site in the vicinity of Bath in the UK. These figures were calculated using data from CCIR (1964) and ITU (1999) models. In the figure the dotted and the top dashed curves show the behavior of manmade noise (rural and urban), the lower dashed curve shows

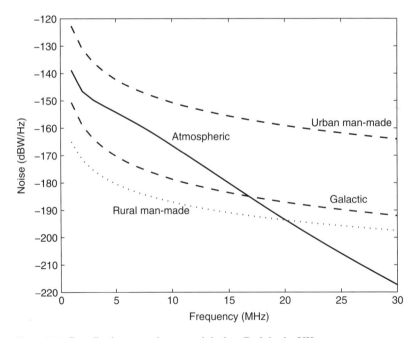

Figure 8.33 Contributions to noise around dusk at Bath in the UK.

galactic noise and the solid curve atmospheric noise. Except for the immediate vicinity of a thunderstorm, most lightning noise will arrive from remote locations by means of propagation in the ionospheric duct. The atmospheric noise drops off significantly with frequency due to the reduction in propagation support as frequency rises. Atmospheric noise varies considerably with time, frequency and location due to spatial and temporal variations in the ionosphere and hence the propagation. Additionally, the rate of lightning strikes can vary considerably over the globe and this can further add to the variability. Figure 8.34 shows the global distribution of lightning strike rates at 12 UT in December (Kotaki and Kato, 1984). Using a simple virtual height propagation model, Kato (1984) has shown how the distribution of lightning strikes can translate into an effective model of the distribution of atmospheric noise. The distribution provided by the ITU and Kato models is in terms of total noise that is received by a standard monopole antenna. It is often assumed that noise is isotropic and hence that this noise is the same for all antennas. This is a questionable assumption that we will consider further.

From the considerations of this chapter, it is clear that HF propagation will not take place for all elevations, and, as a consequence, noise will be elevation dependent. Further, due to the global variation in the propagation medium, and the distribution of lightning, it will also be azimuth dependent. As a consequence, different antennas will collect different amounts of noise and the consideration of the directionality of noise will be important for the accurate prediction of the performance of a radio system. Based on ray tracing through realistic ionospheric models, Coleman (2002) has developed

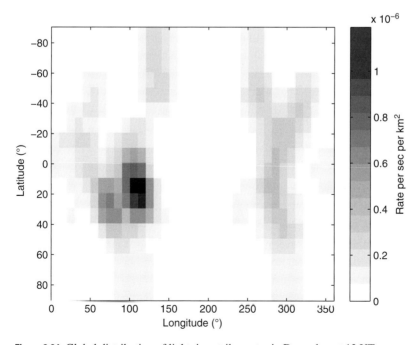

Figure 8.34 Global distribution of lightning strikes rates in December at 12 UT.

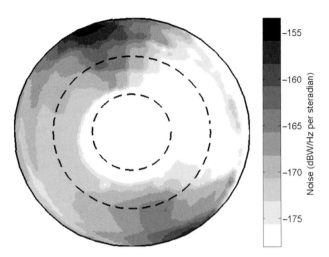

Figure 8.35 Alice Springs angular distribution of noise at dusk for a frequency of 10 MHz.

a directional model of atmospheric noise. At a particular location, the noise density (W/Hz/sr) for a given bearing θ and elevation ϕ is given by

$$N_a(\theta, \phi) = \sum WS \left(\frac{\lambda}{4\pi}\right)^2 \frac{1}{L \sin \phi} \qquad (8.53)$$

where W is the radiated energy density (J/Hz) of a lightning strike and S is the strike rate (strikes per second per unit area) at the source of propagation. L is the loss due to ground reflection and ionospheric absorption and λ is the wavelength. Note that the sum in Equation 8.53 is over all propagation that arrives from the bearing θ and elevation ϕ. (Data given in (Jursa, 1985) suggest values of 2×10^{-6} J/Hz and 10^{-5} J/Hz for W at frequencies of 10 and 2.5 MHz, respectively.) Figures 8.35 and 8.36 show the directional noise around dusk calculated from this model for Alice Springs in Australia and Bath in the UK. Both figures show a very strong directionality in the distribution of noise and demonstrate the need to consider noise in choosing an antenna for the best signal to noise ratio.

8.8 Full Wave Solutions

At low frequencies (below about 3 MHz), the geometric optics approach ceases to be valid and so a full wave solution becomes necessary. One approach that has been extensively studied (Wait, 1996) is to treat the region between the ground and ionosphere as a waveguide in which the radio waves are trapped. Due to the low frequency, we assume the ground can be approximated as a PEC material, and, in the simplest model, the ionosphere is also assumed to be a PEC material. A trapped wave in this waveguide will consist of both upward and downward plane waves. We will study the problem in terms of simple modes that are composed of two plane waves traveling at angles $\pm \alpha$ to

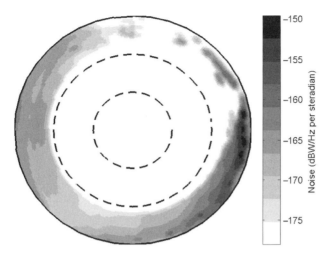

Figure 8.36 Bath angular distribution of noise at dusk for a frequency of 10 MHz.

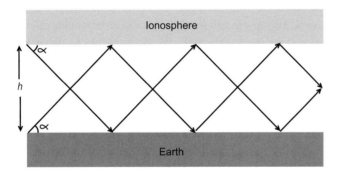

Figure 8.37 Dual plane wave ionospheric mode.

the horizontal (the x direction in this case). More complex problems can be considered to be a combination of such modes (Figure 8.37). The electric field of our simple mode will take the form

$$\mathbf{E} = \mathbf{E}_0^{\text{down}} \exp(-j\beta \left(-z\sin\alpha + x\cos\alpha \right)) + \mathbf{E}_0^{\text{up}} \exp(-j\beta \left(z\sin\alpha + x\cos\alpha \right)) \quad (8.54)$$

where $\mathbf{E}_0^{\text{down}} = E_V \left(\cos\alpha\, \hat{\mathbf{z}} + \sin\alpha\, \hat{\mathbf{x}} \right) + E_H \hat{\mathbf{y}}$ and $\mathbf{E}_0^{\text{up}} = E_V \left(\cos\alpha\, \hat{\mathbf{z}} - \sin\alpha\, \hat{\mathbf{x}} \right) - E_H \hat{\mathbf{y}}$. This field can be rearranged into the form

$$\mathbf{E} = 2E_V \left(\cos\alpha\, \hat{\mathbf{z}} \cos(\beta z \sin\alpha) + j\sin\alpha\, \hat{\mathbf{x}} \sin(\beta z \sin\alpha) \right) \exp(-j\beta x \cos\alpha)$$
$$+ 2jE_H \hat{\mathbf{y}} \sin(\beta z \sin\alpha) \exp(-j\beta x \cos\alpha) \quad (8.55)$$

We note that $\mathbf{E}_0^{\text{down}} + \mathbf{E}_0^{\text{up}}$ does not have a component parallel to the ground ($z = 0$) and so the PEC boundary condition is satisfied. At the height of the ionosphere ($z = h$), however, the PEC condition requires that $h\beta \sin\alpha = n\pi$ where n is an arbitrary positive integer. Consequently, propagation will only be possible on modes for which $\beta \cos\alpha = \beta_n = \sqrt{\beta^2 - n^2\pi^2/h^2}$. It will be noted that there is a *cut-off frequency*

$\omega_n = n\pi c/h$ below which the propagation constant β_n will become imaginary and hence not allow a propagating wave. Below the cut-off frequency, the fields will be exponentially decaying in the propagation direction and are said to be *evanescent*. The phase speed of the composite wave is given by

$$v_p = \frac{\omega}{\beta_n} = \frac{c}{\sqrt{1 - \frac{\omega_n^2}{\omega^2}}} \qquad (8.56)$$

and from which it is seen to be dispersive.

A more realistic approximation for the ionosphere is to represent it as PMC material (Davies, 1990) and this implies that $h\beta \sin\alpha = (n - 1/2)\pi$, where n is an arbitrary positive integer. Consequently, propagation will only be possible on modes for which $\beta \cos\alpha = \beta_n = \sqrt{\beta^2 - (2n-1)^2 \pi^2 / 4h^2}$ and the cut-off frequency will now be given by $\omega_n = (2n-1)\pi c/2h$. It is important to note that the phase speed is dependent on both wave frequency and the height of the ionosphere. At low frequencies, the D layer is the dominant reflecting layer of the ionosphere and this occurs at a height of around 80 km. The dynamics of this layer are fairly simple, but it does have a diurnal variation that satisfies Equation 8.6. This in turn results in a diurnal variation of phase propagation speed for the modes in the ionospheric duct. Observations of this variation have shown the above simple theory to be in accord with the real situation (Davies, 1990). Because of the curvature of Earth, the duct between ionosphere and ground will eventually come back upon itself after a distance $D_S = 2\pi a$ ($a = 6371.135$ km is the radius of the Earth) and this introduces the possibility of resonances. These resonances are known as Schumann resonances (Davies, 1990) and for which the resonance frequency will approximately satisfy $D_S \beta_n = 2m\pi$ where m is an integer. The lowest resonance occurs at approximately 8 kHz and this mode is often excited by lightning strokes.

The above simple model of the Earth–ionosphere duct is useful at very low frequencies, but it ignores the complexity of a real ionosphere and its interaction with Earth's magnetic field. In general, the electric field will satisfy

$$\nabla \times \nabla \times \mathbf{E} - \omega^2 \mu_0 \epsilon_0 \left(1 - \frac{UX}{U^2 - Y^2}\right) \mathbf{E} = \omega^2 \mu_0 \epsilon_0 \frac{X}{U^2 - Y^2} \left(j\mathbf{Y} \times \mathbf{E} + \frac{1}{U} \mathbf{Y} \cdot \mathbf{E} \mathbf{Y}\right) \qquad (8.57)$$

This system of equations, however, is extremely difficult to solve in its full generality and so we seek some simplifications. At frequencies considerably above the gyro frequency (10 MHz and above), we can use the vector Helmholtz equation

$$\nabla^2 \mathbf{E} + \omega^2 \mu_0 \epsilon_0 (1 - X) \mathbf{E} = 0 \qquad (8.58)$$

for all components of the electric field. In this case, one possible approach to its solution is to make use of the paraxial approximation of Chapter 7. In this approximation, we assume that the propagation is predominantly parallel to the horizontal (the x axis) and replace Equation 8.58 by

$$-2j\beta_0 \frac{\partial \mathbf{E}_a}{\partial x} + \nabla_T^2 \mathbf{E}_a + \omega^2 \mu_0 \epsilon_0 \left(\frac{z}{a} - X\right) \mathbf{E}_a = 0 \qquad (8.59)$$

8.8 Full Wave Solutions

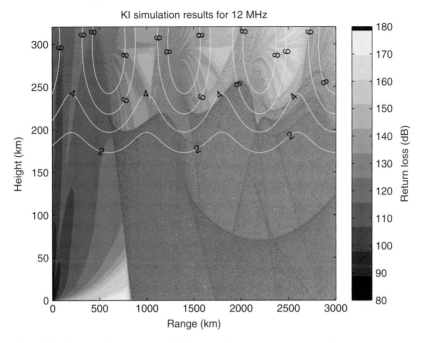

Figure 8.38 Propagation through a TID using the paraxial approximation.

where $\nabla_T^2 = \partial^2/\partial y^2 + \partial^2/\partial z^2$ is the Laplace operator in the transverse direction and $\mathbf{E} = \exp(-j\beta_0 x)\mathbf{E}_a$. (Note that we have used an effective refractive index by including the term z/a and hence the coordinate z is the height above the surface of the Earth and x is the distance along the surface.) We will assume that there is one dominant direction of polarization (vertical or horizontal) and then the electric field will be described by a scalar E_a. We can use the integral equation techniques of Chapter 7 to solve this problem; Figure 8.38 shows an example of its use. The figure shows the power loss for propagation at 12 MHz through an ionosphere that is perturbed by a TID. The plasma frequency (in MHz) is superimposed upon the losses and illustrates how energy can leak out of the ionospheric duct through troughs in the plasma that are formed by the TID.

At frequencies considerably below the gyro frequency, we can use the large Y limit of Equation 8.57, i.e.

$$\nabla \times \nabla \times \mathbf{E} - \omega^2 \mu_0 \epsilon_0 \mathbf{E} = -\omega^2 \mu_0 \epsilon_0 \frac{X}{U} \hat{\mathbf{Y}} \cdot \mathbf{E} \hat{\mathbf{Y}} \qquad (8.60)$$

where $\hat{\mathbf{Y}}$ is a unit vector in the direction of $\mathbf{Y}$. In this limit, it will be noted that the component of electric field that is parallel to Earth's magnetic field behaves as if the Earth had no magnetic field and the components orthogonal to Earth's field behave as if in free space.

When the ionosphere is horizontally stratified (i.e. only z dependence), and the magnetic field is constant, we can take the analysis a bit further. Consider a plane wave that is incident upon the bottom of the ionosphere, then the solution will behave as a combination of upward and downward plane waves outside the ionosphere. In this case, we can assume a solution of the form

$$\mathbf{E}(x, y, z) = \mathbf{E}_0(z) \exp\left(-j\beta_0(xp_x + yp_y)\right) \tag{8.61}$$

where z is the vertical coordinate. If we substitute Equation 8.61 into 8.57 we obtain

$$\frac{d^2\mathbf{E}_0}{dz^2} + \beta_0^2 \left(1 - p_x^2 - p_y^2 - \frac{UX}{U^2 - Y^2}\right) \mathbf{E}_0$$
$$= -\beta_0^2 \frac{X}{U^2 - Y^2} \left(j\mathbf{Y} \times \mathbf{E}_0 + \frac{1}{U}\mathbf{Y} \cdot \mathbf{E}_0\mathbf{Y}\right) \tag{8.62}$$

where we have assumed that the length scales of the medium are much larger than a wavelength. This is a system of three second-order ODEs for the components of field $\mathbf{E}_0$. If we define a new field $\mathbf{W} = d\mathbf{E}_0/dz$, we will have a system of six ordinary differential equations

$$\frac{d\mathbf{W}}{dz} = -\beta_0^2 \left(1 - p_x^2 - p_y^2 - \frac{UX}{U^2 - Y^2}\right) \mathbf{E}_0$$
$$- \beta_0^2 \frac{X}{U^2 - Y^2} \left(j\mathbf{Y} \times \mathbf{E}_0 + \frac{1}{U}\mathbf{Y} \cdot \mathbf{E}_0\mathbf{Y}\right)$$
$$\frac{d\mathbf{E}_0}{dz} = \mathbf{W} \tag{8.63}$$

that can then be solved by the Runge–Kutta techniques of Appendix B. To solve the equations, we will need suitable initial conditions for the ODEs. Budden (1961) has suggested a solution process whereby we start at a level above the ionosphere where the solution must take the form of an upward traveling plane wave. For given p_x and p_y we choose a plane wave solution and then use this to provide initial conditions for the above ODEs. The equations are then solved downward through the ionosphere until, on the other side, the solution becomes a combination of upward and downward plane waves. We split the solution into its upward and downward components, and, if there is a downward component, the upward component will have been "reflected" by the ionosphere. Obviously, in this procedure, we have no choice about the polarization of the wave that is incident upon the ionosphere from below. To overcome this problem, we perform the simulation twice, the second time choosing the plane wave that is incident from above to have polarization that is orthogonal to that of the first simulation. We will now have two different incident waves and can form an arbitrary incident wave by a suitable linear combination of these basic solutions. The "reflected" wave will then be the same linear combination of the reflected components of the basic solutions.

It is instructive to consider the case where Earth's magnetic field is zero and there are no collisions ($\mathbf{Y} = 0$ and $U = 1$), then all field components satisfy the generic equation

$$\frac{d^2\mathbf{E}_0}{dz^2} + \beta_0^2 \left(1 - p_x^2 - p_y^2 - X\right) \mathbf{E}_0 = 0 \tag{8.64}$$

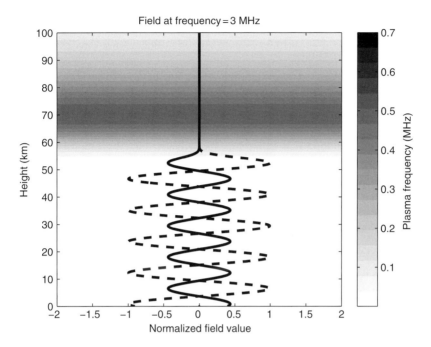

Figure 8.39 The normalized field E_0 of a wave that is reflected by a Chapman layer ionosphere (the real part is the solid curve and the imaginary part is the dashed curve).

We can solve this equation numerically using the procedure outlined in the above paragraph; Figure 8.39 shows the field E_0 calculated in this manner for a wave that is incident upon a Chapman layer ionosphere at an angle of half a degree. In the absence of Earth's magnetic field, it is also possible to obtain some analytic results. Consider the case of an ionosphere with the simple model

$$X = \alpha(z - z_0) \quad z \geq z_0$$
$$= 0 \qquad\qquad z < z_0 \qquad (8.65)$$

for which the ionization starts at a height z_0 above the ground. In this case it is possible to make some headway with an analytic solution (Budden, 1961). The above ionosphere is non-physical since X increases without limit as z increases. It does, however, provide a useful model of the bottom side of an ionospheric layer. We will have

$$\frac{d^2 E_0}{dz^2} + \beta_0^2 \left(C^2 - \alpha(z - z_0)\right) E_0 = 0 \qquad (8.66)$$

for $z \geq z_0$, and

$$\frac{d^2 E_0}{dz^2} + \beta_0^2 C^2 E_0 = 0 \qquad (8.67)$$

for $z < z_0$ where $C^2 = 1 - p_x^2 - p_y^2$ and C is the cosine of the angle between the incident wave and the vertical. We can reduce Equation 8.66 to the form of the Stokes equation by the substitution $\zeta = (\beta_0^2 \alpha)^{1/3} (z - z_0 - C^2/\alpha)$. As a consequence, the solution to Equation 8.66 can be written as $E_0(z) = A_0 Ai(\zeta) + B_0 Bi(\zeta)$. Noting that $Ai(z) \sim \exp(-2z^{3/2}/3)/2\sqrt{\pi}z^{1/4}$ and $Bi(z) \sim \exp(2z^{3/2}/3)/\sqrt{\pi}z^{1/4}$ as $|z| \to \infty$, we see that the Bi part of the solution grows without bound. As a consequence we must set $B_0 = 0$ (i.e. $E_0(z) = A_0 Ai(\zeta)$ for $z > z_0$) for the field to be consistent with the finite energy density of the wave that enters from below the ionosphere. Below the ionosphere, the solution will have the form $E_0(z) = \exp(-j\beta_0 Cz) + R\exp(j\beta_0 Cz)$. (Here we have assumed a plane wave of unit amplitude is incident upon the ionosphere and that a wave of amplitude R is reflected.) We will consider the case where the propagation direction lies in the xz plane (i.e. $p_y = 0$) and the electric field points in the y direction. $E_y = E_0 \exp(-j\beta_0 x p_x)$ will be the only non-zero component of the electric field and so $\mathbf{H} = -(j/\omega\mu_0)(\partial E_y/\partial z)\hat{\mathbf{x}} + (j/\omega\mu_0)(\partial E_y/\partial x)\hat{\mathbf{z}}$ from Maxwell's equations. Across the interface at $z = z_0$, we will need the tangential components of $\mathbf{E}$ and $\mathbf{H}$ to be continuous and so, as a consequence, E_0 and $\partial E_0/\partial z$ will need to be continuous. From these conditions,

$$\exp(-j\beta_0 Cz_0) + R\exp(j\beta_0 Cz_0) = A_0 Ai(\zeta_0) \tag{8.68}$$

and

$$-C\exp(-j\beta_0 Cz_0) + RC\exp(j\beta_0 Cz_0) = -j\left(\frac{\alpha}{\beta_0}\right)^{\frac{1}{3}} A_0 Ai'(\zeta_0) \tag{8.69}$$

where $\zeta_0 = -(\beta_0^2 \alpha)^{1/3} C^2/\alpha$. Eliminating A_0, we obtain

$$R = \exp(-j2\beta_0 Cz_0) \frac{CAi(\zeta_0) - j\left(\frac{\alpha}{\beta_0}\right)^{\frac{1}{3}} Ai'(\zeta_0)}{CAi(\zeta_0) + j\left(\frac{\alpha}{\beta_0}\right)^{\frac{1}{3}} Ai'(\zeta_0)} \tag{8.70}$$

It will be noted that the reflection coefficient R has unit amplitude, consistent with the fact that we have assumed the medium to be lossless (i.e. collision free). In this case, reflection from the ionosphere will merely invoke a phase advance. Noting that ζ_0 is negative, we can, for frequencies consistent with ionospheric reflection, use the asymptotic expressions $Ai(-|\zeta_0|) \approx \sin(2|\zeta_0|^{3/2}/3 + \pi/4)/\sqrt{\pi}|\zeta_0|^{1/4}$ and $Ai'(-|\zeta_0|) \approx -|\zeta_0|^{1/4}\cos(2|\zeta_0|^{3/2}/3 + \pi/4)/\sqrt{\pi}$. From these,

$$R \approx \exp(-j2\beta_0 Cz_0) \frac{\sin\left(\frac{2}{3}|\zeta_0|^{\frac{3}{2}} + \frac{\pi}{4}\right) + j\cos\left(\frac{2}{3}|\zeta_0|^{\frac{3}{2}} + \frac{\pi}{4}\right)}{\sin\left(\frac{2}{3}|\zeta_0|^{\frac{3}{2}} + \frac{\pi}{4}\right) - j\cos\left(\frac{2}{3}|\zeta_0|^{\frac{3}{2}} + \frac{\pi}{4}\right)}$$

$$= j\exp\left(-j2\beta_0 Cz_0 - j\frac{4}{3}C^3\frac{\beta_0}{\alpha}\right) \tag{8.71}$$

The above analytic solution suggests the ansatz $E_0 = A(z)\exp(-j\beta_0 \phi(z))$ for a more general stratified medium. We will assume that X, A and ϕ are all slowly varying on the scale of a wavelength. Equation 8.64 will then imply

$$\frac{d^2 A}{dz^2} - 2j\beta_0 \frac{d\phi}{dz}\frac{dA}{dz} - j\beta_0 A\frac{d^2\phi}{dz^2} - \beta_0^2 A\left(\frac{d\phi}{dz}\right)^2 + \beta_0^2\left(C^2 - X\right)A = 0 \tag{8.72}$$

8.8 Full Wave Solutions

For large β_0, the two highest orders in β_0 will imply

$$\left(\frac{d\phi}{dz}\right)^2 = \left(C^2 - X\right) \tag{8.73}$$

and

$$2\frac{d\phi}{dz}\frac{dA}{dz} = -A\frac{d^2\phi}{dz^2} \tag{8.74}$$

These equations can be formally integrated to yield the solutions

$$E_{\pm}(z) = \left(C^2 - X\right)^{-\frac{1}{4}} \exp\left(\pm j\beta_0 \int_0^z \sqrt{C^2 - X} dz\right) \tag{8.75}$$

from which the general solution will be $E_0(z) = e_+ E_+(x) + e_- E_-(x)$. This is known as the WKB solution (Budden, 1961). From Equation 8.75 we find that $E_{\pm}$ satisfies

$$\frac{d^2 E_{\pm}}{dz^2} + \left(Q + \frac{1}{4}\frac{Q''}{Q} - \frac{5}{16}\left(\frac{Q'}{Q}\right)^2\right) E_{\pm} = 0 \tag{8.76}$$

where $Q = \beta_0^2(C^2 - X)$ and a prime denotes a derivative with respect to z. It is clear that we will need $Q''/4Q - 5\left(Q'/Q\right)^2/16 \ll |Q|$ for $E_{\pm}$ to be approximate solutions to Equation 8.64. Importantly, it will be noted that this condition will break down at a height z_r where $Q(z_r) = 0$. This is the height of ray reflection and also the height where the oscillatory nature of the electric field gives way to a continuously decaying field. Around height z_r, the equation for the electric field will have the form

$$\frac{d^2 E_0}{dz^2} - (z - z_r)\delta E_0 = 0 \tag{8.77}$$

where $\delta = -dQ/dz$ when evaluated at $z = z_r$ and is positive. Equation 8.77 will have the solution $A_0 Ai(\delta^{1/3}(z - z_r)) + B_0 Bi(\delta^{1/3}(z - z_r))$ and we can use this solution to connect the WKB solutions above and below the level $z = z_r$. Above this level, the only physically realistic solution to Equation 8.77 is $A_0 Ai(\delta^{1/3}(z - z_r))$ and so the WKB solution below $z = z_r$ must match up with this. Below $z = z_r$ the WKB solution can be written in the form

$$E_0(z) = \left(C^2 - X\right)^{-\frac{1}{4}} \exp\left(-j\beta_0 \left(\int_0^{z_r} \sqrt{C^2 - X} dz + \Phi(z)\right)\right)$$

$$+ R\left(C^2 - X\right)^{-\frac{1}{4}} \exp\left(j\beta_0 \left(\int_0^{z_r} \sqrt{C^2 - X} dz + \Phi(z)\right)\right) \tag{8.78}$$

where $\Phi(z) = \int_{z_r}^z \sqrt{C^2 - X} dz$. According to the method of matched asymptotic expansions (Nayfeh, 1973), as $z \to z_r$ the behavior of the WKB solution will need to match the behavior of $Ai(\delta^{1/3}(z - z_r))$ as $z - z_r \to -\infty$, i.e. E_0 will need to behave as

$$E_0(z) \approx \frac{A_0}{\sqrt{\pi}\delta^{\frac{1}{12}}|z - z_r|^{\frac{1}{4}}} \sin\left(\frac{2}{3}\delta^{\frac{1}{2}}|z - z_r|^{\frac{3}{2}} + \frac{\pi}{4}\right) \tag{8.79}$$

as $z \to z_r$. For z approaching z_r, $C^2 - X \approx \delta(z_r - z)/\beta_0^2$ and $\Phi(z) \approx (\delta^{1/2}/\beta_0)$ $\int_{z_r}^{z} \sqrt{z_r - z}\, dz = -2\delta^{\frac{1}{2}}/3\beta_0 (z_r - z)^{3/2}$. Consequently, for Equations 8.78 and 8.79 to match up, we will need

$$R = j \exp\left(-j2\beta_0 \int_0^{z_r} \sqrt{C^2 - X}\, dz\right) \quad (8.80)$$

It will be noted that there is a phase advance of $\pi/2$ in addition to that predicted by the ray tracing solutions of Chapter 5 and this advance should be applied to such ray tracing solutions after passage through a caustic surface (Jones, 1994).

8.9 References

L. Barclay (editor), *Propagation of Radio Waves*, Institution of Electrical Engineers, London, 2003.

R.N.E. Baulch and E.C. Butcher, The effect of travelling ionospheric disturbances on the group path, phase path, amplitude and direction of arrival of radio waves reflected from the ionosphere, J. Atmos. Terrest. Phys., vol. 47, pp. 653–662, 1985.

D. Bilitza, International reference ionosphere 1990, Rep. 90-22, Natl. Space Sci. Data Cent., Greenbelt, Md., 1990.

D. Bilitza, D. Altadill, Y. Zhang, C. Mertens, V. Truhlik, P. Richards, L. McKinnell and B. Reinisch, The International Reference Ionosphere 2012: A model of international collaboration, J. Space Weat. Space Clim., vol. 4, 2014.

G. Breit and M.A. Tuve, A test of the existence of the conducting layer, Phys. Rev., vol. 28, pp. 554–575, 1926.

K.G. Budden, *Radio Waves in the Ionosphere*, Cambridge University Press, Cambridge, 1961.

K.G. Budden, *The Propagation of Radio Waves*, Cambridge University Press, Cambridge, 1985.

C.J. Coleman, A direction sensitive model of atmospheric noise and its application to the analysis of HF receiving antennas, Radio Sci., vol. 37, 10.1029/2000RS002567, 2002.

C.J. Coleman, Huygens' principle applied to radio wave propagation, Radio Sci., vol. 37, p. 1105, doi:10.1029/2002RS002712, 2002.

K. Davies, *Ionospheric Radio*, IEE Electomagnetic Waves Series, vol. 31, Peter Peregrinus, London, 1990.

F. Earl, Synthesis of HF ground backscatter spectral characteristics, J. Atmos. Terrest. Phys., vol. 37, pp. 155–1562, 1975.

S.H. Francis, Acoustic-gravity modes and large-scale travelling ionospheric disturbances of a realistic, dissipative atmosphere, J. Geophys. Res., vol. 78, pp. 2278–2301, 1973.

J.P. Friedman, Propagation of internal gravity waves in a thermally stratified atmosphere, J. Geophys. Res., vol. 71, pp. 1033–1054, 1966.

J. Haselgrove, Ray theory and a new method for ray tracing, Proc. Phys. Soc., The Physics of the Ionosphere, pp. 355–364, 1954.

C.O. Hines, Internal atmospheric gravity waves at ionospheric heights, Can. J. Phys., vol. 38, pp. 1441–1481, 1960.

K. Hock and K. Schlegel, A review of atmospheric gravity waves and travelling ionospheric disturbances: 1982–1995, Ann. Geophys., pp. 917–940, 1996.

W.H. Hooke, Ionospheric irregularities produced by internal atmospheric gravity waves, J. Atmos. Terr. Phys., vol. 30, pp. 795–823, 1968.

International Radio Consultative Committee (CCIR), *World Distribution and Characteristics of Atmospheric Radio Noise Data*, Rep. 322, Int. Radio Consult. Comm., Int. Telecommun. Union, Geneva, 1964.

International Telecommunication Union (ITU), *Radio Noise*, ITU-R Rep. P. 372, Geneva, 1999.

D.S. Jones, *Methods in Electromagnetic Wave Propagation*, IEEE Series on Electromagnetic Wave Theory, Oxford and New York, 1994.

A.S. Jursa, (editor), *Handbook of Geophysics and the Space Environment*, Air Force Geophysical Laboratory, Air Force Systems Command, United States Air Force, 1985.

M.C. Kelly, *The Earth's Ionosphere: Plasma Physics and Electrodynamics*, International Geophysics Series, Academic Press, San Diego, CA, 1989.

M. Kotaki, and C. Katoh, The global distribution of thunderstorm activity observed by the ionospheric satellite (ISS-b), J. Atmos. Terr. Phys., vol. 45, pp. 833–850, 1984.

M. Kotaki, Global distribution of atmospheric radio noise derived from thunderstorm activity, J. Atmos. Terr. Phys., vol. 46, pp. 867–877, 1984.

D.F. Martyn, The propagation of medium radio waves in the ionosphere, Proc. Phys. Scc., vol. 47, pp. 323–339, 1935.

A.H. Nayfeh, *Perturbation Methods*, Wiley, New York, 1973.

L.J. Nickisch, Focusing in the stationary phase approximation, Radio Sci., vol. 23, pp. 171–182, 1988.

H. Poeverlein, Strahlwege von Radiowellen in der Ionoshaare., S.B. bayer. Akad. Wiss., Math-nat Klasse, pp. 175–201, 1948.

H. Poeverlein, Strahlwege von Radiowellen in der Ionoshaare. II. Theoretisch Grundlagen., Z. Angew. Phys., vol. 1, pp. 517–525, 1949.

H. Poeverlein, Strahlwege von Radiowellen in der Ionoshaare. III. Bilder theoretisch ermittelter Strahlwege, Z. Angew. Phys., vol. 2, pp. 152–160, 1949.

J.A. Ratcliffe, *An Introduction to the Ionosphere and Magnetosphere*, Cambridge University Press, 1972.

J.A. Secan, WBMOD Ionospheric Scintillation Model, an Abbreviated User's Guide, Rep. NWRA-CR-94 R172/Rev 7, NorthWest Res. Assoc., Inc., Bellevue, Wash.

A.J. Stocker, N.F. Arnold and T.B. Jones, The synthesis of travelling ionospheric disturbance (TID) signatures in HF radar observations using ray tracing, Ann. Geophys., vol. 18, pp. 56–64, 2000.

H.J. Strangeways, Effects of horizontal gradients on ionospherically reflected or transionospheric paths using a precise homing-in method, J. Atmos. Solar-Terr., Phys., vol. 62, pp. 1361–1376, 2000.

J.R. Wait, *Electromagnetic Waves in Stratified Media*, IEEE/OUP Series on Electromagnetic Wave Theory, Institute of Electrical and Electronic Engineers and Oxford University Press, 1996.

9 Propagation in the Lower Atmosphere

In the current chapter we investigate propagation of radio waves close to the ground where the ionosphere has little influence. Under normal circumstances, the optical horizon is a limit to such propagation. However, under suitable meteorological conditions, ducts can form in the lower atmosphere and provide over the horizon propagation. In addition, at low frequencies, ground waves can also provide over the horizon propagation. We will study these mechanisms in the current chapter, but will also investigate the effects of irregular ground, forest and rain on propagation.

9.1 Propagation in Tropospheric Ducts

In a homogenous medium, the propagation direction of an electromagnetic wave will be constant, and this suggests that terrestrial communications will normally be limited by the visual horizon. In reality, the refractive index of air slightly varies with height (under standard conditions $N^2 \approx 1.0006 - 0.00008y$ where y is height above sea level expressed in kilometers) and this will cause the propagation to bend toward the ground. This bending is normally insufficient to cause the propagation to return to the ground, but, under some abnormal weather conditions, the refraction can be sufficiently enhanced so as to cause the electromagnetic waves to return to the ground. The refractive index of the atmosphere is related to the meteorological quantities through the Debye formula

$$N = 1 + \frac{7.76 \times 10^{-5}}{T}\left(P + 4810\frac{e}{T}\right) \tag{9.1}$$

where T is the temperature (in Kelvin), P is the atmospheric pressure (in millibars) and e is the water vapor pressure (in millibars). Pressure will decrease with height ($P \approx P_0 \exp(-y/H)$ where $H \approx 8.5$ km and $P_0 \approx 1010$ mbar) and temperature likewise (the rate is known as the *lapse rate* and a value of about 6.5°C/km is typical). Occasionally, a layer of warm air can form above a layer of cold air (a *temperature inversion*) and this can cause a sufficiently negative gradient of refractive index for the propagation to travel back to the ground and hence result in over the horizon propagation.

9.1 Propagation in Tropospheric Ducts

The effect of the curvature of Earth can be modeled by employing a modified refractive index $\hat{N} = N + y/a$, where a is the radius of Earth ($a = 6371$ km). This additional refractive index will cause the wave to bend upward, but the coordinate y must now be interpreted as the altitude above the surface of Earth and the horizontal coordinate z as the distance along the ground. It is clear that the gradient in N will need to be less than -0.000157 (y in terms of km) for the propagation to return to the ground. The effect of Earth's atmosphere is so slight that it is often more convenient to describe it in terms of M units where $M = (\hat{N} - 1) \times 10^6$, and we will adopt this measure in the rest of this chapter.

Evaporation at the surface of the sea can frequently cause a change in the atmosphere that is sufficient to bend the wave back to the surface. Although the energy will be reflected at the surface, it can once again be bent back and so become captured in a surface layer known as a *duct*. In terms of M units and y in meters, such a duct can be modeled (Levy, 2000) by

$$M(z) = M_0 + 0.125 \, (y - d \ln(y/y_0)) \tag{9.2}$$

where d is the duct thickness and y_0 is the roughness length. Typical values for y_0 and M_0 are 0.00015 m and 330, respectively. The various paraxial techniques of Chapter 7 can be applied to the study of this problem; Figure 9.1 shows an example of the loss incurred in a surface evaporative duct at 10 GHz, simulated using the split step Kirchhoff integral approach (the propagation loss is the power transmitted divided

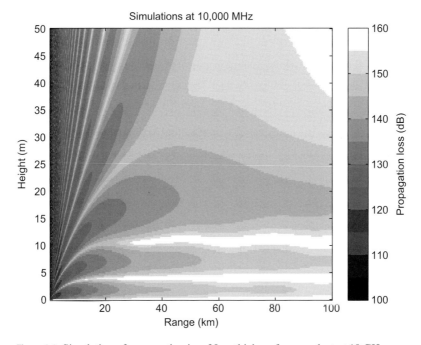

Figure 9.1 Simulation of propagation in a 30-m thick surface sea duct at 10 GHz.

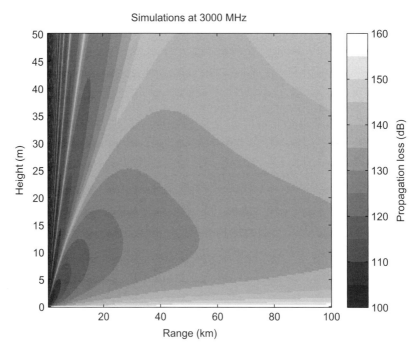

Figure 9.2 Simulation of propagation in a 30-m thick surface sea duct at 3 GHz.

by the power received at a point when translated into dB). The simulation of Figure 9.1 is for a wave frequency of 10 GHz, a duct height of $d = 30$ m and antenna at height 25 m. If we reduce the wave frequency to 3 GHz we obtain the loss behavior shown in Figure 9.2.

The above simulations assume a perfectly flat sea (we assume $\sigma = 5$ S/m and $\epsilon_r = 80$), but this is rarely the case in reality and we need to consider the effect of sea roughness. As discussed in Chapter 2, the effects of roughness can be studied by means of an effective surface impedance. In the case of the integral equation approach, however, a more convenient concept is an effective reflection coefficient (Levy, 2000). Let R_0 be the reflection coefficient for a smooth surface, then the effective reflection coefficient R is given by $R = \rho R_0$ where ρ is known as the roughness reduction coefficient. Miller et al. (1984) have provided a reduction coefficient that gives a good representation of the sea surface,

$$\rho = \exp\left(-\frac{\gamma^2}{2}\right) I_0\left(\frac{\gamma^2}{2}\right) \qquad (9.3)$$

where $\gamma = 2\beta_0 h_{\text{rms}} \cos\theta$ with h_{rms} the rms height of the surface and θ the angle that the propagation direction makes with the vertical. Figure 9.3 shows the same situation as for Figure 9.1, except that the sea is now rough with rms height 1 m. It will be noted that the roughness causes considerable attenuation of the propagation in the duct. Apart

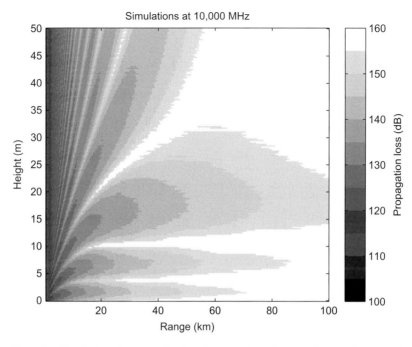

Figure 9.3 Simulation of propagation in a 30-m thick surface sea duct at 10 GHz with rough sea (rms height 1 m).

from this, evaporative ducts can be the cause of considerable anomalous propagation on the sea and this can be problematic for radar.

Other ducts can form at higher levels in the atmosphere due to weather conditions. As mentioned above, a temperature inversion can cause conditions that are conducive to ducting and this can occur in elevated layers of the atmosphere (often during high-pressure conditions). An atmosphere that is typical of such conditions (Levy, 2000) can be described, in terms of M units, by

$$\begin{aligned} M(z) &= 330 + \frac{12}{100}y \quad y < 100 \\ &= 342 - \frac{10}{50}(y - 100) \quad 100 < y < 150 \\ &= 332 + \frac{118}{1000}(y - 150) \quad y > 150 \end{aligned} \quad (9.4)$$

when y is expressed in meters. It will be noted that there is a layer ($100 < y < 150$) in which the gradient of the refractive index is negative and this is sufficient to trap some radio waves. Once again the split step integral approach can be applied to the simulation of such propagation. Figures 9.4 and 9.5 show how the duct captures radio waves at frequencies of 10 and 3 GHz, respectively. It will be noted that most of the radio waves are trapped between the ground and the top of the inversion layer, but that there is also significant leakage into the region above this layer.

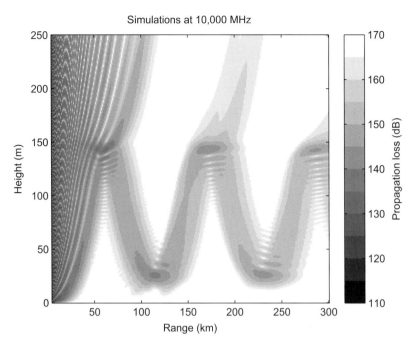

Figure 9.4 Simulation of propagation in an elevated duct at 10 GHz.

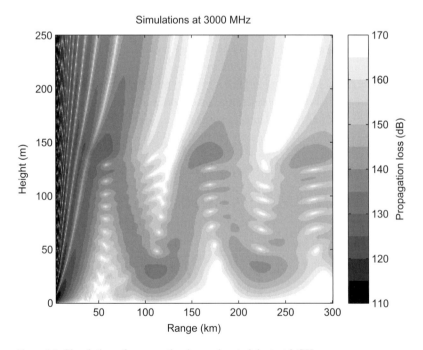

Figure 9.5 Simulation of propagation in an elevated duct at 3 GHz.

9.2 The Effect of Variations in Topography

As we have seen earlier, although terrain can interrupt the propagation of a radio wave, energy can still get into shadow regions by the process of diffraction. Once again, the paraxial techniques of Chapter 7 can be applied to such problems with the terrain treated by a terrain flattening transform. Figure 9.6 shows an example of propagation losses over irregular terrain at a frequency of 500 MHz, simulated using the split step integral algorithm. It will be noted that there is some diffraction into regions that are otherwise masked by the topography. Further, ground reflections cause strong interference with the direct propagation. Figure 9.7 shows the same scenario, but with propagation at the lower frequency of 50 MHz. From these, it is clear that the diffraction into shadow regions becomes increasingly stronger as the frequency reduces.

The techniques of Chapter 7 can also be used to study full 3D propagation, and, in natural terrain, this is important as there can be diffraction around hills as well as over them. Figure 9.8 shows some simulations of propagation losses over irregular terrain that were calculated by the techniques of Chapter 7. The radio frequency is 500 MHz with the antenna (the central cross) at 10 m above the ground. Propagation loss is shown in the figure as a grayscale plot and was sampled at 10 m above the ground. The topography is shown as a contour plot superimposed upon the propagation loss plot. It is clear from this that, although there is shadowing behind hills and in hollows, there is also significant propagation by diffraction into these regions. Figure 9.9 shows the simulations repeated at the frequency of 50 MHz. It can be seen from these simulations

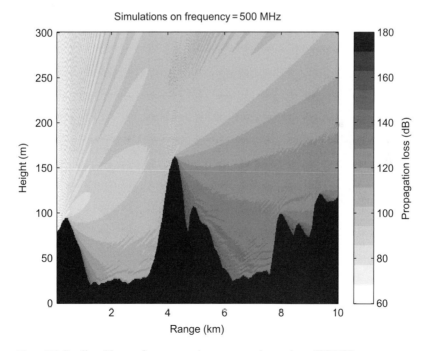

Figure 9.6 Predicted losses for propagation over rough terrain at 500 MHz.

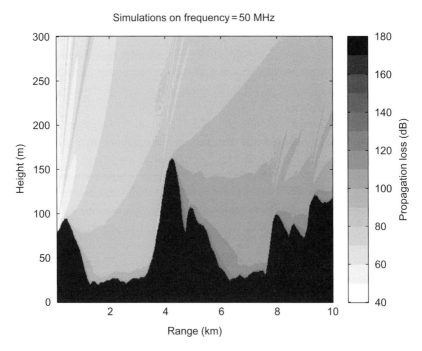

Figure 9.7 Predicted losses for propagation over rough terrain at 50 MHz.

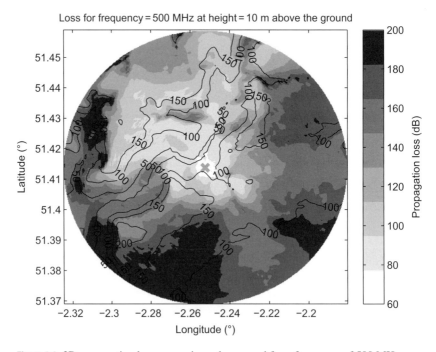

Figure 9.8 3D propagation losses over irregular ground for a frequency of 500 MHz.

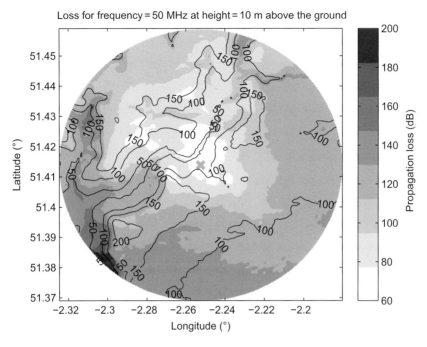

Figure 9.9 3D propagation losses over irregular ground for a frequency of 50 MHz.

that there is much deeper penetration by diffraction for lower frequencies, a fact that needs to be considered in the planning of communications systems. In regions where there is poor line-of-sight coverage, the lower frequencies can often provide better propagation. A major limitation, however, is the size of antenna and the availability of spectrum.

9.3 Surface Wave Propagation

As frequency decreases, it can be seen from the above considerations that there is a significant increase in non-line-of-sight (NLOS) propagation. Figure 9.10 shows that propagation over land at 50 MHz can exhibit considerable over the horizon diffraction. At low enough frequencies, the surface wave will dominate; this can be seen in Figure 9.11, which shows propagation over land ($\sigma = 0.002$ S/m and $\epsilon_r = 10$) on a frequency of 3 MHz. It will be noted that there is considerable NLOS propagation close to the ground. For land, the relative impedance is quite high, but, for sea, the value is quite low and this leads to a significant increase in the strength of the surface wave. Figure 9.12 shows the propagation losses with the ground replaced by sea ($\sigma = 5$ S/m and $\epsilon_r - 80$) and from which it will be noted that there is a large increase in the strength of the surface wave. Figure 9.13 shows propagation over land that is followed by sea (the land to sea interface is located 50 km from the source). It will be noted that the field behaves as for pure ground up until the interface, but after that there is a recovery in

212   Propagation in the Lower Atmosphere

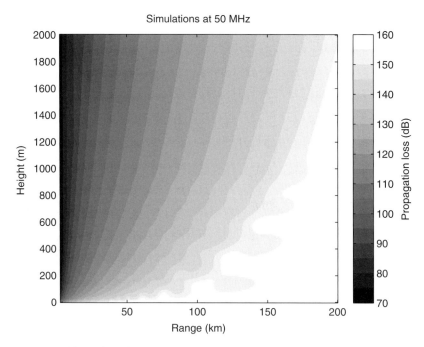

Figure 9.10 Simulation of propagation over land at 50 MHz.

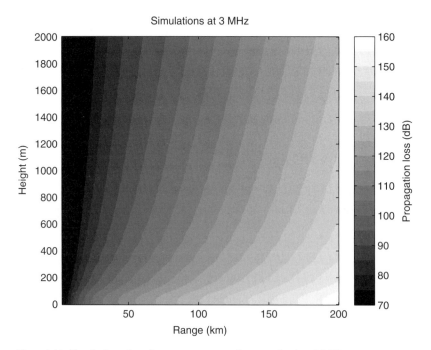

Figure 9.11 Simulation of surface wave propagation over land at 3 MHz.

9.3 Surface Wave Propagation

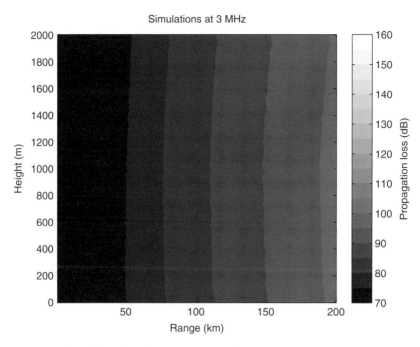

Figure 9.12 Simulation of surface wave propagation over sea at 3 MHz.

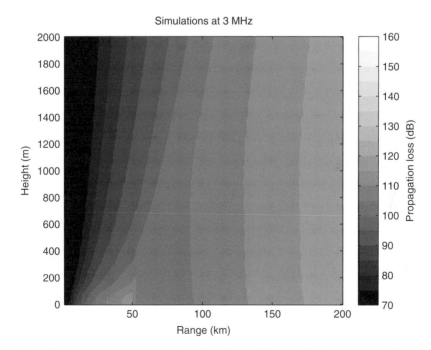

Figure 9.13 Propagation over a land-sea interface at 50 km exhibiting recovery effect.

field strength and the behavior is more like the pure sea case. This is the recovery effect that has been observed in experimental work on surface wave propagation (Millington and Isted, 1950).

When surface waves dominate, and we are primarily interested in the fields close to the ground, we can use the compensation theorem to study them (Monteath, 1973). Consider the field $(\mathbf{H}_A, \mathbf{E}_A)$ of an antenna A and the field $(\mathbf{H}_B, \mathbf{E}_B)$ of an antenna B. These antennas are located on a surface S with the field of antenna A evaluated using surface impedance η and that of antenna B evaluated using the surface impedance $\eta + \delta\eta$. From the considerations of Section 3.4, we have

$$\delta Z_{AB} I_A I_B = \int_S (\mathbf{E}_A \times \mathbf{H}_B - \mathbf{E}_B \times \mathbf{H}_A) \cdot \mathbf{n}\, dS \tag{9.5}$$

where δZ_{AB} is the change in mutual impedance between the antennas when the surface impedance η is changed to $\delta\eta$ (I_A and I_B are the currents that drive antennas A and B, respectively, both antennas assumed to be vertically polarized). We will take the surface S to be a plane R^2 that is perfectly conducting ($\eta = 0$), then

$$\mathbf{E}_A \approx \frac{j\omega\mu I_A}{2\pi} \mathbf{h}_{\text{eff}}^A \frac{\exp(-j\beta\rho_A)}{\rho_A} \tag{9.6}$$

for points close to the surface S and where ρ_A is the horizontal distance from the source of the field. We will take the fields of source B to be G times the value with $\delta\eta = 0$ and so

$$\mathbf{E}_B \approx G(\rho_B) \frac{j\omega\mu I_A}{2\pi} \mathbf{h}_{\text{eff}}^B \frac{\exp(-j\beta\rho_B)}{\rho_B} \tag{9.7}$$

It is assumed that the axis from the source of field B to the source of field A is the z axis with the source B at the origin and with source A a distance Z from B. We can rearrange Equation 9.5 into

$$\delta Z_{AB} I_A I_B = \int_S (\mathbf{H}_B \cdot (\mathbf{n} \times \mathbf{E}_A) - \mathbf{H}_A \cdot (\mathbf{n} \times \mathbf{E}_B))\, dS \tag{9.8}$$

and, from the impedance boundary conditions, we have that $\mathbf{n} \times \mathbf{E}_A = 0$ and $\mathbf{n} \times \mathbf{E}_B = \delta\eta \mathbf{n} \times (\mathbf{n} \times \mathbf{H}_A)$. As a consequence,

$$\delta Z_{AB} I_A I_B = \int_S \delta\eta (\mathbf{n} \times \mathbf{H}_A) \cdot (\mathbf{n} \times \mathbf{H}_B)\, dS \tag{9.9}$$

By stationary phase arguments, the major contribution to the integral in Equation 9.5 will come from around the z axis and so $\mathbf{H}_A \approx -\hat{\mathbf{z}} \times \mathbf{E}_A/\eta_0$ and $\mathbf{H}_B = \hat{\mathbf{z}} \times \mathbf{E}_B/\eta_0$. Then, from Equation 9.5,

$$\delta Z_{AB} I_A I_B = - \int_{R^2} \frac{\delta\eta}{\eta_0^2} \mathbf{E}_A \cdot \mathbf{E}_B\, dS \tag{9.10}$$

Close to the z axis, we can make the approximations

$$\mathbf{E}_A = \frac{j\omega\mu I_A}{2\pi} \mathbf{h}_{\text{eff}}^A \frac{\exp\left(-j\beta\left(Z - z + \frac{x^2}{2(Z-z)}\right)\right)}{Z - z} \tag{9.11}$$

9.3 Surface Wave Propagation

and

$$E_B = G(z) \frac{j\omega \mu I_B}{2\pi} \mathbf{h}^B_{\text{eff}} \frac{\exp\left(-j\beta\left(z + \frac{x^2}{2z}\right)\right)}{z} \quad (9.12)$$

We further note that $\delta Z_{AB} = (G-1)Z^0_{AB}$ where Z^0_{AB} is the mutual impedance with $\delta\eta = 0$, i.e.

$$Z^0_{AB} = \frac{j\omega\mu \, \mathbf{h}^A_{\text{eff}} \cdot \mathbf{h}^B_{\text{eff}}}{2\pi} \frac{\exp(-j\beta Z)}{Z} \quad (9.13)$$

Noting the expressions for Z^0_{AB}, E_A and E_B, we obtain that

$$G(Z) - 1 = -\frac{j\beta}{2\pi} \int_{R^2} \frac{\delta\eta}{\eta_0} G(z) \frac{Z}{(Z-z)z} \exp\left(-j\beta x^2 \frac{Z}{2z(Z-z)}\right) dx dz \quad (9.14)$$

Then, performing the x integral,

$$G(Z) - 1 = -\sqrt{\frac{j\beta}{2\pi}} \int_0^Z \eta_r(z) G(z) \sqrt{\frac{Z}{(Z-z)z}} \, dz \quad (9.15)$$

where $\eta_r = \delta\eta/\eta_0$ is the relative impedance of the surface. Equation 9.15 provides an integral equation that can be solved by standard numerical techniques in order to find $G(Z)$. We consider the case where η_r is constant and define $\hat{G}(Z) = G/\sqrt{Z}$, then $\hat{G}$ satisfies the integral equation

$$\hat{G}(Z) - \frac{1}{\sqrt{Z}} = -\eta_r \sqrt{\frac{j\beta}{2\pi}} \int_0^Z \hat{G}(z) \sqrt{\frac{1}{(Z-z)}} \, dz \quad (9.16)$$

This equation is an Abel integral equation of the second kind and has an analytic solution. From this solution for $\hat{G}$, it can be shown that $G(p) = 1 - j\sqrt{\pi p}\exp(-p)$ erfc($j\sqrt{p}$) where $p = -1/2j\beta\eta_r^2 Z$ is known as the *numerical distance*. This result is the *Sommerfeld formula* (Sommerfeld, 1909) and from which $G(p) \approx -1/2p$ in the limit $Z \to \infty$ (see Appendix F for a further discussion of such solutions).

The above equation can be extended to the case of a slowly undulating ground (see Figure 9.14) if we replace the relative impedance by an effective relative surface

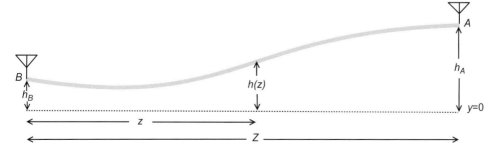

Figure 9.14 Geometry for surface wave equation describing undulating ground.

impedance. We assume the ground to be described by an equation of the form $y = h(z)$ and then, if $dh/dz \ll 1$, $\mathbf{n} \approx (0, 1, -dh/dz)$. The field $\mathbf{E}_A$, however, does not satisfy $\mathbf{n} \times \mathbf{E}_A = 0$ on the now undulating S (it is still the solution for a perfectly conducting plane with $E_{Ax} = E_{Az} = 0$ on $y = 0$) and instead satisfies $\mathbf{n} \times \mathbf{E}_A = (dh/dz)E_{Ay}\hat{\mathbf{x}}$ on S. This means that we cannot discard the first term on the right-hand side of Equation 9.8 but, instead, use the approximation $\mathbf{H}_B \times (\mathbf{n} \times \mathbf{E}_A) \approx -(dh/dz)\mathbf{E}_A \cdot \mathbf{E}_B/\eta_0$. Consequently, we can replace η_r in Equation 9.10 by an effective relative permittivity $\eta'_r = \eta_r + dh/dz$. To incorporate the effects of undulating ground, we therefore replace the relative impedance η_r in Equation 9.15 by effective relative impedance η'_r (note that we also reinterpret the z coordinate as distance along the ground). If the undulations are large, we need to multiply the relative effective permittivity by the phase factor $\exp(-j\beta((h - h_A)^2/2(Z - z) + (h - h_B)^2/2z))$ in order to account for the additional distance between the antennas when connected via a point on the now undulating ground.

9.4 Propagation through Forest

If the ground is covered by forest, there will be a further reduction in the signal as it propagates. The simplest propagation model of a forest consists of a dielectric slab above the ground, and this has been studied by Wait (1967). These ideas have been further developed by Tamir (1977), who has considered propagation through forest that is interrupted by open ground. Let the slab be that shown in Figure 9.15 with region 2 the ground (permittivity μ_0, permeability ϵ_2 and conductivity σ_2), region 1 the forest (permittivity μ_0, permeability ϵ_1 and conductivity σ_1) and region 0 the air (permittivity μ_0, permeability ϵ_0 and conductivity zero). In the case of the forest region, the deviations of the electrical properties from those of air are only small, but nevertheless significant. The solution of Wait extends the solution 9.7 for propagation over an open ground by introducing a *height gain function* $f(y)$ such that

$$\mathbf{E} \approx G(d) \frac{j\omega \mu I}{2\pi} \mathbf{h}_{\text{eff}} \frac{\exp(-j\beta_0 d)}{d} f(y) f(h) \tag{9.17}$$

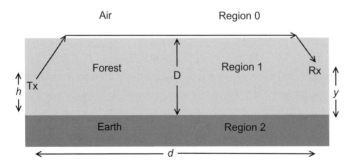

Figure 9.15 Propagation through a forest.

where d is the horizontal distance of the observation point from the source, h is the height of the source above the ground and y is the height of the observation point above the ground. Wait has shown that, for vertical polarization,

$$f(y) \approx 1 + j\beta_0(y - D)\Delta \tag{9.18}$$

when y is above the slab and

$$f(y) \approx \frac{\beta_0^2}{\beta_1^2}\left(\frac{\exp(u_1(y-D)) + R\exp(-u_1(D+y))}{1 + R\exp(-2u_1 D)}\right) \tag{9.19}$$

when y is within the slab (see Appendix F for a derivation of the above solution). In the above expressions $\Delta = Z_1/\eta_0$ where

$$Z_1 = K_1 \frac{K_2 + K_1 \tanh(u_1 D)}{K_1 + K_2 \tanh(u_1 D)} \tag{9.20}$$

and

$$R = \frac{K_2 - K_1}{K_2 + K_1} \tag{9.21}$$

with $K_1 = u_1/(\sigma_1 + j\epsilon_1\omega)$, $K_2 = u_2/(\sigma_2 + j\epsilon_2\omega)$, $u_1 = \sqrt{\beta_0^2 - \beta_1^2}$, $\beta_1^2 = -j\mu_0\omega(\sigma_1 + j\epsilon_1\omega)$, $u_2 = \sqrt{\beta_0^2 - \beta_2^2}$, $\beta_2^2 = -j\mu_0\omega(\sigma_2 + j\epsilon_2\omega)$ and $\beta_0^2 = \omega^2\mu_0\epsilon_0$. For large distances from the source, $G \approx -1/2p$, but with the numerical distance now given by $p = -j\beta_0\Delta^2 d/2$.

Typical values for relative impedance and conductivity of the ground are $\epsilon_2 = 10 \times \epsilon_0$ and $\sigma_2 = 0.01$ S/m and for forest $\epsilon_1 = 1.1 \times \epsilon_0$ and $\sigma_1 = 0.0001$ S/m. Let us first consider both receiver and transmitter to be inside the forest and the forest to have infinite height. In this case the solution of the previous section is still valid, but with air replaced by forest. Although the conductivity of the forest is small, according to Equation 9.7 it will nevertheless cause appreciable attenuation over distances of 100 m or more. This is due to the fact that the propagation constant β is now the propagation constant in the forest. For forest of finite height, the height gain functions of the Wait solution (see Equation 9.17) do not depend on the distance d between transmitter and receiver and so a finite height forest will not induce the same loss. The reason for this is that the energy travels as a *lateral wave* just above the tree tops. The energy travels along the path shown in Figure 9.15 and only a short part of this path is within the lossy forest medium. More general propagation over forest, and built-up terrain, has been considered by Hill (1982) by extending the integral equations for ground wave propagation (discussed in the last section). The paraxial approximation can also be used to study propagation in forest, the forest being treated by means of a height-dependent refractive index.

In the case of forest, the question arises as to what happens when the forest ends. Hill (1982) has applied the methods of Monteath (1973) to this problem. Consider the situation depicted in Figure 9.16, then the major contribution to the signal at the receiver will come from the lateral wave above the forest (the direct signal from the transmitter

Propagation in the Lower Atmosphere

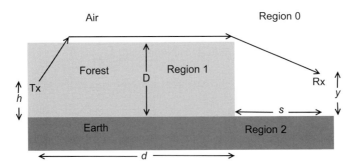

Figure 9.16 Propagation through a forest.

will be absorbed by the forest). From Equation 7.38, the field at the receiver can be expressed in terms of the field in the plane at distance d from the transmitter through

$$\mathbf{E} = -\int_{-\infty}^{\infty}\int_{D}^{\infty} \frac{2}{\eta} E_0 \mathbf{E}(X, Y, d) \, dY dX \qquad (9.22)$$

where

$$E_0 \approx -\frac{j\omega\mu_0}{4\pi} \frac{\exp\left(-j\beta_0\left(s + \frac{X^2 + (y-Y)^2}{2s}\right)\right)}{s} \qquad (9.23)$$

The major contribution to the integral will come from around $X = 0$ and $Y = y$ for $Y > D$. (Note that the major contribution will come from around $Y = D$ when $Y < D$.) As a consequence

$$\mathbf{E} \approx \frac{j\beta_0}{2\pi s} \mathbf{E}(0, \max(y, D), d) \exp(-j\beta_0 s) \int_{-\infty}^{\infty}\int_{D}^{\infty} \exp\left(-j\beta_0 \frac{X^2 + (y-Y)^2}{2s}\right) dY dX \qquad (9.24)$$

We can perform the X integral to obtain

$$\mathbf{E} \approx \sqrt{\frac{j\beta_0}{2\pi s}} \mathbf{E}(0, \max(y, D), d) \exp(-j\beta_0 s) \int_{D}^{\infty} \exp\left(-j\beta_0 \frac{(y-Y)^2}{2s}\right) dY \qquad (9.25)$$

and note that the remaining integral can be evaluated in terms of Fresnel integrals. Above the level of the forest canopy the lateral wave continues unimpeded and below this level some of the lateral wave diffracts into the now open ground.

9.5 Propagation through Water

A further propagation problem that can involve lateral waves is that of propagation through water. We will take the surface of the water to be the $y = 0$ plane, and then, for vertical polarization and deep water, the electrical field takes the form (see Appendix F)

$$\mathbf{E} \approx G(d) \frac{j\omega\mu I}{2\pi} \mathbf{h}_{\text{eff}} \frac{\exp(-j\beta_0 d)}{d} f(y) f(h) \tag{9.26}$$

where $f(y) \approx 1 + j\beta_0 y \Delta$ for $y > 0$ and $f(y) \approx (\beta_0^2/\beta_{\text{water}}^2) \exp\left(j\sqrt{\beta_{\text{water}}^2 - \beta_0^2} y\right)$ for $y < 0$ ($\Delta = (\beta_0/\beta_{\text{water}}^2)\sqrt{\beta_{\text{water}}^2 - \beta_0^2}$ and d is the distance along the surface). At large distances from the source, $G \approx -1/2p$ with the numerical distance p given by $p = -j\beta_0 \Delta^2 d/2$. Typical values for permittivity and conductivity of seawater are $\epsilon_{\text{water}} = 81 \times \epsilon_0$ and $\sigma_1 = 5$ S/m and for fresh water $\epsilon_{\text{water}} = 80 \times \epsilon_0$ and $\sigma_1 = 0.001$ S/m. Direct propagation through water can be heavily attenuated, especially in the case of seawater, and so the lateral wave that travels just above the water can be a very important propagation mechanism.

9.6 Propagation through Rain

Rain (and other hydrometeors such as snow) can cause considerable attenuation of radio waves due to scatter, this being in addition to any loss due to the conductivity of the propagation medium (Goddard, 2003). In Chapter 6 we considered propagation through an irregular medium and derived formula 6.91 for the attenuation of the average field during propagation. Formula 6.91, however, was derived on the basis that the irregularity was much larger than a wavelength, and this, in the case of rain, is certainly not the case for a large amount of the commonly used radio spectrum. With a typical size of around a few millimeters, raindrops can be much smaller than a wavelength and hence scatter radio waves in the Rayleigh sense. At lower frequencies (below 10 GHz), most attenuation is due to absorption in the raindrops themselves, which occurs during forward scatter. We will consider a scatterer in free space that occupies a small volume V with relative dielectric ϵ_r (of the order of 80 for rain). The incident wave will induce an electric dipole in the scatterer and the electric field can be derived from Equation 6.6 of Chapter 6. If E_{inc} is the field incident upon the scatterer, a field

$$E_{\text{scat}} = E_{\text{inc}} \frac{\beta_0^2 V}{4\pi r} \left(\frac{3\epsilon_r - 3}{\epsilon_r + 2} \right) \exp(-j\beta_0 r) \sin\theta \tag{9.27}$$

will be scattered where r is the distance from the scatterer and θ is the angle between the polarization vector of the incident field and the direction of radiation. We now calculate the power that reaches a unit area I of a screen just beyond the scatterer. If we take a coordinate system based on the center of the scatterer, we can approximate the total electric field as

$$E_{\text{tot}} = E_{\text{inc}} + E_{\text{scat}} = E_{\text{inc}} \left(1 + \frac{\beta_0^2 V}{4\pi z} \left(\frac{3\epsilon_r - 3}{\epsilon_r + 2} \right) \exp\left(-j\beta_0 \frac{x^2 + y^2}{2z} \right) \right) \tag{9.28}$$

where the z axis is the propagation direction of E_{inc}, which we have assumed to be approximately plane around the scatterer. The power reaching a unit area I of the screen will be

$$P_{\text{trans}} = \frac{1}{2\eta_0} \int_I \frac{|E_{\text{tot}}|^2}{2\eta_0} dxdy \qquad (9.29)$$

$$\approx \frac{E_{\text{inc}}^2}{2\eta_0} \int_I \left| 1 + \frac{\beta_0^2 V}{4\pi z} \left(\frac{3\epsilon_r - 3}{\epsilon_r + 2} \right) \exp\left(-j\beta_0 \frac{x^2 + y^2}{2z}\right) \right|^2 dxdy$$

We will assume that the screen is far enough away for us to ignore z^{-2} terms in the integrand and that contributions to the integral outside I are negligible, then

$$P_{\text{trans}} \approx \frac{E_{\text{inc}}^2}{2\eta_0} \left(1 + \int_{-\infty}^{\infty} \int_{-\infty}^{\infty} \frac{\beta_0^2 V}{2\pi z} \Re\left\{ \left(\frac{3\epsilon_r - 3}{\epsilon_r + 2} \right) \exp\left(-j\beta_0 \frac{x^2 + y^2}{2z}\right) \right\} dxdy \right)$$

$$= P_{\text{inc}} \left(1 + \beta_0 V \Im \left\{ \frac{3\epsilon_r - 3}{\epsilon_r + 2} \right\} \right) \qquad (9.30)$$

A radio wave will encounter many scatterers, and so, over a unit length of path, the power will be reduced by factor $1 + \beta_0 v \Im \{(3\epsilon_r - 3)/(\epsilon_r + 2)\}$ where v is the volume of scatterer in a unit volume of space ($v = N_s V_s$ with N_s the number of scatterers in a unit volume and V_s a scatterer volume). The attenuation constant will be given by

$$\alpha = (\beta_0 v/2) \Im \{(3\epsilon_r - 3)/(\epsilon_r + 2)\} \qquad (9.31)$$

and, over the total length of a propagation path from A to B, the rain will result in an attenuation $L_{abs} = 8.686 \int_A^B \alpha ds$ in dB terms.

Up until now we have assumed that the scatterers are identical, but micrometeors will usually have a distribution of sizes. For rain (Goddard, 2003), an often quoted empirical distribution of sizes is $N_s(D) = N_0 \exp(-D/D_0)$ where D is the diameter of a raindrop (in mm), $D_0 = 0.122 R^{0.21}$ (in mm), R is the rain rate (in mm hour^{-1}) and $N_0 = 8000$ (in mm^{-1}m^{-3}). The effective scattering volume will now be given by $v = \int_0^\infty N_s(D) V_s(D) dD$ where $V_s(D)$ is the volume of the raindrop with diameter D.

The effect of rain can be incorporated into paraxial techniques by introducing a complex refractive index such that $N_{\text{eff}} = N - j\alpha/\beta_0$. Rain can be an important factor in propagation calculations for microwaves; further techniques for modeling the effect of micrometeors on radio wave propagation can be found in the various International Telecommunication Union (ITU) reports.

9.7 References

C.J. Coleman, An FFT based Kirchhoff integral technique for the simulation of radio waves in complex environments, Radio Sci., vol. 45, RS2002, doi:10.1029/2009RS004197, 2010.

J.W.F. Goddard in L. Barclay (editor), *Propagation of Radio Waves*, Institution of Electrical Engineers, London, 2003.

D.A. Hill, HF ground wave propagation over forested and built-up terrain, NTIA Report 82-114, U.S. Department of Commerce, 1982.

International Telecommunication Union, ITU-R Recommendation 838 [ITU, 838], Specific attenuation model for rain for use in prediction methods, Geneva, 1992.

M. Levy, *Parabolic Equation Methods for Electromagnetic Wave Propagation*, IEE Electromagnetic Wave Series 45, London, 2000.

A.R. Miller, R.M. Brown and E. Vegh, New derivation for the rough surface reflection coefficient and for the distribution of sea-water elevations, IEE Proc., vol. 131, part H, pp. 114–116, 1984.

G.D. Monteath, *Applications of the Electro-magnetic Reciprocity Principle*, Pergamon Press, Oxford, 1973.

S.R. Saunders, *Antenna and Propagation for Wireless Communication Systems*, Wiley, Chichester, 1999.

T.B.A. Senior and J.L. Volakis, *Approximate Boundary Conditions*, IEE Electromagnetic Wave Series 41, The Institution of Electrical Engineers, London, 1995.

A. Sommerfeld, The propagation of waves in wireless telegraphy, Ann. Phys., vol. 28, p. 665, 1909.

T. Tamir, Radio wave propagation along mixed paths in forest environments, IEEE Trans. Ant. Prop. AP-25, pp. 471–477, 1977.

J.R. Wait, Asymptotic theory for dipole radiation in the presence of a lossy slab lying on a conducting half-space, IEEE Trans. Ant. Prop. AP-15, pp. 645–648, 1967.

10 Transionospheric Propagation and Scintillation

We have seen that the ionosphere can provide a duct for radio waves, but most radio waves above a frequency of about 50 MHz will penetrate the ionosphere and escape. Consequently, it is these higher frequencies that are used for satellite communications, radio astronomy and satellite navigation. Although, as frequency rises, the effect of the ionosphere decreases, it can nevertheless have a significant impact on the operation of Earth–space systems. In particular, irregularity in the ionosphere can cause fluctuations in signals (scintillation) that can severely degrade the operations of such systems. The present chapter looks at techniques for analyzing and modeling the impact of the ionosphere on Earth–space systems.

10.1 Propagation through a Benign Ionosphere

When the frequency rises above 50 MHz, most radio waves will penetrate the ionosphere. Figure 10.1 shows what happens to a radio wave that is launched at an angle of 30° on frequencies from 20 to 50 Mhz. It can be seen that by a frequency of 50 MHz there is very little deviation of the ray path from the free space limit. This is important for areas such as radio astronomy where direction needs to be measured accurately. Nevertheless, for satellite navigation and radio astronomy at low frequencies, the effect of the ionosphere still needs to be taken into account. The variational techniques of Chapter 5 are ideal for such studies, the propagation path being that which makes the phase delay stationary (Fermat's principle). For transionospheric propagation, the stationary value of the phase path occurs at a minimum, and so one can use many of the excellent optimization techniques that have been developed in recent years (Press et al., 1992) to solve the discretized form of Fermat's principle. Figure 10.2 shows the propagation between the ground and a low Earth orbiting (LEO) satellite that has been calculated using these techniques (the paths of both ordinary and extraordinary modes are shown). The ionosphere is from around the equatorial regions and the wave frequency is 52 MHz. It will be noted that the ionosphere can still have a substantial effect on the propagation paths and for this reason, most satellite operations occur at frequencies well above 100 MHz. At these frequencies, even though the path is almost that of free space, the ionosphere has a substantial effect on quantities such

10.1 Propagation through a Benign Ionosphere

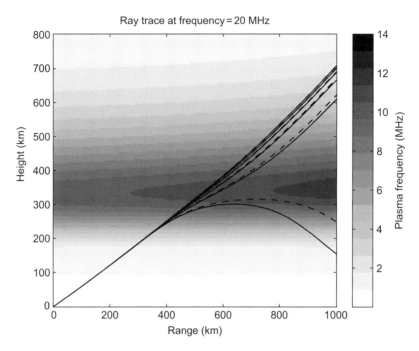

Figure 10.1 Ray paths at an elevation of $40°$ for 20, 25, ..., 50 MHz (both O and X rays shown).

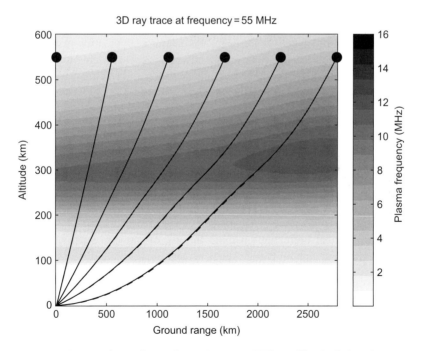

Figure 10.2 Propagation paths to a lower frequency LEO satellite (both O and X rays are shown).

as phase delay, group delay, Doppler and polarization. At frequencies above 100 MHz, the refractive index can be well approximated as

$$N_\pm \approx 1 + \frac{7.76 \times 10^{-5}}{T}\left(P + 4810\frac{e}{T}\right) - \frac{X}{2} \pm \frac{XY_L}{2} \tag{10.1}$$

where the plus sign denotes the index for an ordinary ray and the minus sign for an extraordinary ray. It will be noted that we have also included the effects of the lower atmosphere as these can be significant in some applications (occultation measurements, for example). At higher frequencies ($f \to \infty$), the ray path can therefore be found using Fermat's principle ($\delta\tau_p = 0$) with phase delay approximated by

$$\tau_p = \frac{1}{c_0}\int_A^B \left(1 + \frac{7.76 \times 10^{-5}}{T}\left(P + 4810\frac{e}{T}\right) - \frac{X}{2} \pm \frac{XY_L}{2}\right) ds \tag{10.2}$$

Because the ray path satisfies Fermat's principle, variations in phase will only be second order in variations to path geometry. Consequently, the phase delay can be estimated by evaluating the integral in Equation 10.2 along the free space path between the end points. The group delay can be calculated from

$$\tau_g = \frac{d(f\tau_p)}{df} \tag{10.3}$$

$$= \frac{1}{c_0}\int_A^B \left(1 + \frac{7.76 \times 10^{-5}}{T}\left(P + 4810\frac{e}{T}\right) + \frac{X}{2} \mp XY_L\right) ds$$

In applications, such as the GPS navigation system, timing is very important in order to maintain accuracy. Consequently, we need to be able to estimate corrections for the ionospheric effects. For group delay, such a correction can be calculated using ionospheric models and Equation 10.3, i.e.

$$\Delta\tau_g = \frac{1}{c_0}\int_A^B \left(\frac{7.76 \times 10^{-5}}{T}\left(P + 4810\frac{e}{T}\right) + \frac{X}{2} \mp \frac{XY_L}{2}\right) ds \tag{10.4}$$

The atmospheric and magnetic terms are often ignored and so $\Delta\tau_g = (40.3/c_0 f^2) \times$ TEC (in seconds) where TEC $= \int_A^B N_e ds$ is known as the *total electron content* of the path between A and B. The demand for high accuracy in the data from GPS, and other such systems, has driven the use of dual frequencies, which allow the accurate determination of corrections without the use of models. Such systems use coherent identical modulation on both frequencies and measure the differential group delay $\delta\tau_g$ between the signals. For signals on frequencies f_1 and f_2,

$$\delta\tau_g = 40.3\left(\frac{1}{f_2^2} - \frac{1}{f_1^2}\right) \times \text{TEC} \tag{10.5}$$

from which we can calculate the TEC and hence the delay correction. Using information gleaned from global observations of GPS, one can deduce information about the global distribution of TEC. Maps of the global distribution TEC show a strong similarity to the maps of foF2, as is to be expected. Further, by observation of multiple GPS satellites from each sampling location, it is possible to gain sufficient information to produce a 3D picture of plasma frequency using tomographic techniques (Raymund et al., 1994).

10.2 Faraday Rotation and Doppler Shift

Polarization is another aspect on which the ionosphere can still have a significant effect at higher frequencies. In Chapter 5 we saw that the ionosphere could cause a rotation of the polarization vector during propagation. This *Faraday rotation*, as it is known, will cause a rotation of magnitude

$$\theta = \frac{\beta}{2} \int_A^B XY \cdot d\mathbf{r} \tag{10.6}$$

during propagation between points A and B. Another way of looking at this is to consider a linearly polarized wave as the sum of two circularly polarized waves (see Chapter 2). These circularly polarized waves are the ordinary and extraordinary waves and have different refractive indices, N_+ and N_-. If we take the z axis to be the propagation direction, and the initial polarization to be in the $\hat{\mathbf{x}}$ direction, the electric field at distance z along the propagation direction will take the form

$$\mathbf{E} = \frac{E_0}{2}(\hat{\mathbf{x}} + j\hat{\mathbf{y}}) \exp\left(-j\beta_0 \int_0^z N_+ dz\right) + \frac{E_0}{2}(\hat{\mathbf{x}} - j\hat{\mathbf{y}}) \exp(-j\beta_0 \int_0^z N_- dz) \tag{10.7}$$

and substituting for N_+ and N_- from Equation 10.1, we obtain

$$\mathbf{E} = \frac{E_0}{2} \exp(-j\beta_0 s_0) \tag{10.8}$$
$$\times \left((\hat{\mathbf{x}} + j\hat{\mathbf{y}}) \exp\left(-j\beta_0 \int_0^z \frac{XY_L}{2} dz\right) + (\hat{\mathbf{x}} - j\hat{\mathbf{y}}) \exp\left(j\beta_0 \int_0^z \frac{XY_L}{2} dz\right)\right)$$

where $s_0 = \int_0^z \left(1 + 7.76 \times 10^{-5} \times (P + 4810\frac{e}{T})/T + \frac{X}{2}\right) ds$. Consequently,

$$\mathbf{E} = E_0 \exp(-j\beta_0 s_0) \left(\hat{\mathbf{x}} \cos\left(\beta_0 \int_0^z \frac{XY_L}{2} dz\right) + \hat{\mathbf{y}} \sin\left(\beta_0 \int_0^z \frac{XY_L}{2} dz\right)\right) \tag{10.9}$$

and from which one can see that the polarization vector rotates around the propagation direction in the same manner as predicted by formula 10.6. It is important to note that, as a satellite orbit progresses, there can be changes in the electron density through which a signal must pass before it reaches the ground and this can result in a continually changing polarization at a ground station. Further, this problem will be exacerbated by ionospheric disturbances such as TIDs. For this reason, a single circular polarized mode is often used for Earth–space communication.

A further problem caused by ionospheric motions, such as TIDs, is that of Doppler shift. Most satellites will exhibit a frequency shift as a result of their motion (and possibly the motion of the station with which the satellite communicates), but this can be augmented by a dynamic ionosphere. Doppler shift will result from the time rate of change of the phase distance P between the satellite and communicating station, i.e. $\Delta f = -(f/c)dP/dt$ where f is the wave frequency and Δf is the Doppler shift in frequency. The phase distance is given by

$$P = \int_{Tx}^{Rx} \frac{N}{p} \mathbf{p} \cdot d\mathbf{r} \tag{10.10}$$

and the variation dP of P over time interval dt will consist of three parts: due to the change in path length, due to the change in path bending and due to the change of refractive index. By Fermat's principle, however, the part due to a change in path bending will be zero. Consequently, we will have (Bennett, 1967) that

$$\Delta f = -\frac{f}{c} \left(\int_{Tx}^{Rx} \frac{1}{p} \frac{\partial N}{\partial t} \mathbf{p} \cdot d\mathbf{r} + N_{Rx} v_{Rx} - N_{Tx} v_{Tx} \right) \tag{10.11}$$

where v_{Tx} and v_{Rx} are the transmitter and receiver velocities in the direction of the wave vector. From this, it is clear that the dynamics of the ionosphere might need to be taken into account when evaluating the Doppler experienced in transionospheric propagation.

10.3 Small-Scale Irregularity

Disturbances of the ionosphere, such as gravity waves, can give rise to other more dramatic structures through a mechanism known as the Rayleigh–Taylor instability. This instability is produced by strong density gradients, and such conditions arise quite frequently in the equatorial ionosphere around dusk. After dusk the lower layers of the ionosphere decay rapidly and this gives rise to the necessary strong gradients. In addition, the F2 layer moves upward in altitude and this also enhances the prospect of instability. Large structures, of the order of hundreds of kilometers in height, can form and these are then the source of further instability due to strong density gradients on their sides. This can cause a cascade of irregularity down to quite small scales and the complexity is such that a statistical description is required. After midnight the irregularity starts to decay, but can still remain quite strong until dawn. A simple explanation of the instability is as follows (Kelly, 1989). As shown in Figure 10.3, the strong gradient in the ionospheric density is represented as a jump in plasma density. When a wavelike perturbation occurs on the jump, one half of the wave pushes the plasma into the depleted region and the other pushes the depleted region into the plasma. At equilibrium, the gravity force will be balanced by the magnetic force ($Nm\mathbf{g} + e\mathbf{v} \times \mathbf{B}_0 = 0$, where $\mathbf{v}$ is the plasma velocity and $\mathbf{g}$ is the acceleration due to gravity) and this requires an electron drift to the left and an ion drift to the right. The mass dependence will mean that the ions drift faster than the electrons and so there will be an accumulation of negative charges on the outermost edges of the perturbation and an accumulation of positive charge at the center. This will cause a perturbation electric field $\delta \mathbf{E}$ that will be in opposite

10.4 Scintillation

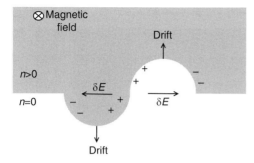

Figure 10.3 Development of the Rayleigh–Taylor instability.

directions in the opposite halves of the perturbation. Importantly, the perturbations will cause a drift velocity $\delta\mathbf{E} \times \mathbf{B}_0/B_0^2$ that is upward in direction on the upward part of the perturbation and downward on the downward part. That is, the changes caused by the perturbation do not restore it to the unperturbed situation but serve to accentuate the perturbation. As a consequence, we have an instability and this will grow until curbed by nonlinear effects. This is known as the Rayleigh–Taylor instability.

The equatorial irregularities are maximum at the equinoxes when the magnetic field lines are aligned with the day-night transition. At other times (certainly around solstices) the field lines in early evening can be shorted out by the daytime E layer and this reduces the potential for instability. The Rayleigh–Taylor instability is one of several mechanisms that can lead to instability and these can act in other zones besides the equatorial zone. In particular, there can be a large amount of irregularity produced in the auroral zone. Irregularity tends to be stretched out along the field lines and so its spectrum can often be non-isotropic. As discussed in Chapter 6, irregularity can cause back scatter, and, as a result of the above anisotropy, this tends to be strongest when the propagation direction is perpendicular to the magnetic field lines. Models of the global distribution of irregularity have been developed (Secan, 2004); Figure 10.4 shows a global distribution of irregular electron density derived from such a model. The figure shows the standard deviation of electron density σ_{N_e} for an equinox (the time of expected maximum irregularity).

10.4 Scintillation

The ionosphere (and/or the atmosphere) between a terrestrial observation point and an extraterrestrial source is almost always disturbed by irregularity to some degree and we need to ascertain the impact of this on our observations. The larger scale irregularities can be handled by geometric optics (i.e. ray tracing) but, at the small scales, the effects of diffraction become important and the GO approximation breaks down. Furthermore, it is often only possible to characterize the small-scale structure in a statistical sense and so we need to employ the techniques of Chapter 6. The effect of irregularity on an observer will mainly consist of fluctuations in the phase and amplitude

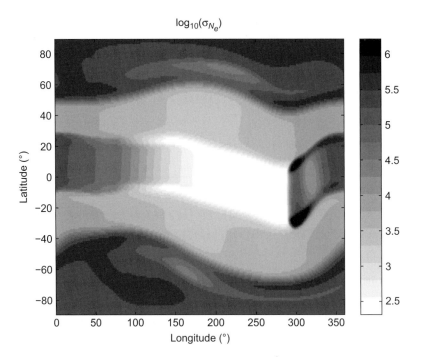

Figure 10.4 Standard deviation of electron density (cm^{-3}) from the local mean for an equinox with $R = 100$ and $K_p = 9$.

of the received signal, this being known as *scintillation*. These fluctuations will often be characterized in terms of the mean square phase fluctuation σ_ϕ^2 and the scintillation index S_4. Parameter S_4 is related to the statistics of the received power P_{Rx} through $S_4^2 = (\langle P_{Rx}^2 \rangle - \langle P_{Rx} \rangle^2)/\langle P_{Rx} \rangle^2$. We would like to find the connection between these statistics and the statistics of the ionospheric irregularity.

The mutual coherence function (MCF) is a major tool for studying the effects of irregularity. As mentioned in Chapter 6, Knepp (1983) has developed an analytic solution for the MCF for a single phase screen (see Figure 10.5) that can be used to study the effect of ionospheric irregularity on satellite signals and radio astronomy observations. In the limit that the screen thickness approaches zero, the MCF has the form $\Gamma = \Gamma_0 \Gamma_a$ where Γ_0 is the MCF with no irregularity present and

$$\Gamma_a(\Delta \mathbf{w}, \omega_d) \approx \exp\left(-\frac{\sigma_\phi^2 \omega_d^2}{2\omega^2}\right) \exp\left(-\frac{\omega_{\text{coh}} |\Delta \mathbf{w}|^2}{l_0^2(\omega_{\text{coh}} + j\omega_d)}\right) \frac{\omega_{\text{coh}}}{\omega_{\text{coh}} + j\omega_d} \quad (10.12)$$

where l_0 is the correlation length of the electric field in the direction transverse to propagation, ω_{coh} is the correlation bandwidth and σ_ϕ is standard deviation of the phase fluctuations. Parameters l_0, ω_{coh} and σ_ϕ can all be derived from the transverse correlation function of the ionospheric irregularity $A(\mathbf{w})$. We consider a simple Gaussian autocorrelation function for the irregularity, i.e. $\langle \delta \epsilon_a \delta \epsilon_b \rangle = \langle |\delta \epsilon|^2 \rangle \exp(-|\mathbf{r}_a - \mathbf{r}_b|^2/L^2)$ where $\langle |\delta \epsilon|^2 \rangle / \epsilon^2 = \langle \delta N_e^2 \rangle / N_e^2$ in the case of the ionosphere (N_e is the electron density

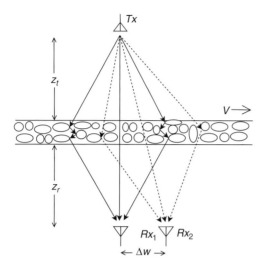

Figure 10.5 Scattering by a phase screen.

and δN_e is the electron density perturbation). From this, the expansion of the transverse autocorrelation $(A(\Delta\mathbf{w}) = \int (\langle \delta\epsilon_a \delta\epsilon_b \rangle / \epsilon^2) dz \approx A_0 + |\Delta\mathbf{w}|^2 A_2 + \cdots)$ has coefficients $A_0 = \sqrt{\pi} L \langle \delta N_e^2 \rangle / N_e^2$ and $A_2 = -(\sqrt{\pi}/L) \langle \delta N_e^2 \rangle / N_e^2$. Then, from Equations 6.120 to 6.121 of Chapter 6, we obtain

$$l_0^2 = \frac{L^2 (z_t + z_r)^2}{z_t^2 \sigma_\phi^2} \tag{10.13}$$

$$\omega_{\text{coh}} = \frac{\pi \omega l_0^2 z_t}{\lambda z_r (z_t + z_r)} \tag{10.14}$$

and

$$\sigma_\phi^2 = \frac{\omega_p^4}{4\omega^2 c_0^2} L L_t \sqrt{\pi} \frac{\langle \delta N_e^2 \rangle}{N_e^2} \tag{10.15}$$

where L_t is the thickness of the screen and $\lambda = 2\pi c/\omega$. The MCF provides us with important information concerning the decorrelation that can occur when the same signal is observed at well-separated points. For example, in astronomical interferometry, antennas can be several kilometers apart and irregularity can cause a significant problem. Strong irregularity will cause the correlation length l_0 to be small and produce an MCF that is delta function in nature. On the other hand, when there is negligible irregularity, the MCF will have a constant value. In the case that there is oblique incidence on the screen, the above theory will still work, but we will need to replace the thickness L_t of the screen by the effective thickness $L_t \sec \chi$, where χ is the angle between the propagation direction and the screen normal.

In order to study S_4, we need to move beyond the MCF and consider equations for the fourth moment of the electric field. For the Rytov approximation, however, it can be shown that $S_4^2 = \Gamma(\mathbf{0})^2 - 1$, where Γ is the MCF. The considerations of Chapter 6 suggest that $S_4^2 = \exp(-L_t \beta^2 A_0/2) - 1$ and so we will have that $S_4^2 \propto L_t$ for small L_t. Consequently, it can be seen that the screen thickness is an important consideration for S_4. The parameters S_4 and σ_ϕ provide important information about the possible damage that can be done by small-scale irregularity to an ionospheric system. A comprehensive global model of scintillation, known as WBMOD (Secan, 2004), is available and from which these parameters can be derived.

When investigating the impact of scintillation on a system, it is often desirable to simulate the signals that will arrive at the receiver. This can be achieved through the methods of Section 6.7. In order to implement these methods, however, we need a suitable generalized power spectral density (GPSD). For a thin screen and isotropic irregularity, there is an analytic expression for this density (Knepp and Wittwer, 1984),

$$S(\mathbf{K}, \tau) = \frac{\omega_{\text{coh}} l_0^2 \alpha}{2^{\frac{5}{2}} \pi^{\frac{3}{2}}} \exp\left(-\frac{|\mathbf{K}|^2 l_0^2}{4} - \frac{1}{2}\alpha^2 \left(\omega_{\text{coh}}\tau - \frac{|\mathbf{K}|^2 l_0^2}{4}\right)\right) \quad (10.16)$$

where $\alpha = \omega/\sigma_\phi \omega_{\text{coh}}$. This expression has proven extremely useful in the simulation of disturbed channels.

The above simulations do not include the Doppler spreading aspect of propagation through irregularity and this can be important when the irregularity is dynamic. In the case of a satellite, even if the irregularity is frozen, the motion of the satellite will cause the propagation path to scan across the irregularity and result in an effective motion. To simulate the channel impulse function, we will need a GPSD that is appropriate to an MCF for temporally varying irregularity. This GPSD has the form (Dana and Wittwer, 1991)

$$S_{\text{GPSD}}(\mathbf{K}, \omega_{\text{Dop}}, \tau)$$
$$= \frac{1}{8\pi^3} \int_{-\infty}^{\infty} \int_{-\infty}^{\infty} \int_{-\infty}^{\infty} \int_{-\infty}^{\infty} \Gamma_a(\mathbf{w}, t_d, \omega_d) \exp\left(-j\mathbf{w} \cdot \mathbf{K} + j\omega_d \tau - j\omega_{\text{Dop}} t_d\right) dw_1 dw_2 d\omega_d dt_d$$

$$(10.17)$$

and the channel function corresponding to this GPSD can be realized through a discrete version of its transform (Nickisch et al., 2012), i.e.

$$H(l\Delta K, m\Delta K, j\Delta\omega_{\text{Dop}}, n\Delta\tau) = \left(\frac{S_{\text{GPSD}}(l\Delta K, m\Delta K, j\Delta\omega_{\text{Dop}}, n\Delta\tau)}{\Delta K \Delta K \Delta\omega_{\text{Dop}} \Delta\tau}\right)^{\frac{1}{2}} r_{lmjn} \quad (10.18)$$

where r_{lmjn} is the complex random variable

$$r_{lmjn} = \sqrt{\frac{1}{2}} \left(a_{lmjn} + jb_{lmjn}\right) \quad (10.19)$$

with a_{lmjn} and b_{lmjn} independent Gaussian random variables that have zero mean and unit variance (l, m, j and n are integers). We can invert this transform (see Section 6.7 for the case of a static medium) through

$$E_a(f\Delta w, g\Delta w, k\Delta T_{\text{ref}}, h\Delta\omega) = \sum_{l=0}^{L-1}\sum_{s=0}^{L-1}\sum_{m=0}^{M-1}\sum_{n=0}^{N-1} H(l\Delta K, s\Delta K, m\Delta\omega_{\text{Dop}}, n\Delta\tau)$$
$$\times \exp\left(jf\, l\Delta K\Delta w + jgs\Delta K\Delta w + jkm\Delta T_{\text{ref}}\Delta\omega_{\text{Dop}} - jhn\Delta\omega\Delta\tau\right)\Delta K\Delta K\Delta\omega_{\text{Dop}}\Delta\tau \quad (10.20)$$

where $\Delta w = D/L$, $\Delta T_{\text{ref}} = T_{\text{ref}}/M$ and $\Delta\tau = T/N$. In order to satisfy the Nyquist limit, we choose the sampling intervals $\Delta K = 2\pi/D$, where D is the extent of lateral coherence; $\Delta\omega_{\text{Dop}} = 2\pi/T_{\text{ref}}$, where T_{ref} is the time scale of changes in the medium and $\Delta\omega = 2\pi/T$, where T is the maximum delay. The MCF derived from the above E_a will then exhibit the desired GPSD. We can now form $\hat{h} = E_0 E_a$, which is the Fourier transform of the impulse response function with E_0 representing $\hat{h}$ in the unperturbed medium.

For a structure function that can be approximated by the quadratic

$$A(\Delta\mathbf{w}, t_d) = A_0 + A_2\left(\frac{\Delta x^2}{L_x^2} + \frac{\Delta y^2}{L_y^2} + \frac{t_d^2}{T_0^2} - 2\frac{C_{xt}\Delta x t_d}{L_x T_0} - 2\frac{C_{yt}\Delta y t_d}{L_y T_0}\right) \quad (10.21)$$

the generalization of Equation 10.12 is given by (Dana and Wittwer, 1991)

$$\Gamma_a(\Delta\mathbf{w}, t_d, \omega_d) = \frac{\exp\left(-\frac{\sigma_\phi^2 \omega_d^2}{2\omega^2}\right)\exp\left(-(1 - C_{xt}^2 - C_{yt}^2)\left(\frac{t_d}{\tau_0}\right)^2\right)}{\sqrt{1 + j\frac{\omega_d\sqrt{2}\ell_x^2}{\omega_{\text{coh}}\sqrt{\ell_x^4 + \ell_y^4}}}\sqrt{1 + j\frac{\omega_d\sqrt{2}\ell_y^2}{\omega_{\text{coh}}\sqrt{\ell_x^4 + \ell_y^4}}}}$$
$$\times \exp\left(-\frac{\left(\frac{\Delta x}{\ell_x} - C_{xt}\frac{t_d}{\tau_0}\right)^2}{1 + j\frac{\omega_d\sqrt{2}\ell_y^2}{\omega_{\text{coh}}\sqrt{\ell_x^4+\ell_y^4}}}\right)\exp\left(-\frac{\left(\frac{\Delta y}{\ell_y} - C_{yt}\frac{t_d}{\tau_0}\right)^2}{1 + j\frac{\omega_d\sqrt{2}\ell_x^2}{\omega_{\text{coh}}\sqrt{\ell_x^4+\ell_y^4}}}\right) \quad (10.22)$$

where the decorrelation distances ℓ_x and ℓ_y are given by

$$\ell_{x,y}^2 = -\frac{A_0(z_t + z_r)^2 L_{x,y}^2}{z_t^2 \sigma_\phi^2 A_2} \quad (10.23)$$

the coherence bandwidth by

$$\omega_{\text{coh}} = -\frac{\omega^2 L_x^2 L_y^2 A_0}{\sqrt{2}c\sqrt{L_x^4 + L_y^4 \sigma_\phi^2 A_2}}\frac{z_t + z_r}{z_t z_r} \quad (10.24)$$

and the decorrelation time by $\tau_0 = (T_0/\sigma_\phi)\sqrt{A_0/A_2}$. The corresponding GPSD (Dana and Wittwer, 1991) is

$$S_{\text{GPSD}}\left(\mathbf{K}, \omega_{\text{Dop}}, \tau\right) = \frac{\pi \tau_o \ell_x \ell_y \alpha \omega_{\text{coh}}}{\sqrt{2\left(1 - C_{xt}^2 - C_{yt}^2\right)}} \exp\left\{-\frac{\left(\tau_o \omega_{\text{Dop}} - C_{xt} K_x \ell_x - C_{yt} K_y \ell_y\right)^2}{4\left(1 - C_{xt}^2 - C_{yt}^2\right)}\right\}$$

$$\times \exp\left\{-\left(\frac{K_x^2 \ell_x^2 + K_y^2 \ell_y^2}{4}\right) - \frac{\alpha^2}{2}\left[\omega_{\text{coh}} \tau - \left(\frac{2\ell_x^4 \ell_y^4}{\ell_x^4 + \ell_y^4}\right)^{1/2} \left(\frac{K_x^2 + K_y^2}{4}\right)\right]^2\right\}$$

(10.25)

where $\alpha = \omega/\sigma_\phi \omega_{\text{coh}}$. We can use this GPSD to generate a realization of the channel impulse function, as described above.

For the above form of irregularity structure function, the parameters C_{xt} and C_{yt} measure the cross correlation between the spatial and temporal aspects of the irregularity. In the case of turbulent irregularity, $C_{xt} \to 0$ and $C_{yt} \to 0$. The structure remaining after turbulence can often take a considerable time to dissipate (particularly in the equatorial regions) and can have the character of fossilized structure that drifts with the background medium. Let the fossilized irregularity have the time-independent structure function $A(\Delta \mathbf{w})$, then, if the irregularity drifts across the propagation path with speed $\mathbf{V}$, the time-dependent structure function will be given by $A(\Delta \mathbf{w}, t) = A(\Delta \mathbf{w} - t\mathbf{V})$. The assumption that the irregularity consists of fossilized structure drifting with the background medium is often known as the *frozen-in assumption*.

The form of GPSD given by Equation 10.25 is fairly general and is useful for simulating the effect of irregularity on a signal that is received by an array. In particular, Nickisch et al. (2012) have used the above GPSD to study the effect of scintillation on radar signals propagating in the ionospheric duct. In this work, representative phase screens were placed in the region of irregularity and, for the purposes of calculating the GPSD, the z coordinate taken to be the distance along the propagation path. The techniques of Section 6.7 can then be used to generate a realization of the channel impulse function. This process requires the impulse function of the background medium and, in the case of ionospheric propagation, this is usually taken to be the GO approximation to this function. Frequently, there will be several GO modes in the ionospheric channel and an impulse function associated with each. In this case, the channel impulse function will be the sum of these separate impulse functions since the different modes will be well decorrelated.

For a radio system with a single fixed antenna, we set $\mathbf{w} = \mathbf{0}$ and define a *channel scattering function* (CSF) through

$$S_{\text{CSF}}(\omega_{\text{Dop}}, \tau) = \frac{1}{2\pi} \int_{-\infty}^{\infty} \int_{-\infty}^{\infty} \Gamma_a(\mathbf{0}, t_d, \omega_d) \exp\left(j\omega_d \tau - j\omega_{\text{Dop}} t_d\right) d\omega_d dt_d \quad (10.26)$$

This function provides us with information concerning the spread of a received signal in both Doppler and delay, such spread being in addition to the Doppler and delay caused by the background ionosphere. We consider the case of the phase screen of Knepp

(1983) (structure function $A = A_0 + |\Delta \mathbf{w}|^2 A_2$) with frozen structure that is moving at speed V across the propagation path, then

$$\Gamma_a(t_d, \omega_d) = \exp\left(-\frac{\sigma_\phi^2 \omega_d^2}{2\omega^2}\right) \exp\left(-\frac{\omega_{\text{coh}} t_d^2}{\tau_0^2 (\omega_{\text{coh}} + j\omega_d)}\right) \frac{\omega_{\text{coh}}}{\omega_{\text{coh}} + j\omega_d} \quad (10.27)$$

where $\tau_0 = \sqrt{A_0/A_2}/\sigma_\phi V$ (see Chapter 6 for the definition of σ_ϕ and ω_{coh}). We can perform the t_d integral in the definition of the CSF to obtain

$$S_{\text{CSF}}(\omega_{\text{Dop}}, \tau) = \frac{\tau_0}{2\sqrt{\pi}} \int_{-\infty}^{\infty} \sqrt{\frac{\omega_{\text{coh}}}{\omega_{\text{coh}} + j\omega_d}} \quad (10.28)$$

$$\times \exp\left(-\frac{\sigma_\phi^2 \omega_d^2}{2\omega^2}\right) \exp\left(-\frac{\omega_{\text{Dop}}^2 \tau_0^2 (\omega_{\text{coh}} + j\omega_d)}{4\omega_{\text{coh}}}\right) \exp(j\omega_d \tau) \, d\omega_d$$

The above integral can be evaluated in the high-frequency limit ($\omega \to \infty$), which effectively encompasses VHF and above. In this case, $\sqrt{\omega_{\text{coh}}/(\omega_{\text{coh}} + j\omega_d)} \approx 1$ in the range of integration where the integrand is significant and, as a consequence,

$$S_{\text{CSF}}(\omega_{\text{Dop}}, \tau) \approx \frac{\tau_0 \omega}{\sqrt{2}\sigma_\phi} \exp\left(-\frac{\omega_{\text{Dop}}^2 \tau_0^2}{4} - \frac{\left(\tau - \frac{\omega_{\text{Dop}}^2 \tau_0^2}{4\omega_{\text{coh}}}\right)^2 \omega^2}{2\sigma_\phi^2}\right) \quad (10.29)$$

Figure 10.6 shows an example of such a CSF for irregularity with an isotropic Gaussian autocorrelation (correlation length is 10 km, $\langle \delta N_e^2 \rangle / N_e^2 = 0.01$, screen thickness is 50 km, $z_r = 500$ km, $z_s = 500$ km, background plasma frequency is 10 MHz, wave frequency is 100 MHz and screen velocity is $V = 100$ m/s).

The CSF is an important element in generating a realization of the channel impulse response function for a single fixed receiver. A discrete transform of the impulse function is generated by

$$H(m\Delta\omega_{\text{Dop}}, n\Delta\tau) = \left(\frac{S_{\text{CSF}}(m\Delta\omega_{\text{Dop}}, n\Delta\tau)}{\Delta\omega_{\text{Dop}} \Delta\tau}\right)^{\frac{1}{2}} r_{mn} \quad (10.30)$$

where r_{mn} is the complex random variable

$$r_{mn} = \sqrt{\frac{1}{2}} (a_{mn} + jb_{mn}) \quad (10.31)$$

with a_{mn} and b_{mn} independent Gaussian random variables that have zero mean and unit variance. We can invert this transform (see Section 6.7 for the case of a static medium) through

$$E_a(k\Delta T_{\text{ref}}, h\Delta\omega) =$$

$$\frac{1}{2\pi} \sum_{m=0}^{M-1} \sum_{n=0}^{N-1} H(m\Delta\omega_{\text{Dop}}, n\Delta\tau) \exp\left(jkm\Delta T_{\text{ref}} \Delta\omega_{\text{Dop}} - jhn\Delta\omega\Delta\tau\right) \Delta\tau \Delta\omega_{\text{Dop}}$$

$$(10.32)$$

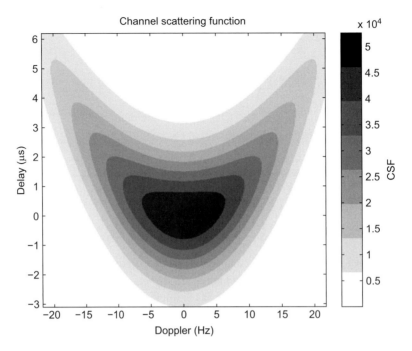

Figure 10.6 Example of a channel scattering function showing parabolic structure.

for $h = 0, \ldots, N-1$ and $k = 0, \ldots, M-1$ where $\Delta\tau = T/N$ and $\Delta T_{\text{ref}} = T_{\text{ref}}/M$. As before, to satisfy the Nyquist limit, we choose $\Delta\omega_{\text{Dop}} = 2\pi/T_{\text{ref}}$ and $\Delta\omega = 2\pi/T$. We can then generate the discretized form of the impulse response function through

$$h(k\Delta T_{\text{ref}}, h\Delta\tau) = \frac{\Delta\omega}{2\pi} \sum_{n=0}^{N-1} \hat{h}(k\Delta T_{\text{ref}}, n\Delta\omega) \exp\left(jhn\Delta\omega\Delta\tau\right) \tag{10.33}$$

where $\hat{h} = E_0 E_a$ with E_0 equal to $\hat{h}$ with no irregularity present. In the case of propagation through the ionospheric duct, there will usually be several propagation modes and the impulse function will need to be the sum of the separate impulse functions for these modes. Providing the signals have narrow bandwidth, the impulse function can be approximated as

$$h(t, \tau) = \sum_i \frac{\hat{H}^i\left(t, \tau - \tau_g^i\right) \exp\left(j\omega_c\left(\tau_g^i - \tau_p^i\right)\right)}{\sqrt{L_i}} \tag{10.34}$$

where, for the ith mode, τ_g^i is the group delay, τ_p^i the phase delay and L_i is the propagation loss. Function $\hat{H}^i$ is defined by

$$\hat{H}^i\left(k\Delta T_{\text{ref}}, n\Delta\tau\right) = \frac{\Delta\omega_{\text{Dop}}}{2\pi} \sum_{m=0}^{M-1} H^i\left(m\Delta\omega_{\text{Dop}}, n\Delta\tau\right) \exp\left(jkm\Delta T_{\text{ref}}\Delta\omega_{\text{Dop}}\right) \tag{10.35}$$

where H^i is function H for the ith mode. It should be noted that τ_p^i and τ_g^i can change with time due to the motion of the background medium and/or the motion of the transmitter and receiver. Such variations with time will result in a Doppler shift in addition to the Doppler spread caused by irregular structure.

10.5 References

J.A. Bennett, The calculation of Doppler shifts due to a changing ionosphere, J. Atmos. Terr. Phys., vol. 29, p. 887, 1967.

R. Dana and L. Wittwer, A general channel model of RF propagation through structured ionisation, Radio Sci., vol. 26, pp. 1059–1068, doi:10.1029/91RS00263, 1991.

K. Davies, *Ionospheric Radio*, IEE Electomagn. Waves Ser., vol. 31, Peter Peregrinus, London, 1990.

A.S. Jursa (editor), *Handbook of Geophysics and the Space Environment*, Air Force Geophysics Laboratory, Air Force Systems Command, United States Air Force, 1985.

J.A. Klobuchar, Ionospheric time delay algorithm for single frequency GPS users, IEEE Transactions, AES-23, p. 325, 1987.

D.L. Knepp, Analytic solution for the two-frequency mutual coherence function for spherical wave propagation, Radio Sci., vol. 18, pp. 535–549, 1983.

D.L. Knepp and L.A. Wittwer, Simulation of wide bandwidth signals that have propagated through random media, Radio Sci., vol. 19, pp. 303–318, 1984.

L.J. Nickisch, Nonuniform motion and extended media effects on the mutual coherence function: An analytic solution for spaced frequency, position and time, Radio Sci., vol. 27, pp. 9–22, 1992.

L.J. Nickisch, G. St. John, S.V. Fridman. M.A. Hausman and C.J. Coleman, HiCIRF: A high-fidelity HF channel simulation, Radio Sci., vol. 47, RS0L11, doi:10.1029/2011RS004928, 2012.

W.H. Press, S. Teukolsky, W. Vetterling and B. Flannery, *Numerical Recipes in FORTRAN: The Art of Scientific Computing*, 2nd edition, Cambridge University Press, Cambridge, 1992.

T.D. Raymund, S.J. Franke and K.C. Yeh, Ionospheric tomography: Its limitations and reconstruction methods, J. Atmos. Terr. Phys., vol. 56, pp. 637–657, 1994.

J.A. Secan, *WBMOD Ionospheric Scintillation Model, an Abbreviated User's Guide*, Rep. NWRA-CR-94 R172/Rev 7, North West Res. Assoc., Inc., Bellevue, Wash., 2004.

Appendix A Some Useful Mathematics

A.1 Vectors

Let $\mathbf{A} = A_x\hat{\mathbf{x}} + A_y\hat{\mathbf{y}} + A_z\hat{\mathbf{z}}$ and $\mathbf{B} = B_x\hat{\mathbf{x}} + B_y\hat{\mathbf{y}} + B_z\hat{\mathbf{z}}$ be vectors where $\hat{\mathbf{x}}$, $\hat{\mathbf{y}}$ and $\hat{\mathbf{z}}$ are unit vectors along the x, y and z axes, respectively. The dot product is a scalar defined by

$$\mathbf{A} \cdot \mathbf{B} = A_x B_x + A_y B_y + A_z B_z \tag{A.1}$$

from which we note that $\mathbf{A} \cdot \mathbf{B} = \mathbf{B} \cdot \mathbf{A}$. The vector product is a product of two vectors that itself is a vector

$$\mathbf{A} \times \mathbf{B} = (A_y B_z - A_z B_y)\hat{\mathbf{x}} + (A_z B_x - A_x B_z)\hat{\mathbf{y}} + (A_x B_y - A_y B_x)\hat{\mathbf{z}} \tag{A.2}$$

from which we note that $\mathbf{A} \times \mathbf{B} = -\mathbf{B} \times \mathbf{A}$. The following identities should also be noted

$$\mathbf{A} \times (\mathbf{B} \times \mathbf{C}) = \mathbf{B}(\mathbf{A} \cdot \mathbf{C}) - \mathbf{C}(\mathbf{A} \cdot \mathbf{B}) \tag{A.3}$$

$$(\mathbf{A} \times \mathbf{B}) \times \mathbf{C} = \mathbf{B}(\mathbf{A} \cdot \mathbf{C}) - \mathbf{A}(\mathbf{B} \cdot \mathbf{C}) \tag{A.4}$$

$$(\mathbf{A} \times \mathbf{B}) \cdot (\mathbf{C} \times \mathbf{D}) = (\mathbf{A} \cdot \mathbf{C})(\mathbf{B} \cdot \mathbf{D}) - (\mathbf{A} \cdot \mathbf{D})(\mathbf{B} \cdot \mathbf{C}) \tag{A.5}$$

$$\mathbf{A} \cdot (\mathbf{B} \times \mathbf{C}) = \mathbf{B} \cdot (\mathbf{C} \times \mathbf{A}) = \mathbf{C} \cdot (\mathbf{A} \times \mathbf{B}) \tag{A.6}$$

A.2 Vector Operators

In terms of Cartesian coordinates, the *gradient*, *divergence*, *curl* and *Laplace* operators are defined as follows. For scalar field ϕ, the gradient operator produces a vector field

$$\nabla \phi = \frac{\partial \phi}{\partial x}\hat{\mathbf{x}} + \frac{\partial \phi}{\partial y}\hat{\mathbf{y}} + \frac{\partial \phi}{\partial z}\hat{\mathbf{z}} \tag{A.7}$$

For vector field $\mathbf{A}$, the divergence operator produces a scalar

$$\nabla \cdot \mathbf{A} = \frac{\partial A_x}{\partial x} + \frac{\partial A_y}{\partial y} + \frac{\partial A_z}{\partial z} \tag{A.8}$$

For vector field **A**, the curl operator produces a vector field

$$\nabla \times \mathbf{A} = \left(\frac{\partial A_z}{\partial y} - \frac{\partial A_y}{\partial z}\right)\hat{\mathbf{x}} + \left(\frac{\partial A_x}{\partial z} - \frac{\partial A_z}{\partial x}\right)\hat{\mathbf{y}} + \left(\frac{\partial A_y}{\partial x} - \frac{\partial A_x}{\partial y}\right)\hat{\mathbf{z}} \quad (A.9)$$

For scalar field ϕ the Laplace operator produces a scalar field

$$\nabla^2 \phi = \frac{\partial^2 \phi}{\partial x^2} + \frac{\partial^2 \phi}{\partial y^2} + \frac{\partial^2 \phi}{\partial z^2} \quad (A.10)$$

The vector operators satisfy the following important identities:

$$\nabla \cdot (\mathbf{X} \times \mathbf{Y}) = \mathbf{Y} \cdot \nabla \times \mathbf{X} - \mathbf{X} \cdot \nabla \times \mathbf{Y} \quad (A.11)$$

$$\nabla \times (\nabla \times \mathbf{G}) = \nabla(\nabla \cdot \mathbf{G}) - \nabla^2 \mathbf{G} \quad (A.12)$$

$$\nabla \times (\nabla \phi) = 0 \quad (A.13)$$

$$\nabla \cdot (\nabla \times \mathbf{A}) = 0 \quad (A.14)$$

Of importance to the current text is the *divergence theorem*, a result that relates the integral of the divergence of a vector field **F** over a volume V to an integral of its normal component over the surface S of V. The relation is of the form

$$\int_V \nabla \cdot \mathbf{F}\, dV = \int_S \mathbf{F} \cdot \mathbf{n}\, dS \quad (A.15)$$

where **n** is the unit normal on the surface S.

A.3 Cylindrical Polar Coordinates

A useful alternative to Cartesian coordinates are cylindrical polar coordinates (ρ, ϕ, z) where $x = \rho \cos \phi$ and $y = \rho \cos \phi$ (Figure A.1). If we define $\hat{\mathbf{e}}_\rho$ to be a unit vector in the direction of coordinate ρ, $\hat{\mathbf{e}}_\phi$ to be a unit vector in the direction of coordinate ϕ and $\hat{\mathbf{e}}_z$ to be the unit vector in the direction of coordinate z,

$$\nabla \phi = \frac{\partial \phi}{\partial \rho} \hat{\mathbf{e}}_\rho + \frac{1}{\rho}\frac{\partial \phi}{\partial \phi} \hat{\mathbf{e}}_\phi + \frac{\partial \phi}{\partial z} \hat{\mathbf{e}}_z \quad (A.16)$$

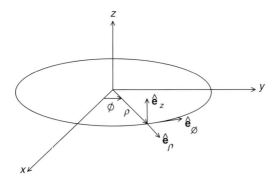

Figure A.1 Cylindrical polar coordinates.

$$\nabla \times \mathbf{A} = \left(\frac{1}{\rho}\frac{\partial A_z}{\partial \phi} - \frac{\partial A_\phi}{\partial z}\right)\hat{\mathbf{e}}_\rho + \left(\frac{\partial A_\rho}{\partial z} - \frac{\partial A_z}{\partial \rho}\right)\hat{\mathbf{e}}_\phi + \frac{1}{\rho}\left(\frac{\partial(\rho A_\phi)}{\partial \rho} - \frac{\partial A_\rho}{\partial \phi}\right)\hat{\mathbf{e}}_z \quad (A.17)$$

$$\nabla \cdot \mathbf{A} = \frac{1}{\rho}\frac{\partial(\rho A_\rho)}{\partial \rho} + \frac{1}{\rho}\frac{\partial A_\phi}{\partial \phi} + \frac{\partial A_z}{\partial z} \quad (A.18)$$

and

$$\nabla^2 \phi = \frac{1}{\rho}\frac{\partial}{\partial}\left(\rho\frac{\partial \phi}{\partial \rho}\right) + \frac{1}{\rho^2}\frac{\partial^2 \phi}{\partial \phi^2} + \frac{\partial^2 \phi}{\partial z^2} \quad (A.19)$$

A.4 Some Useful Integrals

An important integral is given by

$$\int_{-\infty}^{\infty} \exp(-j\alpha x^2)\, dx = \sqrt{\frac{\pi}{j\alpha}} \quad (A.20)$$

Related to this is the integral $\int_v^\infty \exp(-jt^2)\, dt$, which can be evaluated using Fresnel integrals (C and S) via

$$\int_v^\infty \exp(-jt^2)\, dt = C_1(v) - jS_1(v) \quad (A.21)$$

where $C_1(x) = \int_x^\infty \cos(t^2)\, dt$ and $S_1(x) = \int_x^\infty \sin(t^2)\, dt$. For large positive v ($v \to \infty$) we have

$$\int_v^\infty \exp(-jt^2)\, dt \approx \exp(-jv^2)/2jv \quad (A.22)$$

It will also be noted that $\int_v^\infty \exp(-jt^2)\, dt = \frac{1}{2}\sqrt{\pi/j}$ when $v = 0$. Consequently, we can form the rational approximation $\int_v^\infty \exp(-jt^2)\, dt \approx \frac{1}{2}\exp(-jv^2)/(\sqrt{j/\pi} + jv)$ (this is valid in both limits providing $v > 0$). For $v < 0$, it should be noted that $\int_v^\infty \exp(-jt^2)\, dt = \sqrt{\pi/j} - \int_{|v|}^\infty \exp(-jt^2)\, dt$ and this allows us to calculate the integral for $v \to -\infty$. Some other related results are

$$\int_0^\infty t^2 \exp(-jqt^2)\, dt = \frac{1}{2jq}\int_0^\infty \exp(-jqt^2)\, dt \quad (A.23)$$

$$\int_0^\infty \exp(-jqt^2)\, dt = \frac{1}{2}\sqrt{\frac{\pi}{jq}} \quad (A.24)$$

and

$$\int_0^\infty t\exp(-jqt^2)\, dt = \frac{1}{2jq} \quad (A.25)$$

Other related integrals are

$$\int_0^\infty t^2 \exp(-qt^2)\,dt = \frac{\pi^{\frac{1}{2}}}{4q^{\frac{3}{2}}} \qquad (A.26)$$

$$\int_0^\infty t \exp(-qt^2)\,dt = \frac{1}{2q} \qquad (A.27)$$

and

$$\int_0^\infty \exp(-qt^2)\,dt = \sqrt{\frac{\pi}{q}} \qquad (A.28)$$

A multi-dimensional integral that is often useful in propagation calculations is

$$\int_{R^N} \exp\left(-\frac{1}{2}\mathbf{r}^T A \mathbf{r} + \mathbf{b}^T \cdot \mathbf{r}\right) dx_1 dx_2 \ldots dx_N = \frac{(2\pi)^{\frac{N}{2}}}{\det A^{\frac{1}{2}}} \exp\left(\frac{\mathbf{b}^T A^{-1} \mathbf{b}}{2}\right) \qquad (A.29)$$

where $\mathbf{r} = (x_1, x_2, \ldots, x_N)$.

A.5 Trigonometric Identities

$$\exp(j\alpha) = \cos\alpha + j\sin\alpha \qquad (A.30)$$

$$\sin(\theta + \phi) = \sin\theta\cos\phi + \cos\theta\sin\phi \qquad (A.31)$$

$$\cos(\theta + \phi) = \cos\theta\cos\phi - \sin\theta\sin\phi \qquad (A.32)$$

$$\sin\theta + \sin\phi = 2\sin\frac{\theta+\phi}{2}\cos\frac{\theta-\phi}{2} \qquad (A.33)$$

$$\cos\theta + \cos\phi = 2\cos\frac{\theta+\phi}{2}\cos\frac{\theta-\phi}{2} \qquad (A.34)$$

$$\sin\theta - \sin\phi = 2\cos\frac{\theta+\phi}{2}\sin\frac{\theta-\phi}{2} \qquad (A.35)$$

$$\cos\theta - \cos\phi = 2\sin\frac{\theta+\phi}{2}\sin\frac{\phi-\theta}{2} \qquad (A.36)$$

$$\sin\theta\sin\phi = \frac{1}{2}(\cos(\theta-\phi) - \cos(\theta+\phi)) \qquad (A.37)$$

$$\cos\theta\cos\phi = \frac{1}{2}(\cos(\theta-\phi) + \cos(\theta+\phi)) \qquad (A.38)$$

$$\sin\theta\cos\phi = \frac{1}{2}(\sin(\theta-\phi) + \sin(\theta+\phi)) \qquad (A.39)$$

$$\sin\left(\frac{\pi}{2} \pm \theta\right) = \cos\theta \qquad (A.40)$$

$$\cos\left(\frac{\pi}{2} \pm \theta\right) = \mp\sin\theta \qquad (A.41)$$

$$\sin(2\theta) = 2\sin\theta\cos\theta \tag{A.42}$$

$$\cos(2\theta) = \cos^2\theta - \sin^2\theta \tag{A.43}$$

$$\cos^2\theta + \sin^2\theta = 1 \tag{A.44}$$

$$\cos(-\theta) = \cos(\theta) \tag{A.45}$$

$$\sin(-\theta) = -\sin(\theta) \tag{A.46}$$

A.6 Method of Stationary Phase

Consider the integral

$$I = \int_a^b A(x)\exp(-j\beta\phi(x))\,dx \tag{A.47}$$

for the case that $\beta \to \infty$. If $A(x)$ is slowly varying in comparison with the exponential, the maximum contribution to the integral will come from around the point $x = \alpha$ where $\phi'(\alpha) = 0$ (i.e. a point of stationary phase). If the stationary point is not at either of the end points (a and b), then

$$I \approx \int_{-\infty}^{\infty} A(\alpha)\exp\left(-j\beta\left(\phi(\alpha) + (x-\alpha)^2 \frac{\phi''(\alpha)}{2}\right)\right)dx \tag{A.48}$$

and this can be integrated to yield

$$I \approx \sqrt{\frac{2\pi}{j\beta\phi''(\alpha)}} A(\alpha)\exp(-j\beta\phi(\alpha)) \tag{A.49}$$

In the case that the stationary point is one of the end points, we will have half the above value. In the case that there is no point of stationary phase in $[a, b]$, then

$$I \approx \frac{jA(b)\exp(-j\beta\phi(b))}{\beta\phi'(b)} - \frac{jA(a)\exp(-j\beta\phi(a))}{\beta\phi'(a)} \tag{A.50}$$

A.7 Some Expansions

The Taylor expansion about a point α is given by

$$f(x) = f(\alpha) + f'(\alpha)(x-\alpha) + \frac{1}{2}f''(\alpha)(x-\alpha)^2 + \frac{1}{6}f'''(\alpha)(x-\alpha)^3 + \cdots \tag{A.51}$$

As $x \to 0$, we have

$$\sin x = x - \frac{x^3}{6} + \cdots \tag{A.52}$$

$$\cos x = 1 - \frac{x^2}{2} + \cdots \tag{A.53}$$

$$(1+x)^\alpha = 1 + \alpha x + \alpha(\alpha-1)\frac{x^2}{2} + \alpha(\alpha-1)(\alpha-2)\frac{x^3}{6} + \cdots \qquad (A.54)$$

$$\exp x = 1 + x + \frac{x^2}{2} + \frac{x^3}{6} + \cdots \qquad (A.55)$$

and

$$\ln(1+x) = x - \frac{x^2}{2} + \frac{x^3}{3} + \cdots \qquad (A.56)$$

The Taylor expansion about a point (α, β) in two dimensions is given by

$$f(x,y) = f(\alpha,\beta) + f_x(\alpha,\beta)(x-\alpha) + f_y(\alpha,\beta)(y-\beta)$$
$$+ \frac{1}{2}f_{xx}(\alpha,\beta)(x-\alpha)^2 + f_{xy}(\alpha,\beta)(x-\alpha)(y-\beta)$$
$$+ \frac{1}{2}f_{yy}(\alpha,\beta)(y-\beta)^2 + \cdots \qquad (A.57)$$

A.8 The Airy Function

The Stokes equation is given by

$$w'' - zw = 0 \qquad (A.58)$$

and its independent solutions Ai and Bi are known as Airy functions. Airy function Ai is given in series form as

$$Ai(z) = Ai(0)\left(1 + \sum_{j=1}^{\infty} \frac{1.4.7 \cdots 3j-2}{(3j)!} z^{3j}\right)$$
$$+ Ai'(0)\left(z + \sum_{j=1}^{\infty} \frac{2.5.8 \cdots 3j-1}{(3j+1)!} z^{3j+1}\right) \qquad (A.59)$$

where $Ai(0) = \frac{3^{-2/3}}{\Gamma(2/3)} \approx 0.35502805$ and $Ai'(0) = -3^{-\frac{1}{3}}/\Gamma(1/3) \approx -0.25881940$ (note that a prime denotes a derivative). The second airy function, Bi, is given in series form as

$$Bi(z) = \sqrt{3}Ai(0)\left(1 + \sum_{j=1}^{\infty} \frac{1.4.7 \cdots 3j-2}{(3j)!} z^{3j}\right)$$
$$- \sqrt{3}Ai'(0)\left(z + \sum_{j=1}^{\infty} \frac{2.5.8 \cdots 3j-1}{(3j+1)!} z^{3j+1}\right) \qquad (A.60)$$

A general solution to the Stokes equation is given by a linear combination of Ai and Bi. In the limit $|z| \to \infty$, for $|\arg z| < \pi$,

$$Ai(z) \sim \frac{1}{2\sqrt{\pi}z^{\frac{1}{4}}} \exp\left(-\frac{2}{3}z^{\frac{3}{2}}\right) \qquad (A.61)$$

$$Ai'(z) \sim \frac{-z^{\frac{1}{4}}}{2\sqrt{\pi}} \exp\left(-\frac{2}{3}z^{\frac{3}{2}}\right) \qquad (A.62)$$

$$Bi(z) \sim \frac{1}{\sqrt{\pi}z^{\frac{1}{4}}} \exp\left(\frac{2}{3}z^{\frac{3}{2}}\right) \qquad (A.63)$$

$$Bi'(z) \sim \frac{z^{\frac{1}{4}}}{\sqrt{\pi}} \exp\left(\frac{2}{3}z^{\frac{3}{2}}\right) \qquad (A.64)$$

For $|\arg z| < 2\pi/3$,

$$Ai(-z) \sim \frac{1}{\sqrt{\pi}z^{\frac{1}{4}}} \sin\left(\frac{2}{3}z^{\frac{3}{2}} + \frac{\pi}{4}\right) \qquad (A.65)$$

$$Ai'(-z) \sim -\frac{z^{\frac{1}{4}}}{\sqrt{\pi}} \cos\left(\frac{2}{3}z^{\frac{3}{2}} + \frac{\pi}{4}\right) \qquad (A.66)$$

$$Bi(-z) \sim \frac{1}{\sqrt{\pi}z^{\frac{1}{4}}} \cos\left(\frac{2}{3}z^{\frac{3}{2}} + \frac{\pi}{4}\right) \qquad (A.67)$$

$$Bi'(-z) \sim \frac{z^{\frac{1}{4}}}{\sqrt{\pi}} \sin\left(\frac{2}{3}z^{\frac{3}{2}} + \frac{\pi}{4}\right) \qquad (A.68)$$

An integral representation of Ai is given by

$$Ai(z) = \frac{1}{\pi} \int_0^\infty \cos\left(\frac{t^3}{3} + zt\right) dt \qquad (A.69)$$

and it should be noted that Ai has zeros $a_1 \approx -2.33811$, $a_2 \approx -4.08795$, $a_3 \approx -5.52056$ and $a_i \approx -(3\pi(4i-1)/8)^{\frac{2}{3}}$ as $i \to \infty$. The zeros of the derivative of the Airy function Ai' are also negative real and the zero with the smallest magnitude is approximately -1.01879.

A.9 Hankel and Bessel Functions

The solutions to the equation

$$x^2 \frac{d^2y}{dx^2} + x\frac{dy}{dx} + (x^2 - n^2)y = 0 \qquad (A.70)$$

are known as Bessel functions of order n and arise when looking for separable solutions of the 2D wave equation in polar form. The two independent solutions are $J_n(x)$, which

A.9 Hankel and Bessel Functions

is a Bessel function of the first kind, and $Y_n(x)$, which is a Bessel function of the second kind. It should be noted that, for n and integer, $J_{-n}(x) = (-1)^n J_n(x)$ and $Y_{-n}(x) = (-1)^n Y_n(x)$. For the limit $x \to 0$, $J_n(x) \approx x^n/2^n \Gamma(n+1)$ and, providing $n > 0$, $Y_n(x) \approx -2^n(n-1)!/\pi x^n$ ($Y_0(x) \approx 2(\ln(x/2) + \gamma)/\pi$ where $\gamma \approx 0.5772156$ is Euler's constant). In the limit that $|x| \to \infty$

$$J_n \approx \sqrt{\frac{2}{\pi x}} \cos\left(x - \frac{n\pi}{2} - \frac{\pi}{4}\right) \tag{A.71}$$

and

$$Y_n \approx \sqrt{\frac{2}{\pi x}} \sin\left(x - \frac{n\pi}{2} - \frac{\pi}{4}\right) \tag{A.72}$$

The Bessel function can, in general, be represented by a contour integral

$$J_\nu(z) = \frac{1}{2j\pi} \int_C t^{-\nu-1} \exp\left(\frac{z}{2}\left(t - \frac{1}{t}\right)\right) dt \tag{A.73}$$

The integrand has a branch cut along the real axis from $-\infty$ to the origin and the contour C runs from $-\infty$ on the lower side of the branch cut, takes an anticlockwise circuit around the origin and then returns to $-\infty$ along the topside of the branch cut. When ν is an integer, the integrand does not have a branch cut and the integrals along negative real axis cancel. As a consequence, C is simply a closed circuit around the origin (see Figure A.2) and

$$J_n(x) = \frac{1}{\pi} \int_0^\pi \cos(n\theta - x\sin\theta) d\theta \tag{A.74}$$

where n is an integer.

The Hankel functions ($H_n^{(1)}$ of the first kind and $H_n^{(2)}$ of the second kind) are related to the Bessel functions through $H_n^{(1)}(x) = J_n(x) + jY_n(x)$ and $H_n^{(2)}(x) = J_n(x) - jY_n(x)$. They satisfy $H_{-\nu}^{(1)}(z) = \exp(j\nu\pi)H_\nu^{(1)}(z)$, $H_{-\nu}^{(2)}(z) = \exp(-j\nu\pi)H_\nu^{(2)}(z)$, the Wronskian relation

$$H_\nu^{(1)}(x)\frac{d}{dx}H_\nu^{(2)}(x) - H_\nu^{(2)}(x)\frac{d}{dx}H_\nu^{(1)}(x) = -\frac{4j}{\pi x} \tag{A.75}$$

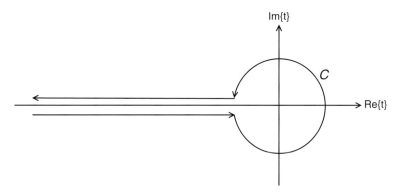

Figure A.2 Integration contour C for the Bessel function.

and the recurrence relation

$$x\frac{d}{dx}H_\nu^{(1,2)}(x) - \nu H_\nu^{(1,2)}(x) = -xH_{\nu+1}^{(1,2)}(x) \tag{A.76}$$

In the limit $x \to \infty$ (ν held constant),

$$H_\nu^{(1)}(x) \sim \sqrt{\frac{2}{x\pi}} \exp\left(j\left(x - \nu\frac{\pi}{2} - \frac{\pi}{4}\right)\right) \tag{A.77}$$

and

$$H_\nu^{(2)}(x) \sim \sqrt{\frac{2}{x\pi}} \exp\left(-j\left(x - \nu\frac{\pi}{2} - \frac{\pi}{4}\right)\right) \tag{A.78}$$

In the limit $\nu \to \infty$ (x held constant), two important results are

$$H_\nu^{(1)}(x) \sim -j\sqrt{\frac{2}{\nu\pi}}\left(\frac{2\nu}{ex}\right)^\nu \tag{A.79}$$

and

$$H_\nu^{(2)}(x) \sim j\sqrt{\frac{2}{\nu\pi}}\left(\frac{2\nu}{ex}\right)^{-\nu} \tag{A.80}$$

All of the above expansions assume that one of ν or x is held constant during the limiting process. More uniform approximations to Hankel's function can be found in terms of Airy functions with

$$H_\nu^{(1)}(x) \sim 2\exp\left(-j\frac{\pi}{3}\right)\left(\frac{4\zeta}{1-(x/\nu)^2}\right)^{\frac{1}{4}} \frac{Ai\left(\exp(j\frac{2\pi}{3})\nu^{\frac{2}{3}}\zeta\right)}{\nu^{\frac{1}{3}}} \tag{A.81}$$

and

$$H_\nu^{(2)}(x) \sim 2\exp\left(j\frac{\pi}{3}\right)\left(\frac{4\zeta}{1-(x/\nu)^2}\right)^{\frac{1}{4}} \frac{Ai\left(\exp(-j\frac{2\pi}{3})\nu^{\frac{2}{3}}\zeta\right)}{\nu^{\frac{1}{3}}} \tag{A.82}$$

where

$$\zeta = -\left(\frac{3}{2}\right)^{\frac{2}{3}}\left(\sqrt{(x/\nu)^2-1} - \sec^{-1}(x/\nu)\right)^{\frac{2}{3}} \tag{A.83}$$

or, alternatively,

$$\zeta = \left(\frac{3}{2}\right)^{\frac{2}{3}}\left(\ln\left(\frac{1+\sqrt{1-(x/\nu)^2}}{x/\nu}\right) - \sqrt{(1-(x/\nu)^2}\right)^{\frac{2}{3}} \tag{A.84}$$

These expressions are valid in the large ν limit, but are uniformly valid in x. In the important limit that $x/\nu \sim 1$, $\zeta \approx 2^{1/3}(1 - \nu/x)^{1/3}$ and so

$$H_\nu^{(1)} \approx 2^{\frac{4}{3}}\nu^{-\frac{1}{3}}\exp\left(-j\frac{\pi}{3}\right)Ai\left(2^{\frac{1}{3}}\exp\left(j\frac{2\pi}{3}\right)x^{-\frac{1}{3}}(x-\nu)\right) \tag{A.85}$$

and

$$H_\nu^{(2)} \approx 2^{\frac{4}{3}}\nu^{-\frac{1}{3}}\exp\left(j\frac{\pi}{3}\right)Ai\left(2^{\frac{1}{3}}\exp\left(-j\frac{2\pi}{3}\right)x^{-\frac{1}{3}}(x-\nu)\right) \tag{A.86}$$

The zeros of Hankel functions are important in many diffraction calculations and, in particular, the zeros of $H_\nu^{(1)}$ and $H_\nu^{(2)}$ in the complex ν plane. For large $|z|$, these can be related to the zeros of the Airy function $a_1, a_2, a_3, \ldots$ through

$$\nu_i \approx \pm \left(x + \left(\frac{x}{2}\right)^{\frac{1}{3}} a_i \exp\left(j\frac{2\pi}{3}\right)\right) \tag{A.87}$$

for $H_\nu^{(2)}(x)$ and

$$\nu_i \approx \pm \left(x + \left(\frac{x}{2}\right)^{\frac{1}{3}} a_i \exp\left(-j\frac{2\pi}{3}\right)\right) \tag{A.88}$$

for $H_\nu^{(1)}(x)$.

Integral expressions for $H_\nu^{(1)}$ and $H_\nu^{(2)}$ (valid for $|\arg z| < \pi/2$) are given by

$$H_\nu^{(1)}(z) = \frac{1}{j\pi} \int_{-\infty}^{\infty + j\pi} \exp(z \sinh t - \nu t) dt \tag{A.89}$$

and

$$H_\nu^{(2)}(z) = -\frac{1}{j\pi} \int_{-\infty}^{\infty - j\pi} \exp(z \sinh t - \nu t) dt \tag{A.90}$$

In the limit that $\nu \to \infty$, it can be seen from these expressions that $\partial H_\nu^{(1,2)}(x)/\partial \nu = -dH_\nu^{(1,2)}(x)/dx$.

A.10 Some Useful Series

Geometric series

$$1 + a + a^2 + a^3 + \cdots + a^{n-1} = \frac{1 - a^n}{1 - a} \tag{A.91}$$

and, if $n \to \infty$, has the limit $1/(1-a)$ for $|a| < 1$.

Trigonometric series

$$1 + a\cos(\theta) + a^2 \cos(2\theta) + a^3 \cos(3\theta) + \cdots = \frac{1 - a\cos(\theta)}{1 - 2a\cos(\theta) + a^2} \tag{A.92}$$

and

$$a\sin(\theta) + a^2 \sin(2\theta) + a^3 \sin(3\theta) + \cdots = \frac{a\sin(\theta)}{1 - 2a\cos(\theta) + a^2} \tag{A.93}$$

converge for $|a| < 1$.

Slowly convergent series can often be converted into fast converging series using the *Poisson summation formula*

$$\sum_{k=-\infty}^{\infty} f(k) = \sum_{m=-\infty}^{\infty} \left(\int_{-\infty}^{\infty} \exp(-2j\pi mx) f(x) dx \right) \tag{A.94}$$

A.11 References

M. Abramowitz and I.A. Stegun, *Handbook of Mathematical Functions*, Dover Publ., New York, 1965.

E.T. Copson, *Asymptotic Expansions*, Cambridge University Press, Cambridge, 1965.

A. Jeffrey, *Handbook of Mathematical Formulas and Integrals*, Academic Press, New York, 1995.

M.R. Spiegel, *Mathematical Handbook of Formulas and Tables*, Schaum's Outline Series, McGraw-Hill, New York, 1999.

G.N. Watson, *A Treatise on the Theory of Bessel Functions*, 2nd edition, Cambridge University Press, Cambridge, 1958.

Appendix B Numerical Methods

B.1 Numerical Differentiation and Integration

Consider a distance interval h, then we can approximate the first derivative of function f by

$$f'(x) \approx \frac{f(x+h) - f(x)}{h} \quad \text{or} \quad \frac{f(x) - f(x-h)}{h} \tag{B.1}$$

with error $O(h)$ (these approximations are known as the *forward difference* and *backward difference* approximations, respectively). A more accurate approximation is given by

$$f'(x) \approx \frac{f(x+h) - f(x-h)}{2h} \tag{B.2}$$

with error $O(h^2)$ (this is known as the *central difference* approximation). The second derivative can be approximated by

$$f''(x) \approx \frac{f(x+h) + f(x-h) - 2f(x)}{h^2} \tag{B.3}$$

with error $O(h^2)$.

For partial derivatives, we have

$$f_x(x, y) \approx \frac{f(x+h, y) - f(x-h, y)}{2h} \tag{B.4}$$

$$f_y(x, y) \approx \frac{f(x, y+h) - f(x, y-h)}{2h} \tag{B.5}$$

$$f_{xx}(x) \approx \frac{f(x+h, y) + f(x-h, y) - 2f(x, y)}{h^2} \tag{B.6}$$

$$f_{yy}(x) \approx \frac{f(x, y+h) + f(x, y-h) - 2f(x, y)}{h^2} \tag{B.7}$$

and

$$f_{xy}(x) \approx \frac{f(x+h, y+h) - f(x+h, y-h) + f(x-h, y+h) - f(x-h, y-h)}{4h^2} \tag{B.8}$$

with error $O(h^2)$.

Integrals can be approximated using a variety of *quadrature rules*. In the lowest order, we have

$$\int_x^{x+h} f(x)dx \approx hf\left(x + \frac{h}{2}\right) \qquad (B.9)$$

with error $O(h^3)$ (this is known as the *midpoint rule*). Another rule with the same order of error is

$$\int_x^{x+h} f(x)dx \approx \frac{h}{2}(f(x) + f(x+h)) \qquad (B.10)$$

which is known as the *trapezoidal rule*. A much more accurate quadrature rule is given by

$$\int_{x-h}^{x+h} f(x)dx \approx \frac{h}{3}(f(x-h) + 4f(x) + f(x+h)) \qquad (B.11)$$

with error $O(h^5)$; this is known as *Simpson's rule*.

B.2 Zeros of a Function

If we need to find a value of x for which $f(x) = 0$ (i.e. a root of f), we will first need an initial guess x_0. This can be found by searching a sequence of x to find consecutive points for which f experiences a change in sign (we could then use the average of these points for x_0). Given an estimate x_i, we can expand the function in Taylor series about this point $f(x) = f(x_i) + (x - x_i)f'(x_i) + \cdots$ and then an approximate root can be found from $f(x_i) + (x - x_i)f'(x_i) = 0$. Consequently, we can find an improved estimate x_{i+1} through *Newton's method*, i.e.

$$x_{i+1} = x_i - \frac{f(x_i)}{f'(x_i)} \qquad (B.12)$$

We continue this refinement process until the sequence of estimates has suitably converged.

If we do not want to calculate f', an alternative is the *secant method*. This essentially uses a finite difference approximation to the derivative. In this case, the refinement is given by

$$x_{i+1} = x_i - \frac{f(x_i)(x_i - x_{i-1})}{f(x_i) - f(x_{i-1})} \qquad (B.13)$$

To make this scheme work, we need two starting values x_0 and x_1. If we search a sequence of points to find where f changes sign, x_0 and x_1 could be the consecutive points with differing signs in f.

Newton's method can be extended to a technique for solving a system of nonlinear equations $f_i(x_1, \ldots, x_n) = 0$ for $i = 1, \ldots, n$.

If we have an initial estimate $\mathbf{x}^I$, we can expand the above equations around this estimate to obtain the equation

$$f_i + \sum_{j=1}^{N} \left(x_j^{I+1} - x_j^I\right) \frac{\partial f_i}{\partial x_j} = 0 \quad i = 1, 2, \ldots, n \tag{B.14}$$

for the next approximation $\mathbf{x}^{I+1}$ to the stationary point (note that the various derivatives are evaluated at point $\mathbf{x}^I$). Starting from an initial estimate $\mathbf{x}^0$, we can use the above algorithm to improve the estimate until there is sufficiently small change in $|\mathbf{x}^{I+1} - \mathbf{x}^I|$ between iterations. It will be noted that, at each stage of iteration, we will need to solve a linear system of equations; this can be achieved by any number of standard algorithms (Gaussian elimination, for example).

B.3 Numerical Solution of Ordinary Differential Equations

Given the initial condition that $y = y_0$ when $t = t_0$, the ordinary differential equation (ODE)

$$\frac{dy}{dt} = f(t, y) \tag{B.15}$$

can be formally integrated to yield

$$y_1 = \int_{t_0}^{t_1} f(t, y)\, dt \tag{B.16}$$

where $t_1 = t_0 + h$, $y_1 = y(t_1)$ and h is the integration step. A first approximation to the integral is to assume that f is approximately constant throughout the interval, then

$$y_1 \approx y_0 + f(t_0, y_0)h \tag{B.17}$$

which has error $O(h^2)$. Once we have found y_1, we may proceed in a similar fashion to find an approximation to $y_2 = y(t_2)$ where $t_2 = t_1 + h$ and so on (note that step length h does not have to be the same at each step). This approach is known as the Euler method and is the simplest numerical approach to the solution of ODEs. An improvement can be gained if we use a more sophisticated approximation to the integral. If we use Simpson's rule

$$y_1 \approx y_0 + \frac{1}{2}(f(t_0, y_0) + f(t_1, y_1))h \tag{B.18}$$

Unfortunately, in general, this will become a transcendental equation for y_1 that requires numerical solution. An alternative is to use the Euler approximation for the value of y_1 on the right-hand side. This is a Runge–Kutta (RK) method of the second order and has error $O(h^3)$ at each step. The approach can be used for systems of equations. Consider the system

$$\frac{d\mathbf{y}}{dt} = \mathbf{f}(t, \mathbf{y}) \tag{B.19}$$

with initial condition $\mathbf{y} = \mathbf{y}_0$ when $t = t_0$. We form

$$\mathbf{k}_1 = h\mathbf{f}(t_0, \mathbf{y}_0) \tag{B.20}$$
$$\mathbf{k}_2 = h\mathbf{f}(t_0 + h, \mathbf{y}_0 + \mathbf{k}_1) \tag{B.21}$$

from which

$$\mathbf{y}_1 = \mathbf{y}_0 + \frac{1}{2}(\mathbf{k}_1 + \mathbf{k}_2) \tag{B.22}$$

A popular form of the RK method is the fourth order scheme that has error $O(h^5)$ at each step. We form

$$\mathbf{k}_1 = h\mathbf{f}(t_0, \mathbf{y}_0) \tag{B.23}$$
$$\mathbf{k}_2 = h\mathbf{f}\left(t_0 + \frac{1}{2}h, \mathbf{y}_0 + \frac{1}{2}\mathbf{k}_1\right) \tag{B.24}$$
$$\mathbf{k}_3 = h\mathbf{f}\left(t_0 + \frac{1}{2}h, \mathbf{y}_0 + \frac{1}{2}\mathbf{k}_2\right) \tag{B.25}$$
$$\mathbf{k}_4 = h\mathbf{f}(t_0 + h, \mathbf{y}_0 + \mathbf{k}_3) \tag{B.26}$$

from which

$$\mathbf{y}_1 = \mathbf{y}_0 + \frac{1}{6}(\mathbf{k}_1 + 2\mathbf{k}_2 + 2\mathbf{k}_3 + \mathbf{k}_4) \tag{B.27}$$

We can exploit the ability to change step length at each step through the Runge–Kutta–Fehlberg (RKF) method (Press et al., 1982; Mathews, 1987). Form the vectors $\mathbf{k}_1, \mathbf{k}_2, \mathbf{k}_3, \mathbf{k}_4, \mathbf{k}_5$ and $\mathbf{k}_6$ through the process

$$\mathbf{k}_1 = h\mathbf{f}(t_0, \mathbf{y}_0) \tag{B.28}$$
$$\mathbf{k}_2 = h\mathbf{f}(t_0 + h/4, \mathbf{y}_0 + \mathbf{k}_1/4) \tag{B.29}$$
$$\mathbf{k}_3 = h\mathbf{f}(t_0 + 3h/8, \mathbf{y}_0 + 3\mathbf{k}_1/32 + 9\mathbf{k}_2/32) \tag{B.30}$$
$$\mathbf{k}_4 = h\mathbf{f}(t_0 + 12h/13, \mathbf{y}_0 + 1932\mathbf{k}_1/2197 - 7200\mathbf{k}_2/2197 + 7296\mathbf{k}_3/2197) \tag{B.31}$$
$$\mathbf{k}_5 = h\mathbf{f}(t_0 + h, \mathbf{y}_0 + 439\mathbf{k}_1/216 - 8\mathbf{k}_2 + 3680\mathbf{k}_3/513 - 845\mathbf{k}_4/4104) \tag{B.32}$$
$$\mathbf{k}_6 = h\mathbf{f}(t_0 + h/2, \mathbf{y}_0 - 8\mathbf{k}_1/27 + 2\mathbf{k}_2 - 3544\mathbf{k}_3/2565 + 1859\mathbf{k}_4/4104 - 11\mathbf{k}_5/40) \tag{B.33}$$

The approximate value of $\mathbf{y}$ at t_1 is obtained using

$$\hat{\mathbf{y}}_1 = \mathbf{y}_0 + 25\mathbf{k}_1/216 + 1408\mathbf{k}_3/2565 + 2197\mathbf{k}_4/4104 - \mathbf{k}_5/5 \tag{B.34}$$

for the fourth-order RK method and by

$$\tilde{\mathbf{y}}_1 = \mathbf{y}_0 + 16\mathbf{k}_1/135 + 6656\mathbf{k}_3/12825 + 28561\mathbf{k}_4/56430 - 9\mathbf{k}_5/50 + 2\mathbf{k}_6/55 \tag{B.35}$$

for the fifth-order method. The truncation error at each step will be $O(h^5)$ for the fourth-order method and $O(h^6)$ for the fifth-order method. The magnitude of the difference between the fourth- ($\hat{\mathbf{y}}_1$) and fifth- ($\tilde{\mathbf{y}}_1$) order estimates $\Delta = |\tilde{\mathbf{y}}_1 - \hat{\mathbf{y}}_1|$ gives an estimate of the truncation error in the fourth-order method. The RKF method proceeds by adjusting the step at each stage so that the error remains close to a pre-assigned value δ. That is, at each stage, we adjust h so that

$$h_{\text{new}} = h_{\text{old}} \left(\frac{\delta}{\Delta}\right)^{\frac{1}{5}} \tag{B.36}$$

B.4 Multidimensional Integration

Consider the multidimensional integral

$$I = \int_0^1 \int_0^1 \cdots \int_0^1 f(x_1, x_2, \ldots, x_N) dx_1 dx_2 \ldots dx_N \tag{B.37}$$

The integration can be regarded as an average of f over an N-dimensional unit cube, and, on this basis, Korobov (1959) has introduced an algorithm of the form

$$I \approx \frac{1}{M} \sum_{i=1}^{M} f\left(\left\{\frac{i}{M}\mathbf{a}(M)\right\}\right) \tag{B.38}$$

where $\mathbf{a}(M)$ is a suitably chosen N-dimensional vector of integers that depends on the number of quadrature points M and $\{x\}$ represents the non-integer part of x. On the assumption that f is periodic of period 1, Korobov has shown that $\mathbf{a}(M)$ can be chosen so that the algorithm converges with increasing N, the rate of convergence depending on the smoothness of function f. If the function f is not periodic, we can instead use

$$F(x_1, \ldots, x_N) = \frac{1}{2}(f(x_1, \ldots, x_N) + f(1 - x_1, \ldots, 1 - x_N)) \tag{B.39}$$

(with the usual periodic extension) since

$$\int_0^1 \cdots \int_0^1 f(x_1, \ldots, x_N) dx_1 \ldots dx_N = \int_0^1 \cdots \int_0^1 F(x_1, \ldots, x_N) dx_1 \ldots dx_N \tag{B.40}$$

We consider the Fourier expansion of f

$$f(x) = \sum_{\mathbf{m}} c_{\mathbf{m}} \exp(2\pi j \mathbf{m} \cdot \mathbf{x}) \tag{B.41}$$

where the sum is over all integer N-tuples $\mathbf{m}$. It turns out that the only Fourier coefficients that contribute to error are those that satisfy the Diophantine equation $\mathbf{a}(M) \cdot \mathbf{m} = 0 \pmod{M}$. For a convergent Fourier series, coefficients $c_{\mathbf{m}} \to 0$ as $|\mathbf{m}| \to \infty$ and so, for good accuracy, we will need to choose $\mathbf{a}(M)$ such that solutions to the Diophantine equation are only possible for large $|\mathbf{m}|$.

The choice of an effective $\mathbf{a}(M)$ can be quite difficult, but one of the simplest approaches is to choose $\mathbf{a}(M)$ such that the approximate integral is the most accurate possible for some "worst" function f_W with known integral I_W (Korobov, 1959; Conroy, 1967). Korobov uses the function $f(x_1, \ldots, x_N) = \prod_{i=1}^{N}(1 - 2x_i)^2$ in $[0, 1]$ with periodic extension outside this interval ($I_W = (1/3)^N$). The problem can now be considered as that of finding the $\mathbf{a}(M)$ that minimizes function $E(\mathbf{a}(M))$ given by

$$E(\mathbf{a}(M)) = \left| \sum_{i=1}^{M} f_W \left(\left\{ \frac{i}{M} \mathbf{a}(M) \right\} \right) - I_W \right| \qquad (B.42)$$

Korobov simplifies this problem by choosing $a_1 = 1$ and then setting $a_i = a^{i-1}$ (modulo N) for some parameter a and N prime. In this case, the minimization will only take place with respect to the single parameter a. Other useful references concerning numerical approaches to multiple integration are Boersma (1978) and Haselgrove (1961).

In general, the integral

$$I = \int_{A_1}^{B_1} \int_{A_2}^{B_2} \cdots \int_{A_N}^{B_N} f(x_1, x_2, \ldots, x_N) dx_1 dx_2 \ldots dx_N \qquad (B.43)$$

can be reduced to the above form by introducing the new variables $y_i = (x_i - A_i)/(B_i - A_i)$. When $B_i = \infty$, we can use a transform of the form $y_i = \exp(A_i - x_i)$, but caution must be exercised due to the artificial singularities that are introduced at $y_i = 0$. A possible solution to the problem, however, is to window the integrand at this boundary, a suitable window being $W(y_i) = ((1 - \cos(\pi y_i))/2)^\alpha$ where α has a small positive number.

B.5 References

J. Boersma, On certain multiple integrals occurring in a waveguide scattering problem, SIAM J. Math. Anal., vol. 9, pp. 377–393, 1978.

H. Conroy, Molecular Schrodinger equation. VII: A new method for the evaluation of multidimensional integrals, J. Chem. Phys., vol. 47, pp. 5307–5318, 1967.

C.B. Haselgrove, A method for numerical integration, Mat. Comp., vol. 15, pp. 323–337, 1961.

N.M. Korobov, The approximate computation of multiple integrals, Doklady Akademii Nauk SSSR, vol. 124, pp. 1207–1210, 1959.

J.H. Mathews, *Numerical Methods for Computer Science, Engineering and Mathematics*, Prentice Hall, Englewood Cliffs, NJ, 1987.

W.H. Press, S. Teukolsky, W. Vetterling and B. Flannery, *Numerical Recipes in FORTRAN: The Art of Scientific Computing*, 2nd edition, Cambridge University Press, Cambridge, 1982.

Appendix C Variational Calculus

Consider the functional

$$I(y) = \int_{t_1}^{t_2} L(t, y, y')\,dt \qquad (C.1)$$

where y is an arbitrary function of t and y' is the first derivative of y with respect to t. We consider a variation δy in y and expand the functional in a functional Taylor series, i.e.

$$I(y + \delta y) = I + \delta I + \frac{1}{2}\delta^2 I + \cdots \qquad (C.2)$$

$$= \int_{t_1}^{t_2} L\,dt + \int_{t_1}^{t_2} \left(\delta y \frac{\partial L}{\partial y} + \delta y' \frac{\partial L}{\partial y'}\right) dt$$

$$+ \frac{1}{2} \int_{t_1}^{t_2} \left(\delta y^2 \frac{\partial^2 L}{\partial y^2} + 2\delta y \delta y' \frac{\partial^2 L}{\partial y \partial y'} + \delta y'^2 \frac{\partial^2 L}{\partial y'^2}\right) dt + \cdots$$

Integrating by parts, we have

$$\delta I = \int_{t_1}^{t_2} \delta y \left(\frac{\partial L}{\partial y} - \frac{d}{dt}\left(\frac{\partial L}{\partial y'}\right)\right) dt + \left[\delta y \frac{\partial L}{\partial y'}\right]_{t_1}^{t_2} \qquad (C.3)$$

and

$$\delta^2 I = \int_{t_1}^{t_2} \left(\delta y^2 L_{00} - \delta y^2 \frac{dL_{01}}{dt} - \delta y \frac{d}{dt}(\delta y' L_{11})\right) dt + \frac{1}{2}\left[\delta y^2 L_{01} + \delta y \delta y' L_{11}\right]_{t_1}^{t_2} \qquad (C.4)$$

where subscript 0 on L denotes a partial derivative with respect to y and subscript 1 denotes a partial derivative with respect to y'. We now consider the variational equation $\delta I = 0$, i.e. the condition that I be stationary with respect to variations in y. For variations in y with end points fixed (i.e. $\delta y(t_1) = \delta y(t_2) = 0$), Equation C.3 will imply that

$$\frac{\partial L}{\partial y} - \frac{d}{dt}\left(\frac{\partial L}{\partial y'}\right) = 0 \qquad (C.5)$$

Equation C.5 is known as the Euler–Lagrange equations and must be satisfied by the function $y(t)$ that makes functional I stationary subject to the fixed values of y at t_1 and t_2 (y is known as an *extremal*). In the case that L is independent of t (i.e. $L = L(y, y')$), the Euler–Lagrange equation has the first integral

$$L - y'\frac{\partial L}{\partial y'} = C \tag{C.6}$$

where C is a constant of integration.

We now need to consider the nature of the stationary value and so we will need to consider the second variation $\delta^2 I$. We first consider the equation

$$L_{00}u - \frac{dL_{01}}{dt}u - \frac{d}{dt}\left(L_{11}\frac{du}{dt}\right) = 0 \tag{C.7}$$

where the $y(t)$ that is used to evaluate L_{00}, L_{01} and L_{11} on the extremal of interest, an equation that is known as Jacobi's accessory equation. Let u be a solution to the Jacobi's equation, then it can be shown (Fox, 1987) that

$$\delta^2 I = \int_{t_1}^{t_2} L_{11}\left(\delta y' - y\frac{u'}{u}\right)^2 dt \tag{C.8}$$

when δy is zero at the end points t_1 and t_2. The result suggests that the stationary value of I is a minimum if $L_{11} > 0$ throughout the interval $[t_1, t_2]$ and maximum if $L_{11} < 0$. This is known as the *Legendre test*. If L_{11} changes sign in the interval $[t_1, t_2]$, the stationary value is neither a maximum nor a minimum.

Unfortunately, although the Legendre test is a necessary condition for a minimum or a maximum, it is not sufficient. To study this, we need to consider the Jacobi equation further. Consider the general solution to the Euler–Lagrange equation $y = y(t, c^A, c^B)$ where c^A and c^B are constants of integration. The constants of integration are determined by the boundary values of y (i.e. $y(t_1)$ and $y(t_2)$). We will consider functions $u^A = \partial y/\partial c^A$ and $u^B = \partial y/\partial c^B$, where y is a solution of the Euler–Lagrange equation

$$L_0 - \frac{dL_1}{dt} = 0 \tag{C.9}$$

Taking the partial derivative of the Euler–Lagrange equation with respect to either c^A or c^B, we find that both u^A and u^B satisfy the Jacobi accessory equation

$$L_{00}u - \frac{dL_{01}}{dt}u - \frac{d}{dt}\left(L_{11}\frac{du}{dt}\right) = 0 \tag{C.10}$$

If quantities L_{ij} are evaluated at the solution to the Euler–Lagrange equation, then $\delta y = \delta c_A u^A + \delta c_B u^B$ is the deviation in y that occurs when the constants of integration are varied by amounts δc_1 and δc_2, respectively. The properties of the solution to the accessory equation are important as the nature of extremum in I is critically dependent on them. We note that the Legendre test will only be definitive if $\delta y' - yu'/u \neq 0$ except at isolated points. Consequently, we need to check for the possibility that a variation δy can be chosen such that $\delta y' - yu'/u$ is zero. Since u is a given, this equation implies that $\delta y \propto u$. Consider a solution to Equation C.10 that vanishes at $t = t_1$, then the next point on this curve ($t = t_C$) at which u vanishes is known as a *conjugate point*. If this point lies beyond $t = t_2$ then we cannot have $\delta y \propto u$ since $\delta y = 0$ at $t = t_2$. Consequently, providing that there is a solution to the accessory equation with $u(t_1) = 0$ and no other zero in the range $[t_1, t_2]$, the stationary value will be a minimum if $L_{11} > 0$ for all points in $[t_1, t_2]$ and maximum if $L_{11} > 0$. This is the *Jacobi test*.

In all the above, we have considered a functional with arguments that consist of a single scalar function of t. However, the above theory also carries over to the situation where the argument is a vector function of t, i.e.

$$I(y) = \int_{t_1}^{t_2} L(t, \mathbf{y}, \mathbf{y}')\,dt \tag{C.11}$$

In this case, the Euler–Lagrange equations become

$$\frac{\partial L}{\partial y_i} - \frac{d}{dt}\left(\frac{\partial L}{\partial y_i'}\right) = 0 \tag{C.12}$$

with first integral

$$L - y_j'\frac{\partial L}{\partial y_j'} = C \tag{C.13}$$

in the case that L is independent of t. (Note that we have assumed the Einstein summation convention, i.e. that repeated indices denote a sum over those indices.) The accessory equation becomes

$$u_j \frac{\partial^2 L}{\partial y_i \partial y_j} - u_j \frac{d}{dt}\frac{\partial^2 L}{\partial y_i' \partial y_j} - \frac{d}{dt}\left(\frac{\partial^2 L}{\partial y_i' \partial y_j'}\frac{du_j}{dt}\right) = 0 \tag{C.14}$$

where the various derivatives of L are evaluated at the appropriate solution to the Euler–Lagrange equation. If we solve the above equation with $\mathbf{u}(t_1) = 0$ and $\mathbf{u}(t_2) = \delta \mathbf{y}$, the solution $\mathbf{u}$ describes the deviation between two solutions to the Euler–Lagrange equations that start with the same value at $t = t_1$ and deviate by $\delta\mathbf{y}$ when they reach $t = t_2$.

For the scalar function y, we have used the solutions of the accessory equation to introduce the idea of a conjugate point, this being the point where the solution to the accessory equation becomes zero again. The conjugate point is, in fact, the point where two nearby solutions to the Euler–Lagrange equations cross over (i.e. they have a point in common). This concept clearly generalizes to the vector function $\mathbf{y}$ since we can regard $\mathbf{y}(t)$ for $t \in [t_1, t_2]$ as a curve in a multidimensional space, the conjugate point being where two curves cross over again. The generalization of the Jacobi test is then as follows. Given an extremal, let there be no nearby solution to the Euler–Lagrange equations with the same start point that crosses over the extremal within the interval $(t_1, t_2]$. Further, assume that

$$\left(\frac{\partial^2 L}{\partial y_i'^2}\right)\left(\frac{\partial^2 L}{\partial y_j'^2}\right) > \left(\frac{\partial^2 L}{\partial y_i' \partial y_j'}\right)^2 \tag{C.15}$$

on the extremal. Then, if

$$\left(\frac{\partial^2 L}{\partial y_i'^2}\right) > 0 \tag{C.16}$$

for all $t \in [t_1, t_2]$ we have a minimum and if

$$\left(\frac{\partial^2 L}{\partial y_i'^2}\right) < 0 \qquad (C.17)$$

for all $t \in [t_1, t_2]$ we have a maximum.

We now consider the situation where $L = L(\mathbf{y}, \mathbf{y}')$ is homogeneous in the first degree, that is, $L(\mathbf{y}, \lambda \mathbf{y}') = \lambda L(\mathbf{y}, \mathbf{y}')$. For such an L, the functional $I(y) = \int_{t_1}^{t_2} L(\mathbf{y}, \mathbf{y}')dt$ is *parameter invariant*, i.e.

$$I(y) = \int_{t_1}^{t_2} L\left(\mathbf{y}, \frac{d\mathbf{y}}{dt}\right) dt = \int_{t_1}^{t_2} L\left(\mathbf{y}, \frac{d\mathbf{y}}{d\tau}\right) d\tau \qquad (C.18)$$

for $t = t(\tau)$ that is smooth (parameter invariance also implies that L is homogeneous in $\mathbf{y}'$). In particular, we can choose τ to be one of the components of $\mathbf{y}$. Consider a vector $\mathbf{y}$ that has three components, i.e.

$$I(x, y, x) = \int_{t_1}^{t_2} L\left(x, y, z, \frac{dx}{dt}, \frac{dy}{dt}, \frac{dz}{dt}\right) dt \qquad (C.19)$$

Let $\lambda = \frac{dx}{dt}$, then, since L is homogeneous in the first degree,

$$I(x, y, x) = \int_{t_1}^{t_2} L\left(x, y, z, \frac{dx}{dt}, \frac{dy}{dt}, \frac{dz}{dt}\right) dt$$

$$= \int_{t_1}^{t_2} \frac{dx}{dt} L\left(x, y, z, 1, \frac{dt}{dx}\frac{dy}{dt}, \frac{dt}{dx}\frac{dz}{dt}\right) dt$$

$$= \int_{t_1}^{t_2} L\left(x, y, z, 1, \frac{dy}{dx}, \frac{dz}{dx}\right) dx \qquad (C.20)$$

We now have a functional that depends on only two functions (y and z) with x an independent parameter. It is clear that, for homogeneous L, there is a freedom of choice in independent parameter and this can be exploited to simplify the problem. For an L that is homogeneous in the first degree, a property worth noting is the Euler identity

$$y_i' \frac{\partial L}{\partial y_i'} = L \qquad (C.21)$$

This identity, however, is a first integral of the Euler–Lagrange equations. Essentially, when L is homogeneous, the Euler–Lagrange equations are linearly dependent and we must add a constraint to complete the system. As we have seen above, we can do this by choosing an independent parameter τ to be related to $\mathbf{y}$ in some way.

The question now arises as to how we actually solve the variational equation $\delta I = 0$ where

$$I = \int_{t_A}^{t_B} L(t, y, y')dt \qquad (C.22)$$

We could obviously turn the problem into one of solving ODEs by deriving the appropriate Euler–Lagrange equations. It is, however, also possible to find approximate solutions directly without the intermediate step of ODEs. This can be done through what is known as the *Rayleigh–Ritz technique*. Consider a function $\phi(t, \alpha_1, \alpha_2, \ldots, \alpha_N)$ such that $\phi(t, \alpha_1, \alpha_2, \ldots, \alpha_N) = y(t_A)$ and $\phi(t, \alpha_1, \alpha_2, \ldots, \alpha_N) = y(t_B)$ where $y(t_A)$ and $y(t_B)$ are the given values of $y(t)$ at the end points. The function ϕ is chosen such that variations of the N parameters $\alpha_1, \alpha_2, \ldots, \alpha_N$ provide a set of functions that covers the range of expected behavior in $y(t)$. We replace y by ϕ in Equation C.22 and then I becomes a function of $\alpha_1, \alpha_2, \ldots, \alpha_N$. We can now find appropriate α_i's by finding the values of these parameters that make I stationary; i.e. we need to solve the equations

$$\frac{\partial I}{\partial \alpha_i} = 0 \quad i = 1, 2, \ldots, N \tag{C.23}$$

These are normally coupled nonlinear equations and hence require techniques such as that of Newton and Raphson. Although ϕ can be a nonlinear function of $\alpha_1, \alpha_2, \ldots, \alpha_N$, we more often than not consider ϕ to be a linear combination of basis functions $\phi_1(t), \phi_2(t), \ldots, \phi_N(t)$, i.e.

$$y(t) \approx \phi(t, \alpha_1, \alpha_2, \ldots, \alpha_N) = \sum_{i=0}^{N+1} \alpha_i \phi_i(t) \tag{C.24}$$

where α_0 and α_{N+1} are chosen such that ϕ attains the correct values at the end points. Possible choices for the basis functions are those used in Chebyshev and Fourier series expansions. For our current purposes, however, it is more convenient to use basis functions with bounded support (i.e. they are only nonzero over a finite interval). This leads to what is commonly known as the finite element method (Strang and Fix, 1973). As an example, we will consider the simple basis functions of the *hat* variety. Let the interval from t_A to t_B divided by points $t_1, t_2, \ldots, t_N$ such that $t_0 < t_1 < t_2 < \cdots < t_{N-1} < t_N < t_{N+1}$ where $t_0 = t_A$ and $t_{N+1} = t_B$. We define the hat basis function ϕ_i by

$$\phi_i(t) = \frac{t - t_{i-1}}{t_i - t_{i-1}} \quad t_{i-1} \leq t \leq t_i$$

$$\phi_i(t) = \frac{t_{i+1} - t}{t_{i+1} - t_i} \quad t_i \leq t \leq t_{i+1}$$

$$\phi_i(t) = 0 \quad \text{otherwise} \tag{C.25}$$

If $y_0, y_1, y_2, \ldots, y_N, y_{N+1}$ are the values of y at the points $t_0, t_1, t_2, \ldots, t_N, t_{N+1}$, we can approximate y by

$$y(t) \approx \phi(t, \alpha_1, \alpha_2, \ldots, \alpha_N) = \sum_{i=0}^{N+1} y_i \phi_i(t) \tag{C.26}$$

It will be noted that

$$\phi'_i(t) = \frac{1}{t_i - t_{i-1}} \quad t_{i-1} < t < t_i$$

$$\phi'_i(t) = \frac{-1}{t_{i+1} - t_i} \quad t_i < t < t_{i+1}$$

$$\phi'_i(t) = 0 \quad \text{otherwise} \tag{C.27}$$

and so we can approximate y' by

$$y'(t) \approx \frac{y_{i+1} - y_i}{t_{i+1} - t_i} \quad t_i \leq t \leq t_{i+1} \qquad (C.28)$$

Consistent with the order of approximation that is afforded by the hat basis functions, we can approximate the integral in Equation C.22 by means of the trapezoidal rule and obtain

$$I \approx \frac{1}{2} \sum_{i=0}^{N} \left(L\left(t_i, y_i, \frac{y_{i+1} - y_i}{t_{i+1} - t_i}\right) + L\left(t_{i+1}, y_{i+1}, \frac{y_{i+1} - y_i}{t_{i+1} - t_i}\right) \right)(t_{i+1} - t_i) \qquad (C.29)$$

On noting that y_0 and y_{i+1} are given, I is now a function of the arbitrary parameters y_1 to y_N and these can be found by solving the coupled equations $\partial I/\partial y_i = 0$ for $i = 1, \ldots, N$.

An important special case is the functional I that arises from Fermat's principle for isotropic media

$$I = \int_{x_A}^{x_B} N(x, y) \sqrt{\left(\frac{dy}{dx}\right)^2 + 1} \; dx \qquad (C.30)$$

where $y_A = y(x_A)$ and $y_B = y(x_B)$ are given. Let the interval from x_A to x_B be divided by points $x_1, x_2, \ldots, x_N$ such that $x_0 < x_1 < x_2 < \cdots < x_{N-1} < x_N < x_{N+1}$ where $x_0 = x_A$ and $x_{N+1} = x_B$. An approximate form of I is then given by

$$I \approx \frac{1}{2} \sum_{i=0}^{N} (N(x_i, y_i) + N(x_{i+1}, y_{i+1})) \sqrt{(y_{i+1} - y_i)^2 + (x_{i+1} - x_i)^2} \qquad (C.31)$$

On noting that $y_0 = y(x_A)$ and $y_{N+1} = y(x_B)$, this will be a function of the N arbitrary parameters $y_1, y_2, \ldots, y_N$ and these can be found by solving the coupled equations $\frac{\partial I}{\partial y_i} = 0$ for $i = 1, \ldots, N$. Some useful references concerning variational calculus are Sagan (1969), Fox (1980) and Gelfand and Fomin (2000).

C.1 References

A.M. Arthurs, *Complementary Variational Principles*, Clarendon Press, Oxford, 1980.
C. Fox, *An Introduction to Calculus of Variations*, Dover Publ., New York, 2010.
I.M. Gelfand and S.V. Fomin, *Calculus of Variations*, Dover Publ., New York, 2000.
H. Sagan, *An Introduction to Calculus of Variations*, Dover Publ., New York, 1969.
G. Strang and G. Fix, *An Analysis of the Finite Element Method*, Prentice Hall, Englewood Cliffs, NJ, 1973.

Appendix D The Fourier Transform

The Fourier transform is given by

$$F(\alpha) = \mathcal{F}\{f(x)\} = \int_{-\infty}^{\infty} \exp(-j\alpha x) f(x)\, dx \tag{D.1}$$

(note that this transform is sometimes defined with a multiplicative factor of $1/\sqrt{2\pi}$). Fourier's integral theorem states that

$$f(x) = \frac{1}{2\pi} \int_{-\infty}^{\infty} \exp(j\alpha x) \int_{-\infty}^{\infty} \exp(-j\alpha u) f(u)\, du\, d\alpha \tag{D.2}$$

and this provides a means of inverting the transform. From Fourier's integral theorem, we have the important result that

$$\frac{\delta(x)}{A} = \frac{1}{2\pi} \int_{-\infty}^{\infty} \exp(jA\alpha x)\, d\alpha \tag{D.3}$$

where $\delta(x)$ is the Dirac delta function. Four other important results are the convolution theorem

$$\frac{1}{2\pi} \int_{-\infty}^{\infty} F(\alpha) G(\alpha) \exp(j\alpha x)\, d\alpha = \int_{-\infty}^{\infty} f(u) g(x - u)\, du \tag{D.4}$$

(this is an important result for solving integral equations with displacement invariance) Paserval's identity

$$\int_{-\infty}^{\infty} |f(x)|^2\, dx = \frac{1}{2\pi} \int_{-\infty}^{\infty} |F(\alpha)|^2\, d\alpha \tag{D.5}$$

the transformation of derivatives

$$\int_{-\infty}^{\infty} \exp(-j\alpha x) f^{(n)}(x)\, dx = (j\alpha)^n F(\alpha) \tag{D.6}$$

and the property

$$\mathcal{F}\{f(\beta x) \exp(j\gamma x)\} = \frac{1}{\beta} F\left(\frac{\alpha - \gamma}{\beta}\right) \tag{D.7}$$

An important Fourier transform is (Jeffreys, 1995)

$$\mathcal{F}\{\exp(-\beta x^2)\} = \sqrt{\frac{\pi}{\beta}} \exp\left(-\frac{\alpha^2}{4\beta}\right) \tag{D.8}$$

Consider function f to be of bounded support (f is zero outside the interval $[0, X]$) and consider the $N + 1$ sample points $x_0 = 0, x_1 = h, x_2 = 2h, x_3 = 3h, \ldots, x_N = X$ where $h = X/N$. Let $f_0 = f(x_0), f_1 = f(x_1), f_2 = f(x_2), f_3 = f(x_3), \ldots, f_N = f(x_N)$ be the values of f at these sample points. We will assume that $f_0 = f_N$ and so, using the trapezoidal rule, we can approximate the Fourier transform by

$$F(\alpha) \approx h \sum_{i=0}^{N-1} f_i \exp(-j\alpha ih) \tag{D.9}$$

and from which it will be noted that the approximation is a periodic function in α. According to the Nyquist sampling theorem, we only need to sample the frequency at intervals of $2\pi/Nh$ in order to gain sufficient information to completely reconstruct the transform. Consequently, we produce a discrete Fourier transform by sampling the above approximation at the sample points $0, 2\pi/Nh, 4\pi/Nh, \ldots, 2(N-1)\pi/Nh$. The discrete Fourier transform (DFT) is defined to be (Mathews, 1987)

$$F_k = \sum_{i=0}^{N-1} f_i \exp\left(-2\pi j \frac{ik}{N}\right) \quad \text{for } k = 0, 1, \ldots, N-1 \tag{D.10}$$

and its inverse will be

$$f_k = \frac{1}{N} \sum_{i=0}^{N-1} F_i \exp\left(2\pi j \frac{ik}{N}\right) \quad \text{for } k = 0, 1, \ldots, N-1 \tag{D.11}$$

Due to the convolution theorem, Fourier transform techniques have proven extremely useful in the solution of integral equations with kernels of the displacement invariant variety. Further, by means of the DFT, such an approach can be turned into an effective numerical procedure. Unfortunately, a DFT can require a large number of arithmetic operations (N^2 multiplications) and this can make it quite slow computationally. If, however, N is a power of two ($N = 2^M$ where M is an integer) an extremely fast DFT can be made through what is known as the fast Fourier transform (FFT) (Press et al., 1982). Let $W = \exp(-j2\pi/N)$, then Equation D.10 can be rewritten as

$$F_k = \sum_{i=0}^{N-1} f_i W^{ik} \quad \text{for } k = 0, 1, \ldots, N-1 \tag{D.12}$$

and this can be factorized to the form

$$F_k = \sum_{i=0}^{N/2-1} f_{2i} W^{ik} + W^k \sum_{i=0}^{N/2-1} f_{2i+1} W^{ik} \quad \text{for } k = 0, 1, \ldots, N-1 \tag{D.13}$$

i.e. $F_k = F_k^{\text{even}} + W^k F_k^{\text{odd}}$ where F_k^{even} and F_k^{odd} are the DFTs of the even and odd samples of f, respectively. (It will be noted that both F_k^{even} and F_k^{odd} are periodic in k with period $N/2$. Consequently, $F_{k+N/2}^{\text{even}} = F_k^{\text{even}}$ and $F_{k+N/2}^{\text{odd}} = F_k^{\text{odd}}$.) This new process for evaluating the DFT is faster, requiring only $N^2/2 + N$ multiplications. Obviously we could also split the two DFTs above in the same manner and so on in a recursive fashion. Effectively, for N a power of 2, we can build up a long DFT from a sequence of shorter DFTs. If we do this in the correct sequence, we can greatly reduce the number of arithmetic operations required to reach the long DFT. In the limit of this process, we obtain a DFT for which the number of multiplications is now only $O(N \log_2 N)$. The use of this highly efficient algorithm for calculating DFTs can lead to extremely efficient numerical techniques for solving propagation problems (see Chapter 7).

For a computer language that allows recursion, we can use Equation D.13 to implement the FFT algorithm. The following is a FORTRAN subroutine that achieves this:

```
recursive subroutine FFT(f)
complex*16 :: f(:)
complex*16, allocatable :: feven(:)
complex*16, allocatable :: fodd(:)
complex*16 WK,W
real*8 pi
N=size(f)
if(N.eq.1) return
pi=4.d0*atan(1.d0)
allocate(feven(N/2))
allocate(fodd(N/2))
do i=1,N/2
fodd(i)=f(2*i-1)
feven(i)=f(2*i)
enddo
call FFT(fodd)
call FFT(feven)
W=cmplx(cos(pi/dble(N/2)),sin(pi/dble(N/2)),8)
WK=1.d0
do i=1,N/2
WK=WK*W
f(i)=feven(i) + WK*fodd(i)
f(i+N/2)=feven(i) - WK*fodd(i)
enddo
deallocate(feven)
deallocate(fodd)
return
end subroutine
```

It will be noted that the subroutine replaces the function samples f_i with their FFT samples F_i and this can be useful if storage is at a premium. (Note that only very minor changes to the above code are required in order to produce an inverse FFT algorithm.)

D.1 References

A. Jeffreys, *Handbook of Mathematical Formulas and Integrals*, Academic Press, San Diego, CA, 1995.

J.H. Mathews, *Numerical Methods for Computer Science, Engineering and Mathematics*, Prentice Hall, Englewood Cliffs, NJ, 1987.

W.H. Press, S. Teukolsky, W. Vetterling and B. Flannery, *Numerical Recipes in FORTRAN: The Art of Scientific Computing*, 2nd edition, Cambridge University Press, Cambridge, 1982.

Appendix E Finding Stationary Values

E.1 Newton–Raphson Approach

Consider function $f = f(x_1, \ldots, x_n)$ and the problem of finding the stationary value of f (including maxima and minima) with respect to x_1 to x_n. The values of $\mathbf{x} = (x_1, \ldots, x_n)$ that make f stationary are found as the solutions to

$$\frac{\partial f}{\partial x_i} = 0 \quad i = 1, 2, \ldots, n \tag{E.1}$$

If we have an estimate to the stationary point $\mathbf{x}^I$, we can expand the above equation around this estimate to obtain the equation

$$\frac{\partial f}{\partial x_i} + \sum_{j=1}^{N} \left(x_j^{I+1} - x_j^I \right) \frac{\partial^2 f}{\partial x_i \partial x_j} = 0 \quad i = 1, 2, \ldots, n \tag{E.2}$$

for the next approximation $\mathbf{x}^{I+1}$ to the stationary point (note that the various derivatives are evaluated at point $\mathbf{x}^I$). Starting from an initial estimate $\mathbf{x}^0$, we can use the above algorithm to improve the estimate until there is sufficiently small change in $|\mathbf{x}^{I+1} - \mathbf{x}^I|$ between iterations. It will be noted that, at each stage of iteration, we will need to solve a linear system of equations, and this can be achieved by any of a number of standard algorithms (Gaussian elimination, for example) (Press et al.,1982).

E.2 Nelder–Mead Method

The Nelder–Mead method provides an algorithm for minimizing the function $f(x_1, x_2, \ldots, x_n)$ without the use of derivatives (Mathews, 1987). The algorithm starts with an n-dimensional simplex ($n + 1$ vertices) that roughly defines the region of solution (a triangle in the case of function of two variables, i.e. $n = 2$). It then refines this simplex so that it converges to a point that makes the function minimum. The algorithm proceeds as follows:

(I) Label the $n + 1$ vertices $\mathbf{x}^1, \mathbf{x}^2, \ldots, \mathbf{x}^n, \mathbf{x}^{n+1}$ such that $f(\mathbf{x}^1) \leq f(\mathbf{x}^2) \leq \cdots \leq f(\mathbf{x}^n) \leq f(\mathbf{x}^{n+1})$ and calculate the centroid $\mathbf{x}^0 = \left(\mathbf{x}^1 + \mathbf{x}^2 + \cdots + \mathbf{x}^n + \mathbf{x}^{n+1} \right)/n$

(II) Now proceed as follows:

$$\text{form } \mathbf{x}^a = 2\mathbf{x}^0 - \mathbf{x}^{n+1}$$
$$\text{if } f(\mathbf{x}^a) < f(\mathbf{x}^1)$$
$$\quad \text{set } \mathbf{x}^{n+1} = \mathbf{x}^a \text{ and go to step III}$$
$$\text{else}$$
$$\quad \text{if } f(\mathbf{x}^1) \leq f(\mathbf{x}^a) \leq f(\mathbf{x}^n)$$
$$\quad\quad \text{form } \mathbf{x}^b = 3\mathbf{x}^0 - 2\mathbf{x}^{n+1}$$
$$\quad\quad \text{if } f(\mathbf{x}^b) < f(\mathbf{x}^a)$$
$$\quad\quad\quad \text{set } \mathbf{x}^{n+1} = \mathbf{x}^b \text{ and go to step III}$$
$$\quad\quad \text{else}$$
$$\quad\quad\quad \text{set } \mathbf{x}^{n+1} = \mathbf{x}^a \text{ and go to step III}$$
$$\quad \text{else}$$
$$\quad\quad \text{form } \mathbf{x}_c = \frac{1}{2}\mathbf{x}^0 + \frac{1}{2}\mathbf{x}^{n+1}$$
$$\quad\quad \text{if } f(\mathbf{x}^c) < f(\mathbf{x}^{n+1})$$
$$\quad\quad\quad \text{set } \mathbf{x}^{n+1} = \mathbf{x}_c \text{ and go to step III}$$
$$\quad\quad \text{else}$$
$$\quad\quad\quad \text{set } \mathbf{x}^i = \frac{1}{2}\mathbf{x}^1 + \frac{1}{2}\mathbf{x}^i \text{ for } i = 1 \text{ to } n+1 \text{ and go to step III}$$

(III) Is the simplex small enough? If not, return to step I.

In essence, the algorithm searches the line passing through the simplex centroid $\mathbf{x}^0$ and the vertex $\mathbf{x}^{n+1}$ with maximum value of f. When it finds the point of minimum f on this line, it makes this the new value of $\mathbf{x}^{n+1}$ and hence forms a new simplex. If there is no minimum, the new simplex is formed by shrinking the old simplex toward the vertex of minimum f. This process is repeated until the algorithm has suitably converged.

E.3 References

J.H. Mathews, *Numerical Methods for Computer Science, Engineering and Mathematics*, Prentice Hall, Englewood Cliffs, NJ, 1987.

W.H. Press, S. Teukolsky, W. Vetterling and B. Flannery, *Numerical Recipes in FORTRAN: The Art of Scientific Computing*, 2nd edition, Cambridge University Press, Cambridge, 1982.

Appendix F Stratified Media

F.1 Two-Layer Medium

In this appendix, we will study the effect of horizontally stratified media (see Figure F.1) on a vertical dipole using the methods developed by Wait (1967, 1996). (These solutions are used extensively in Chapter 9.) Consider the dipole to be located on the vertical axis (the z axis in this case) and then we will have cylindrical symmetry about this axis. We can study electromagnetic–magnetic fields through a magnetic vector potential $\mathbf{A}$ for which

$$\mathbf{H} = \frac{1}{\mu} \nabla \times \mathbf{A} \tag{F.1}$$

We define $\mathbf{A}$ such that it satisfies a vector Helmholtz equation with current as its source

$$\nabla^2 \mathbf{A} + \omega^2 \mu \epsilon \mathbf{A} = -\mu \mathbf{J} \tag{F.2}$$

Then, if the electric field is $\mathbf{E}$ defined by

$$\mathbf{E} = -j\omega \mathbf{A} + \frac{\nabla(\nabla \cdot \mathbf{A})}{j\omega \epsilon \mu} \tag{F.3}$$

the time harmonic Maxwell equations will be satisfied. For a bounded source

$$\mathbf{A}(\mathbf{r}) = \mu \int_V \frac{\mathbf{J}(\mathbf{r}') \exp(-j\beta |\mathbf{r} - \mathbf{r}'|)}{4\pi |\mathbf{r} - \mathbf{r}'|} dV' \tag{F.4}$$

where $dV' = dx' dy' dz'$ in Cartesian coordinates and V is a volume that contains the sources.

In the case of a vertical dipole located on the z axis, A_z is the only nonzero component of the vector potential and so we only need to consider the Helmholtz equation

$$\frac{\partial^2 A_z}{\partial \rho^2} + \frac{1}{\rho} \frac{\partial A_z}{\partial \rho} + \frac{\partial^2 A_z}{\partial z^2} + \beta^2 A_z = -\mu J_z \tag{F.5}$$

where $\beta^2 = \omega^2 \mu \epsilon$. For a z directed dipole in a uniform medium,

$$\mathbf{A} = -\frac{\mu I h_{\text{eff}}}{4\pi R} \hat{\mathbf{z}} \exp(-j\beta R) \tag{F.6}$$

where I is the current in the dipole, h_{eff} is its effective antenna length (taken to be a constant for current purposes) and $R = \sqrt{\rho^2 + (z - h)^2}$ where h is the height of the dipole above the origin. Away from the dipole the field will satisfy the source-free Helmholtz

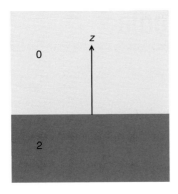
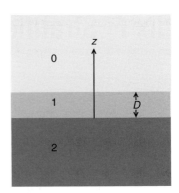

Figure F.1 Two- and three-layer slabs.

equation. Solutions to Helmholtz equation will consist of linear combinations of the basic solutions $\exp(\pm j\gamma z)H_0^{(1)}(\alpha\rho)$ and $\exp(\pm j\gamma z)H_0^{(2)}(\alpha\rho)$ where α is arbitrary and $\gamma^2 = \beta^2 - \alpha^2$. We can, however, discard the $H_0^{(1)}$ part of this solution as we will only have outgoing waves for a bounded source. Consequently, we can write the general solution for the source-free Helmholtz equation as a sum of these solutions, i.e.

$$A_z = \int_{-\infty}^{\infty} (A\exp(+j\gamma z) + B\exp(-j\gamma z)) H_0^{(2)}(\alpha\rho) d\alpha \tag{F.7}$$

where A and B are functions of α. We will first consider the situation where the $z = 0$ plane divides a space with $\epsilon = \epsilon_0$ and $\mu = \mu_0$ in the upper half-space and $\epsilon = \epsilon_2$ and $\mu = \mu_0$ in the lower half-space. The continuity of $\mathbf{H} \times \mathbf{n}$ and $\mathbf{E} \times \mathbf{n}$ across the interface ($z = 0$) will imply that A_z and $\beta^{-2}\partial A_z/\partial z$ are both continuous across this interface. In the upper half-space, the wave must be an upgoing wave as $z \to \infty$ and so

$$A_z = \int_{-\infty}^{\infty} B\exp(-j\gamma_0 z) H_0^{(2)}(\alpha\rho) d\alpha \tag{F.8}$$

and in the lower half-space the wave must be downgoing as $z \to -\infty$ and so

$$A_z = \int_{-\infty}^{\infty} A\exp(+j\gamma_2 z) H_0^{(2)}(\alpha\rho) d\alpha \tag{F.9}$$

(Note that $\gamma_0^2 = \beta_0^2 - \alpha^2$ and $\gamma_2^2 = \beta_2^2 - \alpha^2$.) In addition, we must add the dipole field to the half-space in which the dipole lies. The dipole solution can be expanded in terms of Hankel functions since (Wait, 1996)

$$\frac{\exp(-j\beta R)}{R} = -\frac{j}{2}\int_{-\infty}^{\infty} \exp(-j\gamma|z-h|) H_0^{(2)}(\alpha\rho) \frac{\alpha d\alpha}{\gamma} \tag{F.10}$$

If we take the dipole to be in the upper half-plane, the field for $z \geq 0$ will take the form

$$A_z = \int_{-\infty}^{\infty} B\exp(-j\gamma_0 z) H_0^{(2)}(\alpha\rho) d\alpha + \frac{j\mu I h_{\text{eff}}}{8\pi}\int_{-\infty}^{\infty} \exp(-j\gamma_0|z-h|) H_0^{(2)}(\alpha\rho)\frac{\alpha}{\gamma_0} d\alpha \tag{F.11}$$

From the conditions at the interface, we obtain that

$$B + \frac{j\mu I h_{\text{eff}}}{8\pi} \exp(-j\gamma_0 h) \frac{\alpha}{\gamma_0} = A \qquad (F.12)$$

and

$$-jB\gamma_0 + \frac{j\mu I h_{\text{eff}}}{8\pi} \exp(-j\gamma_0 h) j\alpha = jA\gamma_2 \frac{\beta_0^2}{\beta_2^2} \qquad (F.13)$$

Then, eliminating A between these equations,

$$\left(\frac{\beta_0^2}{\beta_2^2} + \frac{\gamma_0}{\gamma_2}\right) B = -j\frac{\mu I h_{\text{eff}}}{8\pi} \left(\frac{\alpha \beta_0^2}{\gamma_0 \beta_2^2} - \frac{\alpha}{\gamma_2}\right) \exp(-j\gamma_0 h) \qquad (F.14)$$

from which

$$B = -j\frac{\mu I h_{\text{eff}}}{8\pi} \frac{\alpha}{\gamma_0} \frac{\gamma_2 \beta_0^2 - \gamma_0 \beta_2^2}{\gamma_2 \beta_0^2 + \gamma_0 \beta_2^2} \exp(-j\gamma_0 h) \qquad (F.15)$$

We will now have that

$$A_z = -\frac{j\mu I h_{\text{eff}}}{8\pi} \int_{-\infty}^{\infty} \frac{\gamma_2 \beta_0^2 - \gamma_0 \beta_2^2}{\gamma_2 \beta_0^2 + \gamma_0 \beta_2^2} \exp(-j\gamma_0(z+h)) H_0^{(2)}(\alpha \rho) \frac{\alpha}{\gamma_0} d\alpha$$

$$+ \frac{j\mu I h_{\text{eff}}}{8\pi} \int_{-\infty}^{\infty} \exp(-j\gamma_0 |z-h|) H_0^{(2)}(\alpha \rho) \frac{\alpha}{\gamma_0} d\alpha \qquad (F.16)$$

The above integral can be evaluated by contour integral methods (Carrier et al., 1966), but the calculations are quite complex. We now note that

$$A_z = -\frac{\mu I h_{\text{eff}}}{4\pi} \left(\frac{\exp(-j\beta R)}{R} - \frac{\exp(-j\beta \tilde{R})}{\tilde{R}}\right)$$

$$+ \frac{j\mu I h_{\text{eff}}}{8\pi} \int_{-\infty}^{\infty} \frac{2\alpha \beta_2^2}{\gamma_2 \beta_0^2 + \gamma_0 \beta_2^2} \exp(-j\gamma_0(z+h)) H_0^{(2)}(\alpha \rho) d\alpha \qquad (F.17)$$

where $\tilde{R} = \sqrt{\rho^2 + (z+h)^2}$. Regarding the integrand as a function in the complex plane, we can deform the integration contour (see Figure F.2) to one in the lower half-plane. The integrand has a branch point at $\alpha = \beta_0$ (arising from the term γ_0) and so we can deform the contour around a branch cut that runs from $\alpha = \beta_0$ to $\alpha = \beta_0 - j\infty$. This branch point is the dominant singularity. Although there are other singularities, their effect is negligible when the lower medium has significant conductivity (the conductivity damps out their contribution). The integral will now consist of a part that runs from $\beta_0 - i\infty$ to β_0 on the left-hand side of the branch cut and another that runs from β_0 to $\beta_0 - i\infty$ on the right-hand side. On either side of the branch cut, the integrands are almost identical except that we choose different branches of γ_0 on either side. We consider the limit $\rho \to \infty$ and so maximum contribution will come from around $\alpha = \beta_0$ due to the exponential decay of $H_0^{(2)}$ in this limit ($H_0^{(2)}(\alpha \rho) \approx \sqrt{2j/\pi \alpha \rho} \exp(-j\alpha \rho)$ in this limit). We now parameterize the integral by $\alpha = \beta_0 - js^2$ where s is a parameter that runs from ∞ to 0 on the left-hand side and 0 to ∞ on the right-hand side. Now

$\gamma_0 = \sqrt{\beta_0 - \alpha}\sqrt{\beta_0 + \alpha}$ and so, around $\alpha = \beta_0$, we will have $\gamma_0 \approx s\sqrt{2j\beta_0}$ on the right-hand side of the branch cut and $\gamma_0 \approx -s\sqrt{2j\beta_0}$ on the left. Consequently,

$$A_z \approx -\frac{\mu I h_{\text{eff}}}{4\pi}\left(\frac{\exp(-j\beta R)}{R} - \frac{\exp(-j\beta \tilde{R})}{\tilde{R}}\right)$$

$$-\frac{j\mu I h_{\text{eff}}}{8\pi}\int_0^\infty 2\beta_0 \left(\frac{\exp(js\sqrt{2j\beta_0}(z+h))}{\Delta\beta_0 - s\sqrt{2j\beta_0}} - \frac{\exp(-js\sqrt{2j\beta_0}(z+h))}{\Delta\beta_0 + s\sqrt{2j\beta_0}}\right)$$

$$\times \sqrt{\frac{2j}{\pi\beta_0\rho}}\exp(-j\beta_0\rho)\exp(-\rho s^2)2js\,ds \qquad \text{(F.18)}$$

where $\Delta = \beta_0\sqrt{\beta_2^2 - \beta_0^2}/\beta_2^2$. Expanding the integrand to leading order in s, and noting that $\int_0^\infty s^2 \exp(-\rho s^2)\,ds = \pi^{\frac{1}{2}}/4\rho^{\frac{3}{2}}$, we obtain that

$$A_z \approx -\frac{\mu I h_{\text{eff}}}{4\pi}\left(\frac{\exp(-j\beta R)}{R} - \frac{\exp(-j\beta \tilde{R})}{\tilde{R}}\right) + \frac{\mu I h_{\text{eff}}}{4\pi}\frac{2j}{\Delta^2\beta_0\rho^2}(1 + j(z+h)\Delta\beta_0)$$

(F.19)

for $z > 0$ and $h > 0$.

To find the behavior of A_z for $z < 0$, we need the behavior of coefficient A. From the above expressions, this is found to be

$$A = \frac{j\mu I h_{\text{eff}}}{8\pi}\frac{2\alpha\beta_0^2}{\gamma_2\beta_0^2 + \gamma_0\beta_2^2}\exp(-j\gamma_0 h) \qquad \text{(F.20)}$$

and from which

$$A_z = \frac{j\mu I h_{\text{eff}}}{8\pi}\int_{-\infty}^{\infty}\frac{2\alpha\beta_0^2}{\gamma_2\beta_0^2 + \gamma_0\beta_2^2}\exp(j\gamma_2 z - j\gamma_0 h)H_0^{(2)}(\alpha\rho)\,d\alpha \qquad \text{(F.21)}$$

Once again, the dominant contribution will come from the branch cut running from β_0 to $\beta_0 + j\infty$ and can be evaluated in a similar fashion to the previous expression for A_z. This yields

$$A_z \approx -\frac{\mu I h_{\text{eff}}}{4\pi}\left(\frac{\exp(-j\beta R)}{R} - \frac{\exp(-j\beta \tilde{R})}{\tilde{R}}\right)$$

$$+ \frac{\mu I h_{\text{eff}}}{4\pi}\frac{2j}{\Delta^2\beta_0\rho^2}(1 + jh\Delta\beta_0)\exp\left(j\sqrt{\beta_2^2 - \beta_0^2}z\right) \qquad \text{(F.22)}$$

In the case that the source is located in the lower layer ($h < 0$), we have

$$A_z = \int_{-\infty}^{\infty} B\exp(-j\gamma_0 z)H_0^{(2)}(\alpha\rho)\,d\alpha \qquad \text{(F.23)}$$

for $z > 0$ and

$$A_z = \int_{-\infty}^{\infty} A\exp(+j\gamma_2 z)H_0^{(2)}(\alpha\rho)\,d\alpha + \frac{j\mu I h_{\text{eff}}}{8\pi}\int_{-\infty}^{\infty}\exp(-j\gamma_2|z-h|)H_0^{(2)}(\alpha\rho)\frac{\alpha}{\gamma_2}\,d\alpha$$

(F.24)

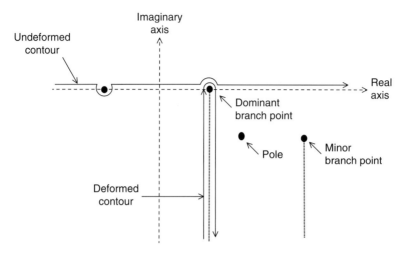

Figure F.2 Integration contour and deformed contour.

for $z < 0$. From the boundary conditions on the interface, we have

$$-B\frac{\beta_2^2}{\beta_0^2} = \frac{\gamma_2}{\gamma_0}A - \frac{j\mu I h_{\text{eff}}}{8\pi}\exp(j\gamma_2 h)\frac{\alpha}{\gamma_0} \tag{F.25}$$

and

$$B = A + \frac{j\mu I h_{\text{eff}}}{8\pi}\exp(j\gamma_2 h)\frac{\alpha}{\gamma_2} \tag{F.26}$$

Eliminating B between the above expressions

$$A = \frac{\gamma_2\beta_0^2 - \gamma_0\beta_2^2}{\gamma_2\beta_0^2 + \gamma_0\beta_2^2}\frac{j\mu I h_{\text{eff}}}{8\pi}\exp(j\gamma_2 h)\frac{\alpha}{\gamma_2} \tag{F.27}$$

and from which

$$A_z = \frac{j\mu I h_{\text{eff}}}{8\pi}\int_{-\infty}^{\infty}\frac{\gamma_2\beta_0^2 - \gamma_0\beta_2^2}{\gamma_2\beta_0^2 + \gamma_0\beta_2^2}\exp(+j\gamma_2(z+h))H_0^{(2)}(\alpha\rho)\frac{\alpha}{\gamma_2}d\alpha - \frac{\mu I h_{\text{eff}}}{4\pi}\frac{\exp(-j\beta_2 R)}{R} \tag{F.28}$$

This can be reinterpreted as

$$A_z = -\frac{\mu I h_{\text{eff}}}{4\pi}\left(\frac{\exp(-j\beta_2 R)}{R} - \frac{\exp(-j\beta_2 \tilde{R})}{\tilde{R}}\right)$$

$$+ \frac{j\mu I h_{\text{eff}}}{8\pi}\int_{-\infty}^{\infty}\frac{2\alpha\beta_0^2}{\gamma_2\beta_0^2 + \gamma_0\beta_2^2}\exp(+j\gamma_2(z+h))H_0^{(2)}(\alpha\rho)d\alpha \tag{F.29}$$

Once again, the dominant contribution will come from the branch cut running from β_0 to $\beta_0 + j\infty$ and can be evaluated in a similar fashion to the previous expressions for A_z. We find that

$$A_z \approx -\frac{\mu I h_{\text{eff}}}{4\pi}\left(\frac{\exp(-j\beta R)}{R} - \frac{\exp(-j\beta \tilde{R})}{\tilde{R}}\right)$$

$$+ \frac{\mu I h_{\text{eff}}}{4\pi}\frac{2j}{\Delta^2 \beta_0 \rho^2}\frac{\beta_0^2}{\beta_2^2}\exp\left(j\sqrt{\beta_2^2 - \beta_0^2}\,(z+h)\right) \quad \text{(F.30)}$$

when both y and h are below the interface.

Once we have expressions for A_z, $\mathbf{E}$ can be calculated from Equation F.3.

F.2 Three-Layer Medium

We now consider the case of three layers with $\epsilon = \epsilon_0$ for $z > D$, $\epsilon = \epsilon_2$ for $z < 0$ and $\epsilon = \epsilon_1$ elsewhere ($\mu = \mu_0$ everywhere). On assuming the source to be located in the uppermost region ($h > D$), we will have

$$A_z = \int_{-\infty}^{\infty} B \exp(-j\gamma_0 z) H_0^{(2)}(\alpha \rho)\, d\alpha + \frac{j\mu I h_{\text{eff}}}{8\pi} \int_{-\infty}^{\infty} \exp(-j\gamma_0|z-h|) H_0^{(2)}(\alpha \rho)\frac{\alpha}{\gamma_0}\, d\alpha \quad \text{(F.31)}$$

for $z > D$,

$$A_z = \int_{-\infty}^{\infty} A \exp(+j\gamma_2 z) H_0^{(2)}(\alpha \rho)\, d\alpha \quad \text{(F.32)}$$

for $z < 0$ and

$$A_z = \int_{-\infty}^{\infty} (E \exp(+j\gamma_1 z) + F \exp(-j\gamma_1 z)) H_0^{(2)}(\alpha \rho)\, d\alpha \quad \text{(F.33)}$$

elsewhere (note that $\gamma_1 = \sqrt{\beta_1^2 - \alpha^2}$). Once again, we will need A_z and $\beta^{-2}\partial A_z/\partial z$ to be continuous across the interfaces ($z = 0$ and $z = D$). From the interface at $z = D$ we have

$$E \exp(j(\gamma_0+\gamma_1)D) - F \exp(j(\gamma_0-\gamma_1)D) = -\frac{\gamma_0}{\gamma_1}\frac{\beta_1^2}{\beta_0^2}B + \frac{j\mu I h_{\text{eff}}}{8\pi}\frac{\beta_1^2}{\beta_0^2}\exp(-j\gamma_0(h-2D))\frac{\alpha}{\gamma_1} \quad \text{(F.34)}$$

$$E \exp(j(\gamma_0+\gamma_1)D) + F \exp(j(\gamma_0-\gamma_1)D) = B + \frac{j\mu I h_{\text{eff}}}{8\pi}\exp(-j\gamma_0(h-2D))\frac{\alpha}{\gamma_0} \quad \text{(F.35)}$$

From the interface at $z = 0$, we have $E - F = A(\beta_1^2/\beta_2^2)(\gamma_2/\gamma_1)$ and $E + F = A$ and these can be solved to yield $2E = L_2^+ A$ and $2F = L_2^- A$ where $L_2^+ = 1 + (\beta_1^2/\beta_2^2)(\gamma_2/\gamma_1)$ and $L_2^- = 1 - (\beta_1^2/\beta_2^2)(\gamma_2/\gamma_1)$. Consequently, from Equation F.34,

$$L_2^+ \exp(j(\gamma_0+\gamma_1)D)A = L_1^- B + L_1^+ \frac{j\mu I h_{\text{eff}}}{8\pi}\exp(-j\gamma_0(h-2D))\frac{\alpha}{\gamma_0} \quad \text{(F.36)}$$

and, from Equation F.35,

$$L_2^- \exp(j(\gamma_0-\gamma_1)D)A = L_1^+ B + L_1^- \frac{j\mu I h_{\text{eff}}}{8\pi}\exp(-j\gamma_0(h-2D))\frac{\alpha}{\gamma_0} \quad \text{(F.37)}$$

where $L_1^+ = 1 + (\beta_1^2/\beta_0^2)(\gamma_0/\gamma_1)$ and $L_1^- = 1 - (\beta_1^2/\beta_0^2)(\gamma_0/\gamma_1)$. Eliminating A,

$$\left(\frac{L_1^-}{L_2^+}\exp(-j\gamma_1 D) - \frac{L_1^+}{L_2^-}\exp(j\gamma_1 D)\right) B$$

$$= -\left(\frac{L_1^+}{L_2^+}\exp(-j\gamma_1 D) - \frac{L_1^-}{L_2^-}\exp(j\gamma_1 D)\right)\frac{j\mu I h_{\text{eff}}}{8\pi}\exp(-j\gamma_0(h-2D))\frac{\alpha}{\gamma_0} \quad \text{(F.38)}$$

Consequently, for $z > D$,

$$A_z = -\frac{j\mu I h_{\text{eff}}}{8\pi}\int_{-\infty}^{\infty} K(\alpha)\exp(-j\gamma_0(z+h-2D))H_0^{(2)}(\alpha\rho)\frac{\alpha}{\gamma_0}d\alpha$$

$$+ \frac{j\mu I h_{\text{eff}}}{8\pi}\int_{-\infty}^{\infty}\exp(-j\gamma_0|z-h|)H_0^{(2)}(\alpha\rho)\frac{\alpha}{\gamma_0}d\alpha \quad \text{(F.39)}$$

where

$$K(\alpha) = \frac{L_1^+ L_2^- \exp(-j\gamma_1 D) - L_1^- L_2^+ \exp(j\gamma_1 D)}{L_1^- L_2^- \exp(-j\gamma_1 D) - L_1^+ L_2^+ \exp(j\gamma_1 D)} \quad \text{(F.40)}$$

We can rearrange the above expression into

$$A_z = -\frac{\mu I h_{\text{eff}}}{4\pi}\left(\frac{\exp(-j\beta_0 R)}{R} - \frac{\exp(-j\beta_0 \tilde{R})}{\tilde{R}}\right)$$

$$- \frac{j\mu I h_{\text{eff}}}{8\pi}\int_{-\infty}^{\infty}\frac{2\alpha}{\gamma_0}\hat{K}(\alpha)\exp(-j\gamma_0(z+h-2D))H_0^{(2)}(\alpha\rho)d\alpha \quad \text{(F.41)}$$

where

$$\hat{K}(\alpha) = \frac{L_2^- \exp(-j\gamma_1 D) + L_2^+ \exp(j\gamma_1 D)}{L_1^- L_2^- \exp(-j\gamma_1 D) - L_1^+ L_2^+ \exp(j\gamma_1 D)} \quad \text{(F.42)}$$

and $\tilde{R} = \sqrt{\rho^2 + (z+h-2D)^2}$. The above integral can be evaluated by contour integral techniques and, as with the two-layer case, the dominant contribution will come from around the branch cut from β_0 to $\beta_0 - i\infty$. (There are other singularities, but their effect is negligible when the lower two layers have significant conductivity.) As with the two-layer case, we parameterize the integral by $\alpha = \beta_0 - js^2$ and expand all functions of α around the branch point $\alpha = \beta_0$. For $z < D$, we can calculate A_z by similar contour integral methods, but we will need to calculate the coefficients A, E and F. For $h < D$ we will need to move the source term from Equations F.31 to F.32, or Equation F.33, according to whichever layer the source is located in. Once we have calculated A_z for appropriate values of h and z, we can calculate E_z using Equation F.3 and obtain the solutions of Wait (1967, 1996).

F.3 References

G.F. Carrier, M. Krook and C.E. Pearson, *Functions of a Complex Variable*, McGraw-Hill, New York, 1966.

J.R. Wait, Asymptotic theory for dipole radiation in the presence of a lossy slab lying on a conducting half-space, IEEE Trans. Ant. Prop. AP-15, pp. 645–648, 1967.

J.R. Wait, *Electromagnetic Waves in Stratified Media*, IEEE/OUP Series on Electromagnetic Wave Theory, Oxford and New York, 1996.

Appendix G Useful Information

Frequency band designations:

Designation	Band
Extremely low frequencies (ELF)	0–3 kHz
Very low frequencies (VLF)	3–30 kHz
Low frequencies (LF)	30–300 kHz
Medium frequencies (MF)	300–3000 kHz
High frequencies (HF)	3–30 MHz
Very high frequencies (VHF)	30–300 MHz
Ultra high frequencies (UHF)	300–3000 MHz
L-band	1–2 GHz
S-band	2–4 GHz
C-band	4–8 GHz
X-band	8–12 GHz
Ku-band	12–18 GHz
K-band	18–27 GHz
Ka-band	27–40 GHz

Note that microwaves are normally considered to be the frequencies between 3 and 300 GHz.

Some physical constants:
  Velocity of light in free space $c_0 = 2.998 \times 10^8$ m/s
  Charge on an electron $e = -1.602 \times 10^{-19}$ C
  Planck's constant $h = 6.6256 \times 10^{-34}$ Js
  Boltzmann's constant $K = 1.381 \times 10^{-23}$ J/K
  Mass of an electron $m = 9.107 \times 10^{-31}$ kg
  Permittivity of free space $\epsilon_0 = 8.854 \times 10^{-12}$ F/m
  Permeability of free space $\mu_0 = 12.57 \times 10^{-7}$ H/m

Some typical electrical properties:

Material	Relative permittivity	Conductivity (S/m)
Fresh water	80	10^{-3}
Sea water	81	5
Pastoral soil	10	10^{-2}
Sandy soil	10	10^{-3}
Concrete	5	10^{-4}
Bitumen	2.7	10^{-12}
Forest	1.1	10^{-4}

Appendix H A Perfectly Matched Layer

The crucial property of a perfectly matched layer (PML) is that it has non-isotropic electric and magnetic conductivities that are chosen so that the impedance of this medium is the same as a medium with these conductivities set to zero (Berenger, 1994). We consider a two-dimensional field (y and z dependence alone) that is transverse magnetic (TM) (i.e. $\mathcal{E} = (0, \mathcal{E}_y, \mathcal{E}_z)$ and $\mathcal{H} = (\mathcal{H}_x, 0, 0)$). Let the medium be a PML for which we consider $\mathcal{H}_x$ to consist of two components ($\mathcal{H}_x = \mathcal{H}_x^y + \mathcal{H}_x^z$) and likewise the magnetic current $\mathcal{M}_x$. The fields $\mathcal{E}_y, \mathcal{E}_z$, $\mathcal{H}_x^y$ and $\mathcal{H}_x^z$ will satisfy (see Chapter 7)

$$\mu_0 \frac{\partial \mathcal{H}_x^y}{\partial t} = -\frac{\partial \mathcal{E}_z}{\partial y} - \mathcal{M}_x^y \tag{H.1}$$

$$\mu_0 \frac{\partial \mathcal{H}_x^z}{\partial t} = \frac{\partial \mathcal{E}_y}{\partial z} - \mathcal{M}_x^z \tag{H.2}$$

$$\epsilon \frac{\partial \mathcal{E}_y}{\partial t} = \frac{\partial \mathcal{H}_x}{\partial z} - \mathcal{J}_y \tag{H.3}$$

$$\epsilon \frac{\partial \mathcal{E}_z}{\partial t} = -\frac{\partial \mathcal{H}_x}{\partial y} - \mathcal{J}_z \tag{H.4}$$

For a PML layer at an upper horizontal boundary (y = constant), $\mathcal{J}_y = \sigma_e \mathcal{E}_y$, $\mathcal{J}_z = 0$, $\mathcal{M}_x^z = \sigma_m \mathcal{H}_x^z$ and $\mathcal{M}_x^y = 0$.

We now consider a plane wave traveling in the above PML, i.e.

$$\mathcal{H}_x^y = H_0^y \exp(j\omega(t - \alpha y - \beta z)) \tag{H.5}$$

$$\mathcal{H}_x^z = H_0^z \exp(j\omega(t - \alpha y - \beta z)) \tag{H.6}$$

$$\mathcal{E}_y = E_y \exp(j\omega(t - \alpha y - \beta z)) \tag{H.7}$$

$$\mathcal{E}_z = E_z \exp(j\omega(t - \alpha y - \beta z)) \tag{H.8}$$

Substituting into Equations H.1 to H.4, we obtain

$$\mu_0 H_0^y = \alpha E_z \tag{H.9}$$

$$\mu_0 H_0^z = -\beta E_y - \frac{\sigma_m}{j\omega} H_0^z \tag{H.10}$$

$$\epsilon E_y = -\beta H_0 - \frac{\sigma_e}{j\omega} E_y \qquad (H.11)$$

$$\epsilon E_z = \alpha H_0 \qquad (H.12)$$

where $H_0 = H_0^y + H_0^z$. From Equations H.11 and H.12 we obtain that

$$\beta = -\alpha \left(1 + \frac{\sigma_e}{j\omega\epsilon}\right) \frac{E_y}{E_z} \qquad (H.13)$$

and from Equations H.9 and H.10

$$\mu_0 \left(1 + \frac{\sigma_m}{j\omega\mu_0}\right) H_0 = \alpha \left(1 + \frac{\sigma_m}{j\omega\mu_0}\right) E_z - \beta E_y \qquad (H.14)$$

For a PML material we have $\sigma_e/\epsilon = \sigma_m/\mu_0$, and so Equation H.14 will reduce to

$$\mu_0 H_0 = \frac{\alpha}{E_z} E_0^2 \qquad (H.15)$$

on substituting for β using Equation H.13. Then, eliminating α between Equations H.15 and H.12, we obtain

$$H_0^2 = \frac{\epsilon}{\mu_0} E_0^2 \qquad (H.16)$$

i.e. the impedance inside the PML layer is the same as for a layer with $\sigma_e = \sigma_m = 0$. Since there is no impedance change across the PML boundary, there will be no reflections at this boundary.

H.1 Reference

J.B. Berenger, A perfectly matched layer for the absorption of electromagnetic waves, J. Comput. Phys., vol. 114, pp. 185–200, 1994.

Appendix I Equations for TE and TM Fields

Maxwell's equations can be written as

$$\nabla \cdot (\mu \mathcal{H}) = 0 \tag{I.1}$$

$$\nabla \cdot (\epsilon \mathcal{E}) = 0 \tag{I.2}$$

$$\nabla \times \mathcal{E} = -\mu \frac{\partial \mathcal{H}}{\partial t} - \mathcal{M} \tag{I.3}$$

$$\nabla \times \mathcal{H} = \mu \frac{\partial \mathcal{E}}{\partial t} + \mathcal{J} \tag{I.4}$$

when the field is driven by electric and magnetic current alone (ϵ and μ are assumed to be independent of time).

For a two-dimensional field (y and z dependence alone) that is transverse magnetic (TM) (i.e. $\mathcal{E} = (0, \mathcal{E}_y, \mathcal{E}_z)$ and $\mathcal{H} = (\mathcal{H}_x, 0, 0)$), Maxwell's equations imply that

$$\mu \frac{\partial \mathcal{H}_x}{\partial t} = -\frac{\partial \mathcal{E}_z}{\partial y} + \frac{\partial \mathcal{E}_y}{\partial z} - \mathcal{M}_x \tag{I.5}$$

$$\epsilon \frac{\partial \mathcal{E}_y}{\partial t} = \frac{\partial \mathcal{H}_x}{\partial z} - \mathcal{J}_y \tag{I.6}$$

$$\epsilon \frac{\partial \mathcal{E}_z}{\partial t} = -\frac{\partial \mathcal{H}_x}{\partial y} - \mathcal{J}_z \tag{I.7}$$

For a two-dimensional field that is transverse electric (TE) (i.e. $\mathcal{H} = (0, \mathcal{H}_y, \mathcal{H}_z)$ and $\mathcal{E} = (\mathcal{E}_x, 0, 0)$), Maxwell's equations imply that

$$\epsilon \frac{\partial \mathcal{E}_x}{\partial t} = \frac{\partial \mathcal{H}_z}{\partial y} - \frac{\partial \mathcal{H}_y}{\partial z} - \mathcal{J}_x \tag{I.8}$$

$$\mu \frac{\partial \mathcal{H}_y}{\partial t} = -\frac{\partial \mathcal{E}_x}{\partial z} - \mathcal{M}_y \tag{I.9}$$

$$\mu \frac{\partial \mathcal{H}_z}{\partial t} = \frac{\partial \mathcal{E}_x}{\partial y} - \mathcal{M}_z \tag{I.10}$$

For problems with cylindrical symmetry, the Maxwell equations are best described in terms of cylindrical polar coordinates. For a two-dimensional field (ρ and z dependence alone) that is transverse magnetic (TM) (i.e. $\mathcal{E} = (\mathcal{E}_\rho, 0, \mathcal{E}_z)$ and $\mathcal{H} = (0, \mathcal{H}_\phi, 0)$), Maxwell's equations imply

$$\mu \frac{\partial \mathcal{H}_\phi}{\partial t} = -\frac{\partial \mathcal{E}_\rho}{\partial z} + \frac{\partial \mathcal{E}_z}{\partial \rho} - \mathcal{M}_\phi \tag{I.11}$$

$$\epsilon \frac{\partial \mathcal{E}_\rho}{\partial t} = -\frac{\partial \mathcal{H}_\phi}{\partial z} - \mathcal{J}_\rho \tag{I.12}$$

and

$$\epsilon \frac{\partial \mathcal{E}_z}{\partial t} = \frac{1}{\rho} \frac{\partial (\rho \mathcal{H}_\phi)}{\partial \rho} - \mathcal{J}_z \tag{I.13}$$

For a two-dimensional field that is transverse electric (TE) (i.e. $\mathcal{H} = (\mathcal{H}_\rho, 0, \mathcal{H}_z)$ and $\mathcal{E} = (0, \mathcal{E}_\phi, 0)$), Maxwell's equations imply

$$\epsilon \frac{\partial \mathcal{E}_\phi}{\partial t} = \frac{\partial \mathcal{H}_\rho}{\partial z} - \frac{\partial \mathcal{H}_z}{\partial \rho} - \mathcal{J}_\phi \tag{I.14}$$

$$\mu \frac{\partial \mathcal{H}_\rho}{\partial t} = \frac{\partial \mathcal{E}_\phi}{\partial z} - \mathcal{M}_\rho \tag{I.15}$$

and

$$\mu \frac{\partial \mathcal{H}_z}{\partial t} = -\frac{1}{\rho} \frac{\partial (\rho \mathcal{E}_\phi)}{\partial \rho} - \mathcal{M}_z \tag{I.16}$$

We now consider the time harmonic fields $\mathcal{H}(\mathbf{r}, t) = \Re \{\mathbf{H}(\mathbf{r}) \exp(j\omega t)\}$, $\mathcal{E}(\mathbf{r}, t) = \Re \{\mathbf{E}(\mathbf{r}) \exp(j\omega t)\}$, $\mathcal{M}(\mathbf{r}, t) = \Re \{\mathbf{M}(\mathbf{r}) \exp(j\omega t)\}$ and $\mathcal{J}(\mathbf{r}, t) = \Re \{\mathbf{J}(\mathbf{r}) \exp(j\omega t)\}$. For a two-dimensional field (ρ and z dependence alone) that is transverse magnetic (TM) (i.e. $\mathcal{E} = (\mathcal{E}_\rho, 0, \mathcal{E}_z)$ and $\mathcal{H} = (0, \mathcal{H}_\phi, 0)$), Maxwell's equations imply

$$j\omega\mu H_\phi = -\frac{\partial E_\rho}{\partial z} + \frac{\partial E_z}{\partial \rho} - M_\phi \tag{I.17}$$

$$j\omega\epsilon E_\rho = -\frac{\partial H_\phi}{\partial z} - J_\rho \tag{I.18}$$

and

$$j\omega\epsilon E_z = \frac{1}{\rho} \frac{\partial (\rho H_\phi)}{\partial \rho} - J_z \tag{I.19}$$

For a two-dimensional field that is transverse electric (TE) (i.e. $\mathbf{H} = (H_\rho, 0, H_z)$ and $\mathbf{E} = (0, E_\phi, 0)$),

$$j\omega\epsilon E_\phi = \frac{\partial H_\rho}{\partial z} - \frac{\partial H_z}{\partial \rho} - J_\phi \tag{I.20}$$

$$j\omega\mu H_\rho = \frac{\partial E_\phi}{\partial z} - M_\rho \tag{I.21}$$

and

$$j\omega\mu H_z = -\frac{1}{\rho} \frac{\partial (\rho E_\phi)}{\partial \rho} - M_z \tag{I.22}$$

Appendix J Canonical Solutions

We consider a PMC scatterer that is infinite in the z direction and of uniform cross section. We will consider excitation by a TM plane wave with the magnetic field $\mathbf{H} = (0, 0, H_z)$ and will study the problem in terms of cylindrical polar coordinates (ρ, θ, z) (in terms of Cartesian coordinates $x = \rho \cos \theta$ and $x = \rho \sin \theta$). A TM field will imply that $\mathbf{E} = (E_\rho, E_\theta, 0)$ and, assuming no z dependence, the appropriate Maxwell equations will be

$$j\omega\mu H_z = -\frac{1}{\rho}\frac{\partial(\rho E_\rho)}{\partial \rho} + \frac{1}{\rho}\frac{\partial E_\rho}{\partial \theta} \tag{J.1}$$

$$j\omega\epsilon E_\rho = \frac{1}{\rho}\frac{\partial H_z}{\partial \theta} \tag{J.2}$$

and

$$j\omega\epsilon E_\theta = -\frac{\partial H_z}{\partial \rho} \tag{J.3}$$

If μ and ϵ are constant, Equations J.1 to J.3 imply the Helmholtz equation

$$\frac{1}{\rho}\frac{\partial}{\partial \rho}\left(\rho \frac{\partial H_z}{\partial \rho}\right) + \frac{1}{\rho^2}\frac{\partial^2 H_z}{\partial \theta^2} + \omega^2\mu\epsilon H_z = 0 \tag{J.4}$$

We will solve this equation subject to the boundary condition $H_z = 0$ on the surface of the scatterer (the electric fields can then be derived from Equations J.2 and J.3). We will seek separable solutions to Equation J.4 of the form $H_z = R_\nu(\rho)\Theta_\nu(\theta)$, then Φ_ν will satisfy

$$\frac{d^2\Theta_\nu}{d\theta^2} + \nu^2\Theta_\nu = 0 \tag{J.5}$$

and R will satisfy

$$\frac{d}{d\rho}\left(\rho\frac{dR_\nu}{d\rho}\right) + \left(\rho^2\beta^2 - \nu^2\right)R_\nu = 0 \tag{J.6}$$

where ν is an arbitrary constant. This last equation is essentially Bessel's equation, where $\beta^2 = \omega^2\mu\epsilon$. The general solution for H_z is then of the form

$$H_z = \sum_\nu a_\nu R_\nu(\rho)\Theta_\nu(\theta) \tag{J.7}$$

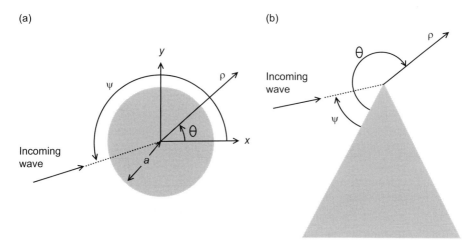

Figure J.1 Diffraction by (a) a cylinder and (b) a wedge.

We first consider the case of a plane wave (see Figure J.1) that is scattered by an infinite cylinder of radius a whose axis is the z axis. The solutions to Equation J.5 will be periodic in θ of period 2π, and so $\Theta_\nu = \exp(jn\theta)$ with $\nu = n$ where n is an integer. The corresponding solutions to Equation J.6 will be Bessel functions $J_n(\beta\rho)$ and $Y_n(\beta\rho)$ or, alternatively, the corresponding Hankel functions ($H_0^{(2)}$ and $H_0^{(1)}$). We will take a scattered field of the form

$$H_z^{\text{scatt}} = \sum_{n=-\infty}^{\infty} j^n a_n H_n^{(2)}(\beta\rho) \exp(jn\theta) \tag{J.8}$$

since the $H_n^{(2)}$ represents the outgoing waves we expect in such fields. The incident field H_z^{inc} can be represented in terms of Bessel functions as (Harrington, 1961)

$$H_z^{\text{inc}} = H_0 \exp(j\beta\rho \cos(\theta - \psi)) = H_0 \sum_{n=-\infty}^{\infty} j^n J_n(\beta\rho) \exp(jn(\theta - \psi)) \tag{J.9}$$

where ψ is the angle of incidence of the incoming plane wave. The boundary condition on the cylinder ($H_z^{\text{inc}} + H_z^{\text{scatt}} = 0$ on $\rho = a$) will imply that

$$a_n = -H_0 \frac{J_n(\beta a)}{H_n^{(2)}(\beta a)} \exp(-jn\psi) \tag{J.10}$$

and so the scattered field will be given by

$$H_z^{\text{scatt}} = -H_0 \sum_{n=-\infty}^{\infty} j^n \frac{J_n(\beta a)}{H_n^{(2)}(\beta a)} H_n^{(2)}(\beta\rho) \exp(jn(\theta - \psi)) \tag{J.11}$$

For the limit of interest in the current context ($\beta a \to \infty$) the above series is only slowly convergent. Jones (1999), however, has used the Poisson summation formula to convert such a series into one that is faster convergent, i.e.

Canonical Solutions

$$H_z^{\text{scatt}} = -\frac{H_0}{2} \sum_{n=-\infty}^{\infty} \int_{-\infty}^{\infty} \left(1 + \frac{H_\nu^{(1)}(\beta a)}{H_\nu^{(2)}(\beta a)}\right) H_\nu^{(2)}(\beta \rho) \exp\left(j\nu\left(\theta - \psi + \frac{\pi}{2}\right)\right) \exp(-j2\nu n\pi) d\nu$$
(J.12)

on noting that $J_n = (H_n^{(1)} + H_n^{(2)})/2$. We now consider the terms of the new series that are defined in terms of an integral. For complex ν, the singularities of the integrand are simple poles that lie in the second and fourth quadrants (they are the zeros of $H_\nu^{(2)}(\beta a)$ in the ν plane). For $\theta - \psi + \pi/2 > 2n\pi$ the contour can be completed in the upper half-plane and for $\theta - \psi + \pi/2 < 2n\pi$ the contour can be completed in the lower half-plane. Contributions to the integral will come from the poles contained by these closed contours (i.e. $\pm\left(\beta a + (\beta a/2)^{1/3} a_i \exp(j2\pi/3)\right)$ for positive integers i). So, for $n < 0$, we complete the contour in the upper half-plane (C_{LUP} the closed contour) and, for $n > 0$, we complete in the lower half-plane (C_{LHP} the closed contour). Appropriately splitting the sum we obtain

$$H_z^{\text{scatt}} = -\frac{H_0}{2} \left\{ \int_{-\infty}^{\infty} \left(1 + \frac{H_\nu^{(1)}(\beta a)}{H_\nu^{(2)}(\beta a)}\right) H_\nu^{(2)}(\beta\rho) \exp\left(j\nu\left(\theta - \psi + \frac{\pi}{2}\right)\right) d\nu \right.$$
$$+ \sum_{n=1}^{\infty} \int_{C_{UHP}} \left(1 + \frac{H_\nu^{(1)}(\beta a)}{H_\nu^{(2)}(\beta a)}\right) H_\nu^{(2)}(\beta\rho) \exp\left(j\nu\left(\theta - \psi + \frac{\pi}{2}\right)\right) \exp(2j\pi n\nu) d\nu$$
$$\left. + \sum_{n=1}^{\infty} \int_{C_{LHP}} \left(1 + \frac{H_\nu^{(1)}(\beta a)}{H_\nu^{(2)}(\beta a)}\right) H_\nu^{(2)}(\beta\rho) \exp\left(j\nu\left(\theta - \psi + \frac{\pi}{2}\right)\right) \exp(-2j\pi n\nu) d\nu \right\}$$
(J.13)

and, converting the integral around C_{UHP} to one around C_{LHP} (note the identities $H_{-\nu}^{(1)}(z) = \exp(j\nu\pi) H_\nu^{(1)}(z)$ and $H_{-\nu}^{(2)}(z) = \exp(-j\nu\pi) H_\nu^{(2)}(z)$),

$$H_z^{\text{scatt}} = -\frac{H_0}{2} \left\{ \int_{-\infty}^{\infty} \left(1 + \frac{H_\nu^{(1)}(\beta a)}{H_\nu^{(2)}(\beta a)}\right) H_\nu^{(2)}(\beta\rho) \exp\left(j\nu\left(\theta - \psi + \frac{\pi}{2}\right)\right) d\nu \right.$$
$$+ \sum_{n=1}^{\infty} \int_{C_{LHP}} \left(1 + \frac{H_\nu^{(1)}(\beta a)}{H_\nu^{(2)}(\beta a)}\right) H_\nu^{(2)}(\beta\rho)$$
$$\left. \times \left(\exp\left(j\nu\left(\theta - \psi - \frac{3\pi}{2}\right)\right) + \exp\left(-j\nu\left(\theta - \psi + \frac{\pi}{2}\right)\right)\right) \exp(-2j\pi n\nu) d\nu \right\}$$
(J.14)

The first integral in the above expression includes the dominant contributions and is the expression that would arise if we simply approximated the sum in Equation J.11 by an integral. We will consider the shadow region $(-3\pi/2 < \theta - \psi < -\pi/2)$ and hence complete this integral in the lower half space. By the methods of residue calculus, the integral is derived from contributions at the poles $\nu_i = \beta a + (\beta a/2)^{1/3} a_i \exp(j2\pi/3)$ and so

$$H_z^{\text{scatt}} \approx j\pi H_0 \sum_{i=1}^{\infty} \frac{H_{\nu_i}^{(1)}(\beta a)}{\frac{\partial H_\nu^{(2)}(\beta a)}{\partial \nu}\Big|_{\nu=\nu_i}} H_{\nu_i}^{(2)}(\beta\rho) \exp\left(j\nu_i\left(\theta - \psi + \frac{\pi}{2}\right)\right) \quad \text{(J.15)}$$

For large βa the pole at $\nu_a = \beta a + (\beta a/2)^{1/3} a_1 \exp(j2\pi/3)$ dominates and we obtain

$$H_z^{\text{scatt}} \approx j\pi H_0 \frac{H_{\nu_a}^{(1)}(\beta a)}{\frac{\partial H_\nu^{(2)}(\beta a)}{\partial \nu}\big|_{\nu=\nu_a}} H_{\nu_a}^{(2)}(\beta \rho) \exp\left(j\nu_i\left(\theta - \psi + \frac{\pi}{2}\right)\right) \quad \text{(J.16)}$$

From the recurrence and Wronskian relations of Appendix A, $H_{\nu_a}^{(1)}(\beta a) H_{\nu_a+1}^{(2)}(\beta a) = 4j/\pi\beta a$. Furthermore, for large ν, $H_{\nu_a+1}^{(1)}(\beta \rho) \approx \partial H_\nu^{(2)}(\beta a)/\partial \nu|_{\nu=\nu_a}$. Consequently,

$$H_z^{\text{scatt}} \approx -\frac{4}{\beta a} H_0 \left(\frac{\partial H_\nu^{(2)}(\beta a)}{\partial \nu}\bigg|_{\nu=\nu_a}\right)^{-2} H_{\nu_a}^{(2)}(\beta \rho) \exp\left(j\nu_i\left(\theta - \psi + \frac{\pi}{2}\right)\right) \quad \text{(J.17)}$$

The Hankel functions in the above expression can be evaluated using the asymptotic results in Appendix A, which are appropriate for $\beta a/\nu_s \sim 1$. From these results we then obtain that $\partial H_\nu^{(2)}(\beta a)/\partial \nu|_{\nu=\nu_a} \approx 2^{5/3}(\beta a)^{-2/3} \exp(-j\pi/3) Ai'(a_1)$.

Next consider scatter from an infinite wedge with z axis along the edge (see Figure J.1) and orientate the coordinate axes so that the θ coordinate is measured clockwise from the illuminated face of the wedge. Let α be the exterior angle of the wedge, then the appropriate solution to the Helmholtz equation will be (Ufimtsev, 2007)

$$H_z^{\text{scatt}} = -H_0 \frac{4\pi}{\alpha} \sum_{n=0}^{\infty} \exp\left(j\frac{\pi}{2}\nu_n\right) J_{\nu_n}(\beta\rho) \sin(\nu_n\theta) \sin(\nu_n\psi) \quad \text{(J.18)}$$

where $\nu_n = \pi n/\alpha$ and n is integer. It is clear from the above expression that this is a solution to the Helmholtz equation that satisfies the correct boundary condition ($H_z = 0$) on the wedge faces. We also need to verify that the solution behaves as an incoming plane wave before it reaches the wedge. Using the identity $2\sin(\nu_n\theta)\sin(\nu_n\phi) = \cos(\nu_n\theta - \nu_n\psi) - \cos(\nu_n\theta + \nu_n\psi)$, we can express Equation J.18 as $H_z = H_0 u(\beta\rho, \theta + \psi) - H_0 u(\beta\rho, \theta - \psi)$ where

$$u(r, \phi) = \frac{2\pi}{\alpha} \sum_{n=0}^{\infty} \exp\left(j\frac{\pi}{2}\nu_n\right) J_{\nu_n}(r) \cos(\nu_n\phi) \quad \text{(J.19)}$$

We now represent the Bessel functions in the integral form (see Appendix A)

$$J_\nu(z) = \frac{1}{2j\pi} \int_C t^{-\nu-1} \exp\left(\frac{z}{2}\left(t - \frac{1}{t}\right)\right) dt \quad \text{(J.20)}$$

and interchange integration and summation (Ufimtsev, 2007) to obtain

$$u(r, \phi) = \frac{1}{j\alpha} \int_C \exp\left(\frac{r}{2}\left(t - \frac{1}{t}\right)\right) \sum_{n=0}^{\infty} \exp\left(j\frac{\pi}{2}\nu_n\right) t^{-\nu_n} \cos(\nu_n\phi) \frac{dt}{t} \quad \text{(J.21)}$$

From the identity $\cos x = (\exp(jx) + \exp(-jx))/2$ and the result $\sum_{n=0}^{\infty} a^n = 1/(1-a)$ ($a \neq 1$), we obtain

$$u(r,\phi) = \frac{1}{2j\alpha} \int_C \exp\left(\frac{r}{2}\left(t - \frac{1}{t}\right)\right)$$
$$\times \left(\frac{1}{1 - t^{-\frac{\pi}{\alpha}}\exp\left(j\left(\phi + \frac{\pi}{2}\right)\frac{\pi}{\alpha}\right)} + \frac{1}{1 - t^{-\frac{\pi}{\alpha}}\exp\left(-j\left(\phi - \frac{\pi}{2}\right)\frac{\pi}{\alpha}\right)}\right)\frac{dt}{t}$$
(J.22)

If we replace the variable t by a new variable w such that $t = \exp(-j(w - \pi/2))$,

$$u(r,\phi) = \frac{-1}{2\alpha}\int_D \exp(jr\cos w)\left(\frac{1}{1 - \exp(j(w+\phi)\frac{\pi}{\alpha})} + \frac{1}{1 - \exp(j(w-\phi)\frac{\pi}{\alpha})}\right)dw$$
(J.23)

where D is now a U-shaped contour from $-\pi/2 + j\infty$ to $-\pi/2 + j\delta$ to $3\pi/2 + j\delta$ to $3\pi/2 + j\infty$ ($\delta > 0$). The above integral can be rearranged into the form

$$u(r,\phi) = \frac{-1}{2\alpha}\int_D \exp(jr\cos w)\left(\frac{1}{1 - \exp(j(w+\phi)\frac{\pi}{\alpha})} - \frac{1}{1 - \exp(-j(w-\phi)\frac{\pi}{\alpha})} + 1\right)dw$$
(J.24)

We can ignore the third term in the integral and change w to $-w$ in the second term. As a consequence

$$u(r,\phi) = \frac{-1}{2\alpha}\int_{D+E} \exp(jr\cos w)\frac{1}{1 - \exp(j(w+\phi)\frac{\pi}{\alpha})}dw \qquad \text{(J.25)}$$

The contour now has two sections consisting of D and the section E running from $\pi/2 - j\infty$ to $\pi/2 - j\delta$ to $-3\pi/2 - j\delta$ to $-3\pi/2 - j\infty$. The contours D and E can now be replaced by the contours F, G and H (see Figure J.2). Contour H will encompass some the poles of the integrand, and the pole at $w = -\phi$ will contribute the GO part of u. This contribution will be $-\exp(jr\cos\phi)$ when $|\phi| < \pi$ and zero when $|\phi| > \pi$. As a consequence

$$H_z^{GO}(r,\phi) = H_0\left(\exp(jr\cos(\theta - \psi)) - \exp(jr\cos(\theta + \psi))\right) \quad 0 \le \theta < \pi - \psi$$
$$= H_0 \exp(jr\cos(\theta - \psi)) \quad \pi - \psi < \theta < \pi + \psi$$
$$= 0 \quad \pi + \psi < \theta \le \alpha \qquad \text{(J.26)}$$

The integrals around F and G constitute the diffracted part u^{diff} of u (note that $u = u^{\text{diff}} + u^{\text{GO}}$). In the integral over contour F, we will now transform to the new parameter $\omega = w + \pi$ and, for that over contour G, transform to the new parameter $\omega = w - \pi$. Consequently, we will now have that the diffracted part of H_z is given by

$$u^{\text{diff}}(r,\phi) = \frac{-1}{2\alpha}\int_I \exp(-jr\cos\omega)$$
$$\times \left(\frac{1}{1 - \exp(j(\omega + \phi - \pi)\frac{\pi}{\alpha})} - \frac{1}{1 - \exp(j(\omega + \phi + \pi)\frac{\pi}{\alpha})}\right)d\omega \quad \text{(J.27)}$$

from which

$$u^{\text{diff}}(r,\phi) = \frac{-1}{2\alpha}\int_I \exp(-jr\cos\omega)\frac{j\sin\left(\pi\frac{\pi}{\alpha}\right)}{\cos\left(\pi\frac{\pi}{\alpha}\right) - \cos\left((\omega + \phi)\frac{\pi}{\alpha}\right)}d\omega$$

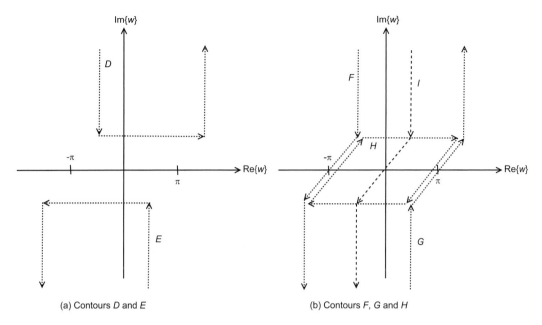

(a) Contours D and E (b) Contours F, G and H

Figure J.2 Contour integration paths for wedge calculations.

where I is a new integration path that goes through the origin (see Figure J.2). For large r, we can use the method of stationary phase to evaluate this integral by deforming contour I such that it runs parallel to the real axis as it passes through the origin. From this we obtain that

$$u^{\text{diff}}(r,\phi) = \frac{j}{2\alpha}\exp(-jr)\sqrt{\frac{2j\pi}{r}}\frac{\sin\left(\pi\frac{\pi}{\alpha}\right)}{\cos\left(\pi\frac{\pi}{\alpha}\right) - \cos\left(\phi\frac{\pi}{\alpha}\right)} \quad (\text{J.28})$$

and hence that

$$H_z^{\text{diff}}(r,\phi) = \frac{H_0}{2\alpha}\exp(-jr)\sqrt{\frac{2\pi}{jr}}$$
$$\times \left(\frac{\sin\left(\pi\frac{\pi}{\alpha}\right)}{\cos\left(\pi\frac{\pi}{\alpha}\right) - \cos\left((\theta+\psi)\frac{\pi}{\alpha}\right)} - \frac{\sin\left(\pi\frac{\pi}{\alpha}\right)}{\cos\left(\pi\frac{\pi}{\alpha}\right) - \cos\left((\theta-\psi)\frac{\pi}{\alpha}\right)}\right) \quad (\text{J.29})$$

J.1 References

G.F. Carrier, M. Krook and C.E. Pearson, *Functions of a Complex Variable*, McGraw-Hill, 1966.

R.F. Harrington, *Time Harmonic Electromagnetic Fields*, McGraw-Hill, 1961.

D.S. Jones, *Methods in Electromagnetic Wave Propagation*, 2nd edition, IEEE/OUP series on Electromagnetic Wave Theory, Oxford University Press, 1999.

J.B. Keller, *Geometrical Theory of Diffraction*, Journal of the Optical Society of America, vol. 52, pp. 116–130, 1962.

P.Y. Ufimtsev, *Fundamentals of the Physical Theory of Diffraction*, John Wiley, Hoboken, NJ, 2007.

Index

2D Helmholtz equation, 130
2D Kirchhoff integral approach, 139
2D ray tracing, 177
3D Kirchhoff integral approach, 148
3D ray tracing, 181

Abel integral equation, 215
absorbing boundary condition, 28, 153
absorption, 189
absorption constant, 190
acoustic cut-off frequency, 166
Airy function, 130, 199, 241
   asymptotic expansion, 242
   integral representation, 242
   zeros, 242
amplitude fading, 192
amplitude modulation (AM), 4
anisotropic media, 23, 80, 87
antenna
   efficiency, 4
   reactance, 5
   reciprocity, 41
Appleton–Hartree formula, 26, 82, 172
atmospheric noise, 193
attenuation, 21
attenuation constant, 21, 111, 220
auroral oval, 164
auroral scatter, 96

back scatter, 12
backward difference, 247
bandwidth, 4
Bessel equation, 242
Bessel function, 242
   asymptotic expansion, 243
   integral representation, 243
Booker scattering formula, 96
Born approximation, 98
Bouger's law, 79
boundary conditions, 27
Breit and Tuve theorem, 169
Brewster angle, 33
Brunt–Vaisala frequency, 166
Bullington method, 57

canonical solutions, 61, 130
carrier signal, 4
caustic surface, 75, 202
cellular radio, 7
central difference, 247
channel impulse response function, 68, 119, 232
channel scattering function, 232
channel simulation, 118
Chapman layer, 161
charge density, 16
Charpit equations, 71, 82, 83
circular polarisation, 20, 225
clutter, 94
coherence bandwidth, 231
coherence
   in the frequency domain, 113
   in the spatial domain, 111
collision loss, 189
compensation theorem, 44, 213
complementary variational principles, 89
conductivity, 16
conformal transformation, 130
conjugate point, 254
cononical solutions, 278
continuity equation, 16, 18
Crank–Nicolson scheme, 138
current density, 16
cut-off frequency, 131, 196
cylindrical polar coordinates, 237

D layer, 161
Debye formula, 204
delay spread, 68
deviative loss, 190
dielectric constant, 16
diffraction, 10, 55
   over a cylinder, 63
   over a screen, 61
   over a wedge, 62
diffraction loss, 57
dipole, 5, 35
dipole antenna, 4
directivity, 4, 35
discrete Fourier transform (DFT), 260

dispersion, 21
displacement invariant, 142, 261
disturbed ionosphere, 166
divergence theorem, 237
domain truncation, 137, 143, 153
Doppler shift, 9, 226
Doppler spread, 68, 98, 235
ducting, 12

E layer, 161
Earth flattening, 133, 148, 197
Earth's curvature, 145
Earth's magnetic field, 163
effective antenna area, 36, 50
effective antenna length, 5, 42
effective permittivity, 20
effective relative surface impedance, 216
eikonal, 70
Einstein summation convention, 254
electric flux density, 16
electric intensity, 16
electromagnetic waves, 2
elevated duct, 207
elliptical polarization, 20
equatorial anomaly, 164
equivalent vertical frequency, 169
Euler Method, 249
Euler–Lagrange equations, 174, 254
evanescence, 131, 196
evaporation duct, 205
external noise, 192
extinction theorem, 46, 122
extremely low frequencies (ELF), 273

F layer, 161
F1 layer, 161
fading, 68, 191
Faraday rotation, 87, 191, 225
fast Fourier transform (FFT), 142, 261
Fermat's principle, 53, 87, 224, 258
finite difference (FD) methods, 134, 136
finite difference time domain (FDTD) methods, 151
finite element (FE) methods, 72, 90, 134, 256
flutter, 68
forward difference, 247
forward scatter, 12
Fourier integral theorem, 259
Fourier transform, 259
Fraunhofer zone, 34
free space, 50
frequency modulation (FM), 4
Fresnel clearance, 57
Fresnel integrals, 57, 66, 67, 218
Friis equation, 6, 9, 50, 52
full wave solutions, 194

gain, 4, 36
galactic noise, 193
generalized parabolic equation, 136
generalized power spectral density (GPSD), 230
generalized power spectrum, 119
generalized Snell's law, 79
geometric optics (GO), 70
geometric series, 245
geometric theory of diffraction, 60
GPS navigation, 224
gravity waves, 166
grazing angle, 142
ground loss, 59
ground reflection loss, 191
ground, effect on a source, 48
group delay, 22, 85, 224
group distance, 22, 71
group refractive index, 85

Hankel function, 243
    asymptotic expansion, 245
    integral representation, 245
    zeros, 245
Hanning window, 142
Haselgrove equations, 83, 182
height gain function, 216
Helmholtz equation, 98, 110, 130
high frequencies (HF), 273
high ray, 178
homogeneous function, 256
horizontally stratified medium, 78
Huygens' principle, 10, 40

impedance boundary condition, 27, 214
impedance of the propagation medium, 18
infinite uniform scatters, 278
integral equation for surface wave, 215
integral equation propagator, 139
internal noise, 192
ionosonde, 167
ionospheric layers, 161
ionospheric propagation, 12, 72
irregular media, 94
irregular terrain, 145, 209

Jacobi equation, 77
Jacobi test, 254, 256

Kirchhoff integral approximation, 139
Kirchhoff scattering theory, 126
knife edge diffraction, 61
Kp index, 164

lapse rate, 204
Legendre test, 254
linear model of ionosphere, 200
linear polarization, 20

Index

Lorentz force, 16
Lorentz reciprocity theorem, 39
loss resistance, 5
low frequencies (LF), 273
low ray, 178

M units, 205
magnetic flux density, 16
magnetic intensity, 16
magneto-ionic effects, 80, 189
manmade noise, 193
Martyn's theorem, 169
  for attenuation, 191
maximum usable frequency (MUF), 171
Maxwell's equations, 15
MCF. *See* mutual coherence function
medium frequencies (MF), 273
micrometeors, 220
midpoint rule, 247
mid-latitude trough, 164
mobile communications, 68
mode expansion, 131
modified Friis equation, 52
modified refractive index, 134, 145
modulation, 3
multipath fading, 68, 192
mutual coherence function (MCF), 102, 228
  for plane wave, 109
  for spherical wave, 109
mutual impedance between antennas, 42, 56, 59, 95

Nelder–Mead method, 182, 263
Newton's method, 248
Newton–Raphson method, 90, 248, 263
noise, 6, 192
noise directionality, 193
non-deviative loss, 190
normal incidence, 28
numerical differentiation, 247
numerical integration, 247
numerical solution of ODEs, 249

O ray, 175
oblique incidence, 30
oblique ionogram, 171
ocean spectrum, 126
Ohm's law, 16
over the horizon propagation, 60, 204

parabolic equation, 135
  for average field, 110
  for MCF, 111
parabolic ionosphere, 73
parameter invariant functional, 256
paraxial approximation, 135
  in ionosphere, 196
peak plasma frequency, 161

perfectly electrically conducting (PEC), 27
perfectly magnetically conducting (PMC), 27
perfectly matched layer (PML), 153
permeability, 16
permittivity, 16
phase delay, 22, 72, 85, 224
phase distance, 22
phase modulation (PM), 4
phase refractive index, 85
phase screen approximation, 114
physical optics, 126
plane harmonic wave, 19
plane to spherical wave transform, 63, 101, 109
plane wave, 17, 23
plane wave Rytov approximation, 101
Poeverlein method, 175
point-to-point ray tracing, 182
Poisson summation formula, 245, 279
polar coordinate ray tracing, 77
polarization, 19
  efficiency, 50
  fading, 192
  vector, 76
power impulse response function, 118
Poynting vector, 16, 18, 26, 75, 84
principal radii of curvature, 54
propagation constant, 19, 220
propagation losses, 187
propagation
  through forest, 216
  through rain, 219
  through water, 219
pseudo Brewster angle, 33
pseudo reciprocity, 42
pulse propagation, 153

quadratic structure function, 231
quadrature rules, 247
quasi parabolic layer, 79
quasi vertical propagation, 170

radar, 8
  cross section, 9
  equation, 8, 9
radiating source, 33
radiation resistance, 5
radiation zone, 34
rain attenuation, 219
ray homing, 180
ray path deviations, 77
Rayleigh criterion, 121
Rayleigh roughness parameter, 121, 207
Rayleigh scattering, 95, 219
Rayleigh–Ritz method, 89, 256
Rayleigh–Taylor instability, 227
rays, 71

Index

reciprocity, 38, 94
recovery effect, 213
recursive FFT algorithm, 261
reflection, 51
  by curved surfaces, 54
  by ground, 53
  coefficient, 32, 142, 147
  coefficient for irregular terrain, 147
refraction, 11
refractive index, 71
refractive index surface, 173
relative impedance, 32
relative surface impedance, 47, 215
rooftop diffraction, 65
rough sea, 207
rough surface boundary condition, 28
rough surface scattering, 121
Runge–Kutta (RK) method, 71, 249
Runge–Kutta–Fehlberg (RKF) method, 71, 251
Rytov approximation, 98, 99, 101, 107, 109, 140

scale height, 160
scatter propagation, 95
scattering, 12
  cross section, 126
  by dielectric anomaly, 95
Schumann resonance, 196
Schwarz–Christoffel transformation, 131
scintillation, 13, 227
sea spectrum, 125
secant law, 169
secant method, 248
second-order Rytov approximation, 107
shadow region, 55
shooting and bouncing rays, 126
shooting method, 180
short dipole, 36
signal to interference ratio (SIR), 8
signal to noise ratio (SNR), 7, 194
Simpson's rule, 248
skip zone, 74, 178
sky wave, 159
small scale irregularity, 226
Snell's law, 11, 72, 78, 79, 170
solar wind, 163
solution of ODEs, 249
Sommerfeld formula, 215
spectrum of the irregularity, 96
specular direction, 121
specular reflection, 121
spherical wave Rytov approximation, 102
spherically stratified medium, 79
Spitze, 176
split step algorithm, 140
spreading loss, 6, 188

stability, 138, 154
  of FD methods, 137
  of FDTD methods, 154
stationary functional, 253
stationary phase, 240
Stokes equation, 241
stratified media, 265
structure function, 104
sunspot cycle, 161
sunspot number, 161
surface impedance, 27, 147
surface roughness, 207
surface waves, 58, 211
  for undulating ground, 216

temperature inversion, 204
three-layer medium, 270
time harmonic field, 18
time harmonic Poynting vector, 19
topography, effect of, 209
total electron content (TEC), 224
total loss, 68, 191
transionospheric propagation, 222
transmission curve, 171
transmission through a screen, 29
transverse electric (TE), 31, 276, 277
transverse magnetic (TM), 31, 276–278
trapezoidal rule, 247
traveling ionospheric disturbance (TID), 166
trigonometric identities, 239
trigonometric series, 245
tropospheric scatter, 96
two-layer medium, 219, 265

ultra high frequencies (UHF), 273
urban propagation, 64

variational calculus, 253
variational principle, 88
vector effective antenna length, 35
vector Helmholtz equation, 129
vector operators, 236
vectors, 236
vertical ionogram, 167
vertical propagation, 167
very high frequencies (VHF), 273
very low frequencies (VLF), 273

wave equation, 17
wavefronts
  distortion by a curved surface, 54
  principal radii of curvature, 188
waveguide model of ionosphere, 194
waves, 1
WBMOD, 230
weakly anisotropic medium, 86
WKB solution for the ionosphere, 202